Plant Cell Culture in Crop Improvement

BASIC LIFE SCIENCES
Alexander Hollaender, General Editor
Associated Universities, Inc., Washington, D.C.

Recent volumes in the series:

A Continuation Order Plan is available for this series. A continuation order will bring delivery of each new volume immediately upon publication. Volumes are billed only upon actual shipment. For further information please contact the publisher.

Plant Cell Culture in Crop Improvement

Edited by

S. K. Sen
Bose Institute
Calcutta, India

and

Kenneth L. Giles
Worcester Polytechnic Institute
Worcester, Massachusetts

SPRINGER SCIENCE+BUSINESS MEDIA, LLC

Library of Congress Cataloging in Publication Data

Main entry under title:

Plant cell culture in crop improvement.

(Basic life sciences; v. 22)
"Proceedings of the International Symposium on Plant Cell Culture in Crop Improvement, held December 6–10, 1981, at Bose Institute, Calcutta, India"—Verso of t.p.
Includes bibliographical references and index.
1. Plant cell culture—Congresses. 2. Plant tissue culture—Congresses. 3. Plant propagation —Congresses. I. Sen, S. K. II. Giles, Kenneth L. III. International Symposium on Plant Cell Culture in Crop Improvement (1981: Bose Institute) IV. Series.
SB123.6.P5 1982 631.5′2 82-16642

ISBN 978-1-4684-4381-3 ISBN 978-1-4684-4379-0 (eBook)
DOI 10.1007/978-1-4684-4379-0

Proceedings of the International Symposium on Plant Cell Culture in Crop Improvement, held December 6–10, 1981, at Bose Institute, Calcutta, India

Originally published by Plenum Press New York in 1983
Softcover reprint of the hardcover 1st edition 1983

PREFACE

The current and potential importance of plant tissue culture techniques in crop improvement is hard to overemphasize. There are few areas where these techniques will have more possible impact than in tropical agriculture, where the availability of high productivity varieties is sadly lacking in many species. The potential for the rapid, clonal propagation of elite individuals and the use of controlled multiline planting could have a major effect on crop yield and disease resistance in many areas of the world.

This volume is a collection of papers presented at the Conference on "Crop Improvement Through Tissue Culture", held at the Base Institute, Calcutta, India in December 1981. It attempts to bring together local research workers, familiar with the agricultural resources of the area and tissue culture and molecular level workers. It was the hope of the conference that the "cross fertilization" of ideas would lead to new approaches and activity in this area.

The editors trust that this collection of papers will stimulate interest and research in the tissue culture and improvement of crop plants everywhere.

ACKNOWLEDGEMENTS

The symposium from which the papers in this book are drawn was held at Bose Institute, Calcutta on December 6 to December 10, 1981. The symposium was made possible through the help and cooperation of the Department of Science and Technology, Government of India and the offices of the Director and Registrar, Bose Institute, Calcutta, and many other organizations. The assistance of each and every member of the Program in Genetical Research, Bose Institute, during the symposium is acknowledged.

The patience and understanding of the Faculty and Administration of the Biology and Biotechnology Department at Worcester Polytechnic Institute during the course of the preparation of this volume is also acknowledged. We are especially thankful to the painstaking and dedicated efforts of Jeanne Burwick, Departmental Secretary, without whose tireless attention to detail the speedy production of this set of papers would not have been possible.

Kenneth L. Giles, Ph.D.
Dept. of Biology and Biotechnology
Worcester Polytechnic Institute
Worcester, Massachusetts, U.S.A.

S.K. Sen, Ph.D.
Bose Institute
Calcutta, India

CONTENTS

PAPERS

POSTERS

PLANT CELL CULTURE TECHNOLOGY IN RELATION TO PLANT BREEDING

Sherret S. Chase

International Plant Research Institute
San Carlos, California, USA

Based upon results realized to date, in research laboratories worldwide, one can predict that means will exist in the near future to set plant breeders free from the historic constraints of season and time, syngamy and meiosis; that is, from the constraints of working within normal sexual cycles, maturation periods and ecological restrictions. The range of relatedness with which plant breeders can work in hybridization and segregation will be widened. The possibilities of inserting genetic material from any species into eukaryotic plant cells will be realized. It will become feasible to further control such foreign deoxyribonucleic acid and its expression in ways appropriate to the plant breeder's requirements and the host genome.

In this talk today I would like to address three subjects: first, briefly, the International Plant Research Institute, of which organization I am a part; second, some old and new research of my own; and third, some lines of study I believe important to the interests of the group assembled here.

I have been a part of the International Plant Research Institute, as head of Plant Breeding, since January of this year. The Institute, a pioneer in the field of commercial plant genetic engineering, was founded in San Carlos, California, in 1978. The primary aim of the organization is to modify the genetic information in higher plants to enhance specific agricultural and industrial uses. In this regard, major efforts are being directed toward improvement of the yield and quality of our primary food plants, including maize, wheat, potatoes, and cassava.

The research of the Institute is divided among five departments: Plant Breeding, Plant Cell Biology, Plant Pathology, Molecular Genetics, and Chemistry. About 80 scientists are presently at work, and of these, 40 are independent scientists at the Ph.D. or equivalent level. Specific plant breeding goals include the development of new plant varieties that can resist diseases and pests, tolerate high levels of soil salinity and extremes of temperature, grow normally with less water, convert more of the sun's energy into carbohydrates, and contain higher levels of better quality protein.

We are very actively concerned with mastering and applying the technologies of plant regeneration from protoplasts and cell masses, "anther" and "ovule" culture, protoplast fusion, transformation and integration, and genetic control of transferred genes. We are also interested in the symbiotic relations of higher plants with rhizobia and mycorrhiza, and with methods for commercial production of useful phytoproducts *in vitro*.

Fundamental to our goals is the development of efficient, specific transformation vectors and transformation systems.

Although the International Plant Research Institute is a "for-profit" organization, it may be several years before the present substantial investment returns profits to the shareholders. However, the pace of reseaerch achievement, both within the Institute and outside, is such that one feels confident that a number of our commercial goals will be achieved in the relatively near future.

For those of you unacquainted with my primary lines of research, I have, since 1946, been interested in spontaneous parthogenesis and related phenomena in maize, the application of the monoploid method in maize breeding, and maize breeding in general, both within the United States and also in other countries, including India.

To recognize spontaneous parthenotes, I have used several marker systems or genetic screens, the primary of which have been a derivative of the old a-tester of Emerson and Randolph (*a B Pl C R lg*), known as the N stock, and a purple embryo, purple aleurone transmitter stock, known as PEM. Both endow their hybrid progeny with seedling or embryo color phenotypes of unusual nature. Putative parthenotes lack the marker colors. Thousands of monoploids have been screened out of marked progenies by use of N and PEM markers as pollinators. A version of the PEM stock carrying the indeterminate gametophyte gene *ig* has special value. When a component of the female genotype, the gene *ig* stimulates high rates of androgenesis. Androgenetic monoploids carry the intact genome of the sperm and the cytoplasm of the egg.

Monoploid maize, as one might expect, is highly sterile. Successful self-pollination depends largely upon formation, through spontaneous doubling of the chromosome complement or colchicine induced doubling, of diploid sectors in the tassel and ears. The rate of successful establishment of homozygous diploid progeny from monoploids is of the order of ten percent of the monoploids isolated. The rate of occurrence of monoploids is of the order of one per thousand marked progeny, the frequency being influenced environmentally and by the genotypes of the two parents of a cross.

Established homozygous diploid lines are equivalent for breeding purposes to highly inbred lines. Most are strikingly uniform. A satisfactory number have proven to be elite parents and have been, and are, currently being used as parents of successful commercial maize hybrids.

Use of the monoploid method in maize breeding has not only been successful directly but has also enabled a more rapid recycling of elite parent stocks. This latter possibility has given the developer of monoploid-derived lines a great advantage over classical breeders.

Today, as plant breeders seek to develop monoploid sporophytes of maize and homozygous diploids by pollen culture, the early work done with spontaneous, parthenogenetically and androgenetically derived monoploids is again of great interest.

Toward the end of the 1980 maize pollination season, while I was still working in New York State, one of my students, Diana Lenard, and I in one afternoon planned and carried out the necessary pollinations for a very simple experiment that was not part of the scheduled work. I think this simple experiment was the most important work of that season. To give you some background, the pollen generation of maize (the microgametophyte) is ordinarily monoploid and, therefore, hemizygous in genotype. Pollen from tetraploid maize is diploid, and, if the tetraploid parent be heterozygous, the pollen also will be heterozygous. Does heterozygous diploid maize pollen exhibit heterosis? That was the question for which we sought an answer.

There is evidence from the potato, an autotetraploid, that more heterozygous male gematophytes have a competitive advantage over less heterozygous gametophytes in delivering sperm to the eggs in fertilization. We had in our maize nursery a series of tetraploid maize inbreds and hybrids that had been planted to make up seed for a trial of heterosis in maize at the tetraploid sporophytic level, as related to the level of heterozygosity of the individuals. These stocks conveniently provided us with the material to test whether heterosis is expressed in diploid pollen; that is, whether heterozygous microgametophytes are more successful in delivering sperm to eggs than are more homozygous microgametophytes. For this study we utilized three tetraploid inbred lines of maize (GGGG, KKKK and NNNN), and the single cross of the first two (GG/KK). Inbred NNNN transmits purple aleurone and plant color; the other stocks are non-purple. A series of three kinds of pollinations were made. In the first, 50% of pollen from line GGGG mixed with 50% from NNNN was used on silks of the single cross hybrid GG/KK. In the second, 50% of pollen from KKKK mixed with 50% from NNNN was used on other plants of the single cross GG/KK. In the third series, 50% of pollen from the single cross hybrid GG/KK mixed with 50% from NNNN was used to pollinate another group of plants of the single cross GG/KK. The results are summarized in Figure 1.

Heterosis was evident; the gain in performance of the pollen from the single cross over that from the inbreds was of the order of 29%. In retrospect we recognize that it would have been better to have used a female tester unrelated to GGGG, KKKK or NNNN. Such a tester might have revealed even higher levels of heterosis.

I suggest that researchers interested in perfecting pollen culture in maize might find it advantageous to obtain and culture anthers from heterozygous tetraploid plants, as the diploid microgametophytes and derived sporophytes should be more vigorous than those obtained from diploid parents.

In addition to the superior performance of the heterozygous diploid pollen, one should note that when compared to pollen from the inbreds GGGG or KKKK, pollen from NNNN was a superior competitor.

This observation has led us to further consideration of the gametophytic generation and the possibilities of doing selective breeding at the gametophytic level, *in vivo* and *in vitro*. The literature is scant, but encouraging.

The competitive relations of pollen grains growing in a single style have a long history of interest among biologists. Charles Darwin in 1876 reported: "We now come to a far more general and, therefore, more important means by which the mutual fertilization of distinct plants is effected, namely the fertilization power of pollen from another variety or individual being greater than that of a plant's own pollen. The simplest and best known case of prepotent action in pollen, though it does not bear directly on our present subject, is that of a plant's own pollen over that from a distinct species. If pollen from a distinct species be placed on the stigma of a castrated flower and then, after the interval of several hours, pollen from the same species be placed on the stigma, the effects of the former are wholly obliterated, except in some rare cases. If the two varieties are treated in the same manner, the result is analogous, though of directly opposite nature; for pollen from any other variety is often or generally prepotent over that from the same flower."

	Purple		Non-Purple	
	Number	%	Number	%
1. 50% KK + 50% NN pollen crossed to GG/KK as female:	714	53	620	47
2. 50% GG + 50% NN pollen crossed to GG/KK as female:	686	56	530	44
Average:		54.5		45.5
3. 50% (G/K) + 50% NN pollen crossed to GG/KK as female:	456	40	677	60
Difference:		14.5		14.5

Heterotic gain = 14.5 x 2 = 29%

GGGG = Maize Coop: 4n, su, 73-2351-5 (0)

KKKK = Maize Coop: 4n, lP-uu, 69-756-1 (2)

NNNN = Maize Coop: 4n, g(ACR), 68-1157-1 (2)

Figure 1.

A number of investigators have studied microgametophyte competition. The results of this work support the conclusion that competition exists among pollen tubes growing concurrently within a single style, that the more vigorous genotypes tend to effect fertilization in higher frequency than less vigorous, and that there may be a positive interaction between the genotype of the pollen and the genotype of the style that favors formation of heterotic progeny.

The observations should encourage both the plant breeder and the cell biologist to utilize these relations in order to select favorable genotypes from among segregating gametophytic individuals, whether these be microgametophytes, megagametophytes, or cells and tissues of the gametophytic generation being grown in vitro. Growth rates in defined environments, whether these be of the culture medium or of the stylar tissue, should enable directed selection among heterogeneous genotypes.

As a plant breeder, I am moving into studies I hope may enable us to use gametophyte competition to screen for gametes efficient in heterosis, in relation to a set stylar genotype. Indeed we may be able to set up a screen that will both run the race and capture the winners as monoploid sporophytes.

This line of thought leads me to urge cell biologists such as yourselves to direct more attention to the alternation of generations in plant species, and to the genetic and environmental controls of this alternation. We need to develop our technology to the level where we can, at will, induce free cells from a sporophytic source to develop into megagametophytes, or as microgametophytes; or induce cells from a microgametophyte or megagametophyte to develop into sporophytes. I recognize that the development of sporophyte from microgametophyte is the outcome of successful anther culture, but suggest that our control over this important phase shift is minimal. Also, we need that level of knowledge and technical skill to enable us, at will, to induce cells to become meiocytes and undergo meiosis. With such abilities in place, along with our present skills in inducing protoplasts to fuse and, in at least some cases, to regenerate progeny cells from fused protoplasts, we would have in hand a complete substitute for the natural sexual process.

The study of meiosis in vivo and in vitro, and of ways to induce meiosis at will in cultured cells and tissues, I consider one of the most exciting lines of research today. Our understanding of the genetic and environmental controls of meiosis is inadequate to our needs. We need a higher level of understanding of meiosis not only to enable us to reduce chromosome complements at will but also to effect integration of foreign DNA into the genetic system of higher plants.

The progressive control -- the restriction, channelization and limitation in time -- of the native propensity of DNA to cross over is one of the grand themes of evolution, from pre-biotic DNA to the most complex of present day organisms. Three decisive pre-adaptive developments, pre-adaptive in the sense that each in turn made possible a vast radiation in life forms, starting from the pre-biotic period of organic evolution, have been (1) the isolation of biologically competent nucleic acid within a self-encoded or self-binding lipo-protein system, giving rise to the protoprokaryote, (2) the further isolation of the hereditary material from the external environment and homologous extra-nuclear DNA by development of the nuclear-cytoplasmic system, giving rise to the protoeukaryote, and (3) the development of an autonomous inhibitory system enabling homologous chromosomes to co-exist within a single diploid nucleus without being subject to crossing over, giving rise to the ancestor of the higher plants and animals and, thereby, to the vast radiation of higher plant and animal species.

It is during the prophase of meiosis that crossing-over between homologous pairs of chromosomes occurs. Therefore, it is at that time in the life cycle that the higher plant nucleus is most vulnerable to genetic integration. What are the changes in the chromosomes that enable crossing-over? What is it in the somatic chromosomal system that ordinarily inhibits crossing-over?

There are many reasons for seeking a better understanding of meiosis and of the controls of that complex process. To give a special example of what might become possible if we had full control of the meiotic process -- in maize breeding populations, such as synthetic varieties, single outstanding heterozygous individuals are observed. Since maize cannot easily be propagated vegetatively, such individuals, though their gametes may be sampled, cannot be regenerated as such. If we could control meiosis so as to enable synapsis to occur without crossing-over taking place, then assure complete disjunction of chromosome pairs, and then isolate and regenerate the immediate daughter cells with their complementary nuclei and from these obtain sporophytes, we would have in hand two "inbred lines" from which the original elite heterotic plant genotype could be regenerated without limit of number.

In conclusion, it is my hope that cell biologists will direct more attention to the genetic and environmental controls of phase change in higher plant cells so that we can at will trigger plant cells to function as sporophytic or gametophytic initials. Further, I urge greater attention to the meiotic process and its control _in vitro_.

TISSUE CULTURE TECHNOLOGY FOR LONG-TERM STORAGE AND PROPAGATION OF POTATO (*SOLANUM TUBEROSUM* L.) GERMPLASMS

L.E. Heszky[1], K. Enyingi[2] and I. Szabo[3]

[1]Department of Scientific Research and Education
Ministry for Agriculture and Food
H-1860 Budapest 55 P.O.B. 1
HUNGARY

[2]Department of Potato Breeding
University of Agriculture
Keszthely
HUNGARY

[3]Potato Research Station
Kisvarda
HUNGARY

INTRODUCTION

The *in vitro* culture of potato (*Solanum tuberosum* L.) shoot meristems began nearly thirty years ago. During this period the basic techniques and culture conditions have been improved upon. They have been partially suitable for virus elimination. However, they were unsuitable for the *in vitro* maintenance, long-term storage and vegetative propagation. The induction of multiple shoot development was successful in 1976 (1), but through a callus phase and, hence, had no practical value. With regard to long-term storage satisfactory results in several species of *Solanum* have been obtained by utilization of meristem and shoot culture techniques (2). In addition cryopreservation methods have also been employed (3, 4). However, until the late seventies: 1) no method was available for increasing the multiplication rate without bringing about some genetical changes; 2) no method was available for unlimited storage without genetical changes; and 3) no culture media were utilizable for a large number of varieties and genetic resources alike without eliminating the genetic and physiological differences.

In Hungary difficulties are encountered in maintenance of germplasm and varieties under adverse environmental and pathological conditions. Hence, studies were initiated with special reference to vegetative propagation of potato varieties. Our aim was to evolve a tissue culture "technology" to be utilized on a large scale in breeding and gene-bank practices to overcome the above described difficulties.

MATERIALS AND METHODS

Six European varieties, Manna and Desire from the Netherlands, Bjeloruskij Rannij from U.S.S.R., Magyar Rozsa, Signal and Somogyi Korai from Hungary were used in the four year study. Meristem cultures were initiated from meristem tips (200-500 µ) of axillary buds (Figure 1A) on semisolid and liquid Murashige and Skoog (17) modi-

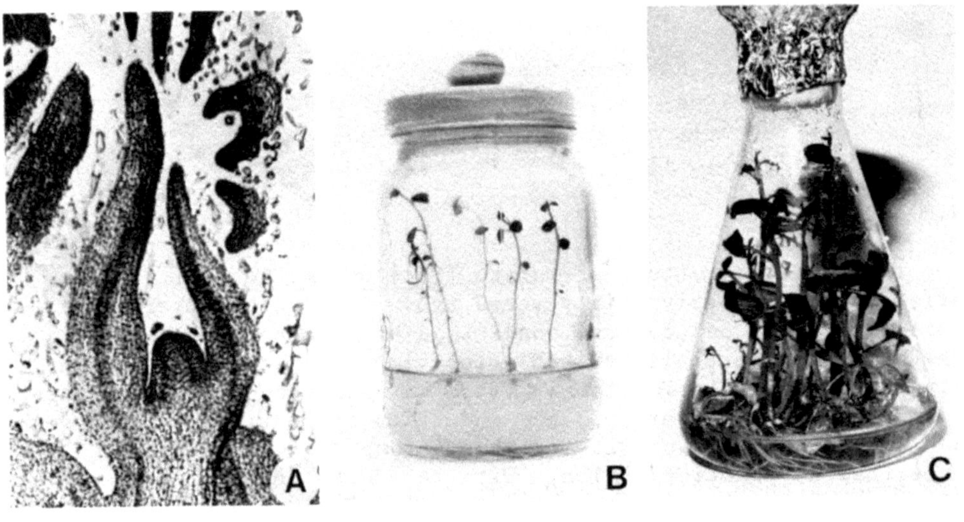

Figure 1. Some cardinal steps of long-term storage and propagation of potato in vitro.
A: meristem tip of axillary bud (marked with white line showing the place of excision);
B: shoot culture used for storage (6 weeks after the nodal segment inoculation);
C: vegetatively propagated shoot culture used for planting (4 weeks after the nodal segment inoculation).

fied MS and Morel and Muller media (16). The media had pH of 5.7. Supplements used in different concentrations and combination included gibberellic acid (GA_3) - 0.025-10 mg/l; indole acetic acid (IAA) - 0.125-10 mg/l; kinetin (Kn) - 0.01-10 mg/l; and benzylaminopurine (BAP) - 0.025-5.0 mg/l.

The liquid media used were either with or without filter paper bridges in test tubes as well as in Erlenmeyer flasks. The culture flasks were kept on a horizontal shaker (90 rpm).

The shoot cultures were raised from nodal segments of test tube grown plants and incubated in artificial illumination varying between 40-200 W/m^2 (16 hr/day) at a temperature range of 15-25°C. Chromosome counts were made of root tips by a method described earlier (5). After the storage and multiplication, the regenerated plants were transferred in the soil, covered for 2-5 days and maintained at 15-25°C air and 12-15°C soil temperatures. The percent relative humidity in the greenhouse was 50-80. Three weeks after planting, two thousand plantlets were transplanted into the field for comparative trials using four treatments and four replicates.

RESULTS AND DISCUSSION

With the aim of preventing gene-erosion of genetic resources and solving problems of disease free maintenance and rapid propagation of breeding material and new varieties, the following techniques were developed during the four year (1977-1980) experiment: 1) tissue culture technology for the long-term storage of germplasm (Figure 2) and 2) tissue culture technology for maintenance and vegetative propagation of breeding materials (Figure 3).

The results published in this field during the last four years (6 - 16), support our work on development and refinement of "industrialized technology" for gene banks and breeders.

Culture Initiation (Figs. 2/1-5; 3/1-6)

The cultures are initiated by the isolation of meristem tips of axillary buds from branches in the rapidly growing plants in vegetative phase and implanting them on the modified MS semisolid culture medium in test tubes. The modified MS (mMS) medium contains no glycine but concentrations of niacin and pyridoxineHCl were doubled (1 mg/l) in addition to the amount of thiamine being present hundred fold (10 mg). Different varieties responded similarly in this culture medium. The mMS could be supplemented on the one hand with GA_3 (0.1-0.5 mg/l) and on the other hand with BAP (0.2-0.5 mg/l) depending upon the need (Figs. 2, 3/3-5). On the GA_3 containing medium, the shoot development is achieved without dedifferentiation in contrast to BAP supplementation. The rooting does not play any role in

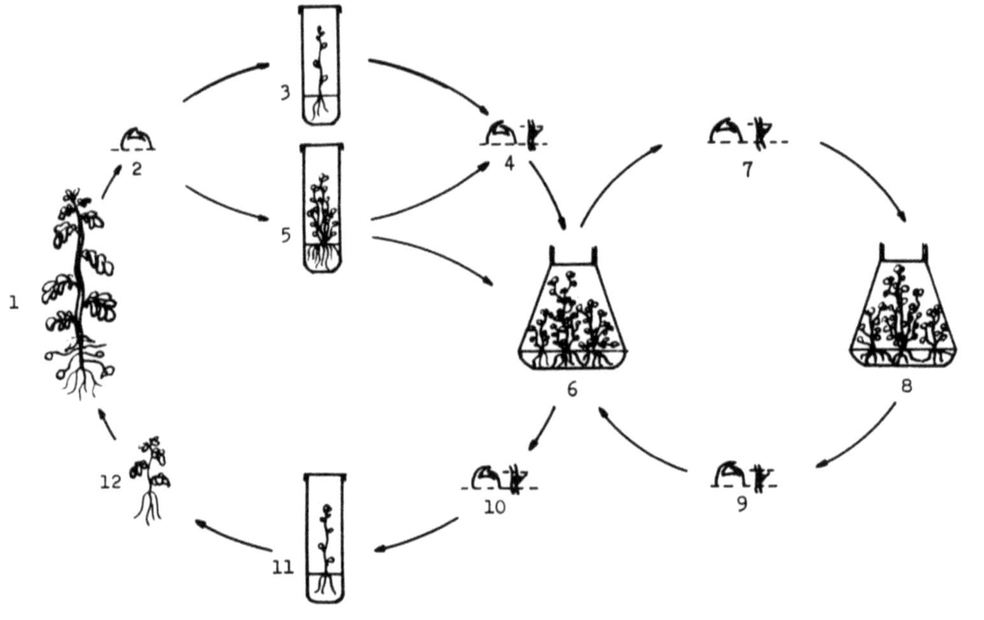

Figure 2. Tissue culture method for long-term storage of potato (Solanum tuberosum L.) germplasms.
1-5: culture initiation;
6-9: long-term storage;
10-11: exchange and transfer;
11-12: planting.

the initiation phase but rapid growth of roots can be achieved by GA_3, IAA and Kn supplementation as indicated in section under maintenance and long-term storage (see below). The cultures are incubated in artificial illumination (200 W/m^2), 16 hr/day at 25°C. The culture initiation can be combined with the methods of virus and disease elimination.

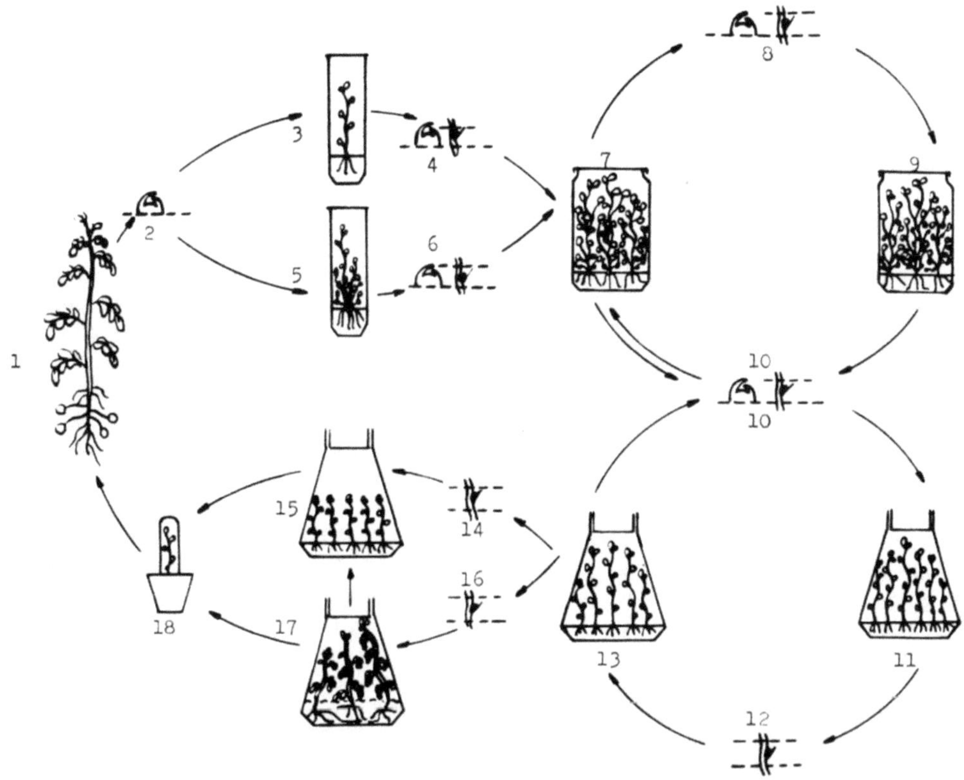

Figure 3. Tissue culture method for maintenance and vegetative propagation of potato (Solanum tuberosum L.) breeding materials and varieties.
1-6: culture initiation;
7-10: maintenance;
11-13: vegetative propagation;
14-18: planting.

Maintenance (Figure 3/7-10)

The nodal segments of test tube developed shoots on the GA_3 containing medium during culture initiation are inoculated on the semisolid mMS in special culture vessels shown in Figure 1B. For reducing the growth rate of shoots and roots the mMS medium is supplemented with GA_3 (0.1 mg/1), IAA (0.5 mg/1) and Kn (0.09 mg/1) and the artificial illumination is maintained at 80 W/m^2 (16 hr/day) at 20°C. Subcultures are carried out by nodal segments of shoots developed in the maintenance cultures, three times a year. Inoculation of 5-10 nodal segments into a vessel is recommended depending on the volume (150 ml, 350 ml or 600 ml).

Three times inoculation is needed annually for keeping the physiological status and the relative quick development of cultures to achieve rapid propagation.

Long-Term Storage (Fig. 2/6-9)

For this purpose the nodal segments of test tube grown shoots developed on the GA_3 containing medium during culture initiation are inoculated on the semisolid mMS medium. For eliminating the differences between varieties, for promotion of root differentiation and for reducing the growing rate of developing shoots and roots, the culture medium is supplemented with GA_3 (0.1 mg/l), IAA (0.5 mg/l) and Kn (0.04 mg/l) in the first year of storage. During the subsequent years there is no need for this change. The culture vessels for storage are the ones depicted in Figure 1B, perforated on the top, having capacity of 150-300 or 600 ml. The number of isolated nodal segments used as explant, generally 1-5 per culture vessel (Figure 1B). Inoculation is carried out once or twice a year. The cultures are maintained in the artificial illumination (80 W/m^2, 16 hr/day) at 15°C.

To enhance safety in the preservation of germplasm, the inoculation of explants from one sample of germplasm was carried out in three culture vessels prepared at time interval of 3-6 months. Thus, the six culture vessels altogether mean that 6-10 clones/varieties or different genotypes can be stored within one square meter. Besides, the difference in age of cultures, 3 or 6 months, has further advantage such as the availability of a large number of nodal segments for propagation or international exchange at any time.

Exchange and Transfer (Fig. 2/10-11)

For this purpose, nodal segments of shoots developed in long-term stored cultures are isolated on the solid mMS medium in special culture tubes. In these tubes within two weeks small plantlets will develop having shoots and roots. The transfer procedure is as recommended (9).

Vegetative Propagation (Fig. 3/10-13)

The vegetative propagation begins with the inoculation of nodal segments of shoots developed in cultures of maintenance or long-term storage on to mMS medium. The number of nodal segments inoculated vary from 5 to 10 on to medium in Erlenmeyer flasks of 100-150 ml capacity. For increasing the growth rate of shoots, the culture medium is supplemented with GA_3 (0.1 to 0.5 mg/l) and the cultures maintained in artificial illumination (200 W/m^2; 16 hr/day) at 25°C. The shoots that developed from the nodal segments had by four weeks 8-12 leaves and axillary buds. The vegetative propagation is carried out by cutting the shoots developed in cultures and by inoculating the nodal segments every four weeks as long as the required number is achieved.

Within four weeks ten times more multiplication can be achieved. Theoretically this means that a hundred billion plants can be produced from one nodal segment by this procedure within one year.

Planting (Figs. 2/11-12; 3/14-18)

The preparation for planting takes place by the change of culture medium and technique in the last multiplication cycle. The nodal segments are inoculated on the solid mMS medium without supplements or on liquid mMS medium with GA_3 (0.1 mg/l). The number of explants per vessel varied form 5 to 25 depending on the capacity of the culture vessel (100 or 250 or 500 ml). The conditions of incubation are similar to those used under vegetative propagation section.

Plants developing in liquid cultures on horizontal shakers are suitable for direct transplantation into soil in greenhouse. The plants developing on semisolid medium are first transplanted into the pots having sterilized soil or non-sterilized soil or 2:1 mixture of sand and manure. The high percent relative humidity necessary is provided by different covering methods during the immediate days after planting. The survival rate by this method varied from 95 to 100%.

Development (Table 1)

After three year long storage and two month vegetative propagation, two thousand plants of variety Desire were regenerated and planted into the field for trials.

The growth and development of plantlets were satisfactory. They all had a dominant shoot and the length of the stem axis was 120-150 cm. There were a number of laterals. The average green mass produced in the experiment was 5.1 kg/m^2. The plants developed rapidly and were to be harvested two weeks earlier (105 days in contrast to 120 days for Desire variety). Plants had 10-12 tubers generally, and the average yield of productivity was about 4.0 kg/m^2. There was no chromosomal variation and also the plant morphology was the same.

Based upon the results of the comparative field trials, the horizontal manner of planting and 40 x 20 cm area of cultivation can

Table 1. Field Trial Data of Two Thousand Plants Derived from Three Year Long Stored and Then Multiplied Cultures.

	Area of Cultivation			
	40 x 40 cm		40 x 20 cm	
	Manner of Planting			
Parameters Examined	Vertical	Horizontal	Vertical	Horizontal
# of tubers per plant	13.5	13.9	10.3	11.5
# of tubers per m^2	77.0	85.9	107.7	112.6
Tuber yield/plant (Kg)	0.64	0.68	0.38	0.42
Tuber yield/m^2	3.7	4.2	3.9	4.2
Avg. tuber weight (g)	48.0	49.0	36.0	37.0
Green mass to tuber ratio	1.28	1.33	1.24	1.23
Tuber yield (1000 Kg/ha)	36.7	41.8	39.4	41.7

be recommended for the further propagation of test tube plants.

CONCLUSIONS

The present study reveals a manner of overcoming the problems mentioned in the INTRODUCTION. The main features of this technology are: 1) unlimited terms of storage; 2) unlimited propagation potential; 3) probability of genetic changes remains at the level of bud mutation; 4) variety independent; 5) very high survival rate; 6) no marked changes in growth, development and productivity; and 7) the procedures are relatively simple, reliable and safe.

This technology has been initiated partially in gene-bank and partially in breeding practice of potato in Hungary.

REFERENCES

1. S. Roest and G.S. Bokelmann, Vegetative propagation of Solanum tuberosum L. in vitro, Potato Res. 19:173-178 (1976).
2. R. Westcott, G.G. Henshaw and W.M. Roca, Tissue culture storage of potato germplasm: culture initiation and plant regeneration, Plant Sci. Lett. 9:309-315 (1977).
3. Y.P.S. Bajaj, Technology and prospects of cryopreservation of germplasm, Euphytica 28:267-285 (1979).
4. Y.P.S. Bajaj, Regeneration of plants from potato meristems freeze-preserved for 24 months, Euphytica 30:141-146 (1981).
5. M.S. Swaminathan and H.W. Howard, Cytology and genetics of potato (Solanum tuberosum L.) and related species, Bibliographia genetica AG:A-192 (1953).
6. I.R. Denton, R.J. Westcott and B.V. Ford-Lloyd, Phenotypic variation of Solanum tuberosum L. cv. Dr. McIntosh regenerated directly from shoot-tip culture, Potato Res. 20:131-137 (1977).
7. F.C. Mellor and . Stace-Smith, Virus-free potatoes by tissue culture, in: "Applied and Fundamental Aspects of Plant Cell, Tissue and Organ Culture," J. Reinert and Y.P.S Bajaj, eds., Springer-Verlag, Heidelberg-Berlin-New York (1977).
8. W.M. Roca, N.O. Espinoza, M.R. Roca and J .E. Bryan, A tissue culture method for the rapid propagation of potatoes, American Potato J. 55:691-706 (1978).
9. W.M. Roca, J.E. Bryan and M.R. Roca, Tissue culture for the international transfer of potato genetic resources, American Potato J. 56:1-11 (1979).
10. P.B. Goodwin, Y.C. Kim and T. Adisarwanto, Propagation of potato by shoot-tip culture. I. Shoot multiplication, Potato Res. 23:9-18 (1980a).
11. P.B. Goodwin, Y.C. Kim and T. Adisarwanto, Propagation of potato by shoot-tip culture. II. Rooting of proliferated shoots, Potato Res. 23:19-24 (1980b).
12. P.B. Goodwin and T. Adisarwanto, Propagation of potato by shoot-tip culture in petri dishes, Potato Res. 23:445-448 (1980c).
13. P B. Goodwin and G. Brown, Field performance of potato shoot-tips proliferated in culture, Potato Res. 23:449-452 (1980d).
14. S. Roest and G.S. Bokelmann, In vitro adventitious bud techniques for vegetative propagation and mutation breeding of potato (Solanum tuberosum L.). I. Vegetative propagation in vitro through adventitious shoot formation, Potato Res. 23:167-181 (1980).
15. F.J. Novak, J. Zadina, V. Horackova and J. Maskova, The effect of growth regulators on meristem tip development and in vitro multiplication of Solanum tuberosum L. plants, Potato Res. 23:155-166 (1980).

16. K.C. Tao, W.T. Yin and H.Y. Cheng, Meristem culture of potatoes and production of virus-free seed potatoes, in: "Plant Tissue Culture," Proc. Symp. on Plant Tissue Culture, Science Press, Peking (1978).
17. G. Morel and J.F. Muller, La culture in vitro du meristeme apical de la pomme de terre, Comptes Rendus 258:5250-5252 (1964).
18. T. Murashige and F. Skoog, A revised medium for rapid growth and bioassays with tobacco tissue cultures, Physiol. Plantarum 15:473-497 (1962).

CRYOPRESERVATION AND INTERNATIONAL EXCHANGE OF GERMPLASM

Y.P.S. Bajaj

Punjab Agricultural University
Ludhiana
Punjab,
INDIA

INTRODUCTION

The rapid development and the increasing pressures of population have led to the indiscriminate clearing of the forest and agricultural land. The natural ecosystem is being interferred with and many plant species are thus threatened with extinction. This has resulted in the depletion of the naturally occurring pools of germplasm of the medicinal, forest and crop plants. Moreover, although green revolution has helped to increase yields, it does not seem to be green any more, since the introduction of some of the exotic germplasm has resulted in the disappearance of the local genetic stocks. This shrinkage of the genetic resources has caused global concern for the conservation of rare germplasms.

Storage of seeds is the traditional method for the preservation of germplasm. However, the seeds of a number of tree species (Table 1) are recalcitrants, i.e., they are sensitive to humidity and temperature and, thus, cannot be stored on a long-term basis (1).

Moreover, for the conservation of germplasm of some of the vegetatively propagated crops, like potato and cassava, the clones have to be maintained through annual propagation in the nurseries. This involves the hazards of pests and pathogens, and instances are not unknown where entire germplasm is lost. At present no satisfactory method is available by which the germplasm of such plants can be conserved on a long-term basis.

International efforts are, therefore, being made to develop methods other than traditional ones for the storage, maintenance,

Table 1. List of Some of the Important Orthodox and Recalcitrant Seeds.

Orthodox seeds	Recalcitrant seeds	
Cereals	Oilpalm Coffee Coco Rubber	Juglans Castanea Taxus Quercus
Grain-Legumes	Coconut	Aesculus

conservation and exchange of germplasm and to explore the possibilities of the creation of "Germplasm Banks" of the rare plant materials. In this regard, tissue culture combined with the technology of cryopreservation are being looked upon as complementary aids to the existing methods (2 - 6).

STORAGE OF CELL AND TISSUE CULTURE

Tissue cultures are generally stored at 23°C-25°C, but for long-term preservation one of the main drawbacks at this temperature is the need for periodic transfer to fresh medium. This can to some extent be delayed by the suppression of growth. For this purpose various methods have been adopted (Table 2). Although growth in the callus cultures can be retarded, the possibility of genetic erosions is not ruled out. The cells undergo endomitosis, mutations, chromosomal aberration, aneuploidy, etc., and, thus, the clones cannot be maintained. This difficulty to some extent can be circumvented by the storage of meristem cultures at lower temperature (5°C-15°C). This has proved to be quite efficient, and the cultures need not be transferred up to 2 years or even more.

Thus, for long-term storage and the maintenance of germplasm of clones, meristems are preferred over cell and callus cultures for a variety of reasons. The meristems are genetically more stable, are easy to regenerate as plants, and yield disease-free stocks.

The meristem tips of _Vitis vinifera_ stored at 9°C with only one transfer a year regenerated plants (7, 8). Similarly, meristem-derived plantlets of 50 cultivars of strawberry were maintained for up to 6 years in cultures at 4°C in darkness (9). The meristem tips of _Lolium multiflorum_ maintained for about 3 years at 2°C-4°C have regenerated into plants (10). In our laboratory capitulum-derived plantlets of _Chrysanthemum morifolium_ maintained in a refrigerator

Table 2. Some *in vitro* Methods for the Long-term Storage of Germplasm.

Growth Suppression	1. Culture on minimal medium
	2. Storage at low temperature
	3. Use of a layer of mineral oil
	4. Partial dehydration of the tissues
	5. Maintenance under low atmospheric pressure
Cryopreservation	1. Cultures of cell, tissue and organs
	2. Plantlets

(4°C-5°C) for 3 1/2 years have flowered on transfer to the soil (Bajaj, unpublished).

Thus, the primary purpose of the suppression of growth at low temperature can be achieved in many plant species, and this delays the time for subculture and helps to store the cultures without transfer to fresh media.

PROGRESS OF WORK ON THE CRYOPRESERVATION OF GERMPLASM

The freeze-preservation of plant cell, tissue and organ cultures is a rather recent development but with abundant potentials (Figure 1) and far-reaching implications for the conservation and international exchange of germplasm. The principle underlying the freeze-preservation basically involves bringing the cells to a state of inactive metabolism or "suspended animation" by subjecting the cultures to super-low temperatures in the presence of cryoprotectants. During the last decade it has been established that entire plants can be regenerated from plant cells (Figure 2), embryos and endosperm (11), anthers and pollen embryos (Tables 3 and 4) and excised meristem (Tables 5 - 7) freeze-preserved in liquid nitrogen for various lengths of time. This has been demonstrated in a number of plant species (Table 8, Figures 1 - 7), some of which include tobacco, carrot, petunia, potato, cassava, strawberry, rice, sugarcane, poplar, date plam, chickpea, pea and peanut. However, the main problem at present seems to be the low percentage of viability,

Table 3. Growth Response of Transversely Cut Halves of 4 Week Old Cultured Anthers of *Nicotiana tabacum* Subjected to Liquid Nitrogen and Recultured in Agitated Liquid Medium (39).

Treatment	No. of anthers cultured	No. of growing anthers	% Survival	Total number of plantlets	No. of plants/ anther
Control (untreated)	45	30	66.6	623	20.7
Frozen at the rate of 20/min to -196°C	60	4	6.6	10	2.5

Table 4. Effect of Various Concentrations of DMSO on Viability of Pollen-Embryos Cooled at the Rate of 20°C/Min (38).

DMSO Concentration %	Viable Pollen-Embryo %
0	2
2	15
5	31
7	36
10	24

Table 5. Survival of Meristems from Tuber Sprouts, Axillary Buds and Shoot Tips of Three Cultivars of Potato Subjected to Sudden Freezing and Stored in Liquid Nitrogen for 31 Days (15).

	cv. Kufri Chandramukhi		cv. Kufri Sindhuri		cv. Kufri Alankar
Cryoprotectant	tuber sprout (3-5 mm)	axillary bud (1-2 mm)	tuber sprout 3-5 mm)	axillary bud (1-2 mm)	shoot tip (0.8-1.5 mm)
DMSO 10%	11	19	12	21	19
DMSO 5% + glycerol 5% + sucrose 5%	13	23	14	26	27

Table 6. The Effect of Freezing (Temperature -196°C; Cryoprotectant Glycerol 10% + Sucrose 5%; Thawing at 35-40°C) on the Excised Meristems of Cassava (16).

Treatment	No. of cultures raised	No. of cultures growing	% Survival*	Survival % of control
Control	103	84	84.5	-
Frozen	76	16	21.0	25

* 13% plantlets; 8% callusing and rooting.

Table 7. Effect of Various Thawing Temperatures on the Survival of Unfrozen and Frozen Excised Meristems of *Cicer arietinum*.

Thawing temperature °C	Material	No. of meristems cultured	No. of meristems showing growth	% showing growth	% of control
25	Unfrozen	29	23	80	-
	Frozen	32	10	31	39
35	Unfrozen	21	16	76	-
	Frozen	23	9	39	51
40	Unfrozen	24	17	70	-
	Frozen	26	9	34	48

For control purposes, before culture the unfrozen meristems were treated with sterile distilled water at the appropriate temperature (25, 35, 40°C).

Table 8. Summary of the Work Done on the Cryopreservation of Plant Cell, Tissue and Organ Culture.

Plant Species	Culture	Results	References
1. Acer pseudoplatanus	Cell suspension	20-30% survival	33, 34, 35
2. A. saccharum	Cell suspension	20% survival	36
3. Arachis hypogaea	Meristems	23-31% survival	20
4. Atropa belladonna	Cell suspension	40% survival	37
	Pollen-embryos	31% survival	2, 38
	Anthers	Occasional callus	39, 40
5. Bromus inermis	Cell suspension	38% survival	41
	Protoplasts	68% survival	
6. Capsicum annuum	Cell suspension	1-25% survival	42
7. Cicer arietinum	Meristems	27-36% survival	20
8. Chrysanthemum morifolium	Callus	8% survival	43
9. Datura stramonium	Cell suspension	40% survival	44
10. Daucus carota	Cell suspension	up to 80% survival	37, 41, 44, 45, 46
	Somatic embryos	Plants	35, 44
	Protoplasts	68-90% survival	41, 42
11. Dianthus caryophyllus	Meristem	Up to 100% survival and plants	47, 48, 49

Plant species	Culture	Results	References
12. *Fragaria* x *ananassa*	Meristem	Plants	17, 50
13. *Glycine max*	Cell suspension	52% survival	44
14. *Gossypium hirsutum*	Anther- and ovule-derived callus	Survival	51
G. arboreum			
15. *Happlopappus ravenii*	Cell suspension	Three times growth	52
16. *Hyoscyamus muticus*	Cell suspension	30% survival and plantlets	53
17. *Ipomoea sp.*	Cell suspension	Growth at high sucrose level	45
18. *Linum usitatissimum*	Cell suspension	14% survival	54
19. *Lycopersicon esculantum*	Seedlings, meristems	Callus, shoots	19
20. *Manihot utilissima*	Meristems	21% shoots, plants	16
21. *M. esculentum*	Meristems	3-8% survival	55
22. *Nicotiana tabacum*	Diploid cell suspension	20% survival 60 % survival	53, 56
	Haploid cell suspension	11% survival	44, 57
	Pollen-embryos	Up to 36% survival	38
	Androgenic anthers	6.6% survival	39

(continued)

Table 8 (continued)

Plant Species	Culture	Results	References
23. *Oryza sativa*	Cell suspension	60-65% survival	58
	Androgenic anthers	Callus, plants	59
	Endosperm	Callus, plants	11
24. *Petunia hybrida*	Callusing anthers	5% callus further	39
25. *Phoenix dactylifera*	Embryogenic callus	Plants	60
26. *Pisum sativum*	Meristem	60% survival	18
27. *Populus euramericana*	Callus	Cold acclimation	61
28. *Primula obconica*	Androgenic anthers	Plants	62
29. *Prunus cerasus*	Callus	Increased hardiness	22
30. *Saccharum officinarum*	Cell suspension	60% survival	63, 64
31. *Solanum tuberosum*	Meristems	Plants	13, 14, 15, 55
32. *S. etuberosum*	Meristems	Plants	65
33. *S. goniocalyx*	Meristems	60% survival	55, 66
34. *Sorghum bicolor*	Cell suspension	25% survival	53
35. *Triticum aestivum*	Cell suspension	Survival	67
36. *Zea mays*	Cell suspension	75% survival	53

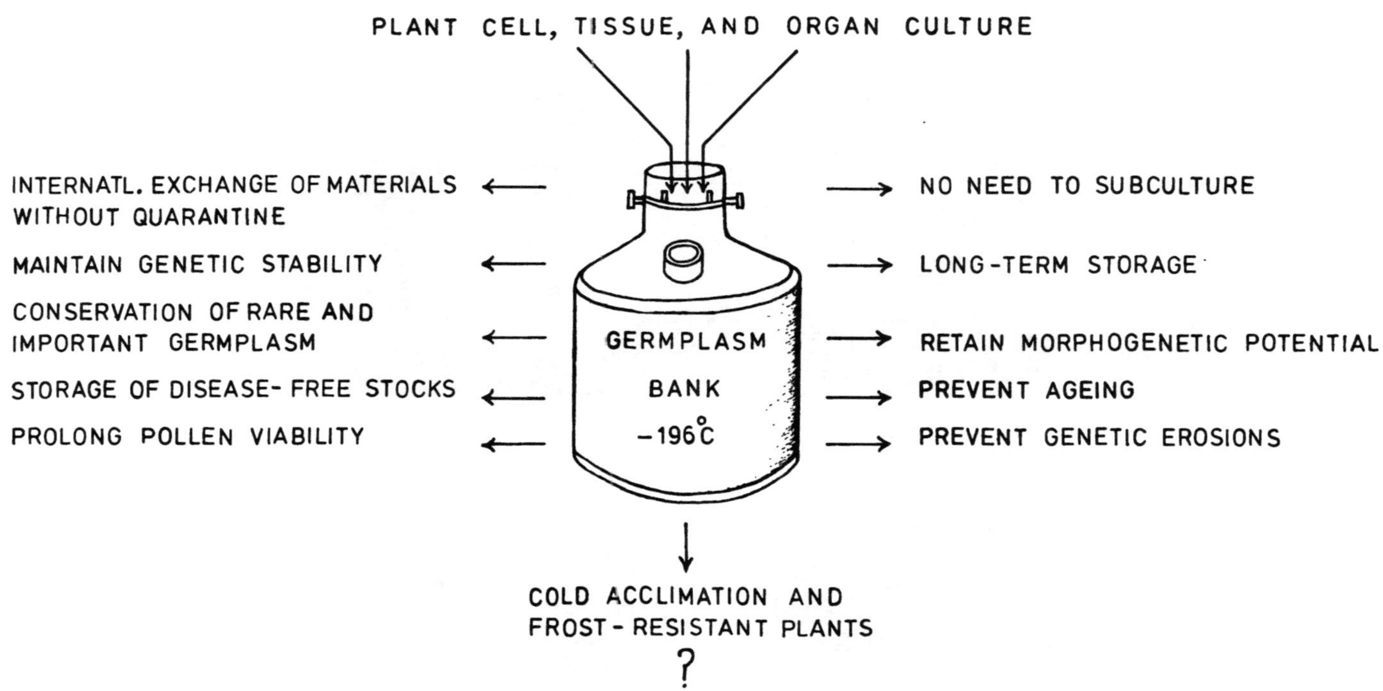

Figure 1. Prospects of the freeze-preservation of plant cell, tissue and organ culture, and the possibility of the conservation of germplasm.

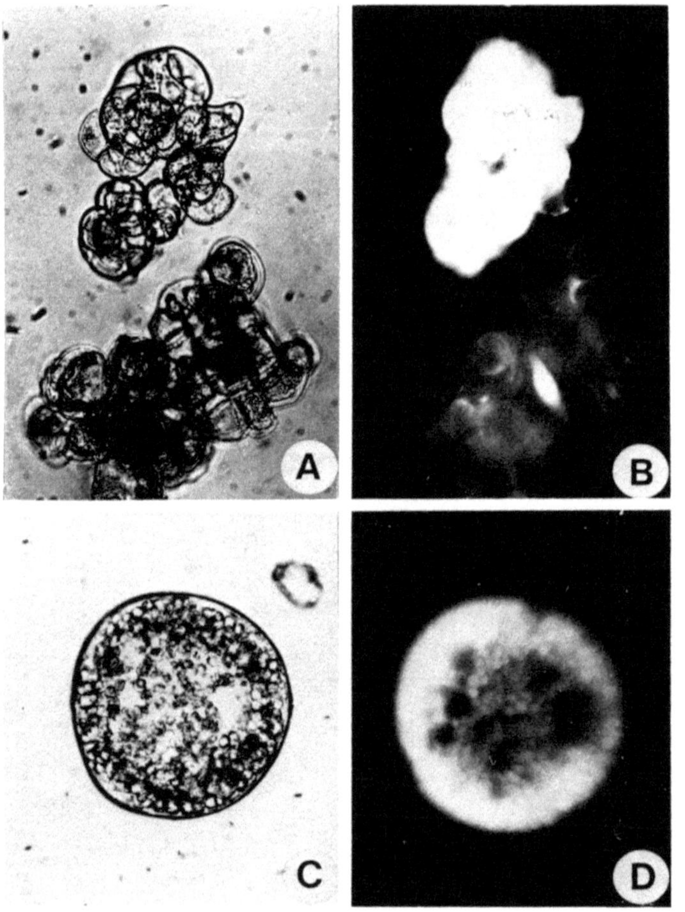

Figure 2 A-D. Survival of callus cell suspensions and protoplasts at -196°C. Figs. A, B: Carrot cell suspension under ordinary tungsten light, and under ultraviolet light respectively; note the complete survival (fluorescence) of a clump of cells. Figs. C, D: Wheat protoplasts under ordinary tungsten and ultraviolet light respectively.

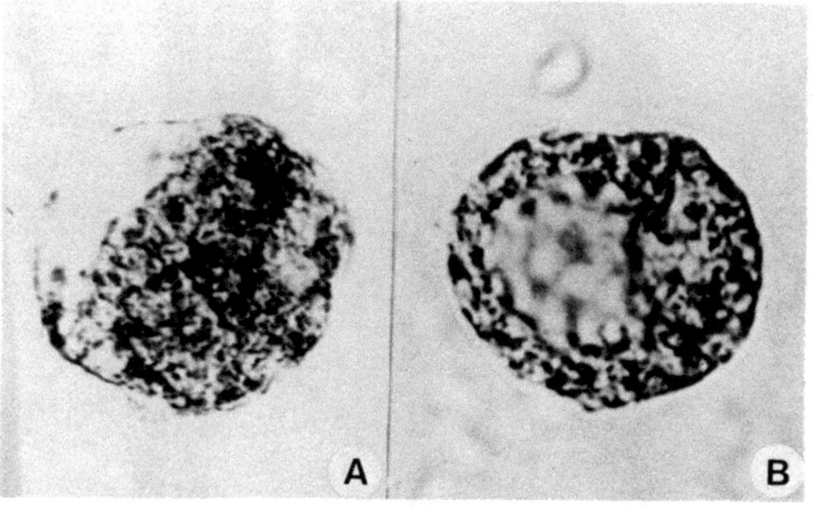

Figure 3 A, B. Survival of somatic hybrid cells (*Triticum aestivum* + *Pisum sativum* = *Tritisum*) subjected to freezing in liquid nitrogen. *Fig. A*: A fused product immediately after thawing at 35°C; *Fig. B*: Same, 21 days after culture in a microdrop; note the mixing of the protoplasts and the healthy nature of the retrieved somatic hybrid cell.

Figure 4 A-H. Formation of tubers in potato plants regenerated from meristems freeze-preserved in liquid nitrogen for various lengths of time. Fig. A: Potato plant as a source of material. Fig. B: The vials containing precultured meristems are maintained in ice during treatment with the cryoprotectant. Fig. C: A container employed for immersing the vials in liquid nitrogen. Fig. D: A liquid nitrogen cylinder (VEB, German Democratic Republic) employed for storage. Fig. E: Induction of growth in a retrieved culture (a meristem precultured for one week followed by freezing). Figs. F-H: Plantlets obtained from meristems freeze-preserved for 14 months; these cultures were maintained in diffuse light.

Figure 4 I-K. Fig. I: A retrieved culture grown under high intensity light. Fig. J: Transfer of a plant from test-tube to a pot. Fig. K: Tuber formation in a plant obtained from a meristem freeze-preserved for seven months (15).

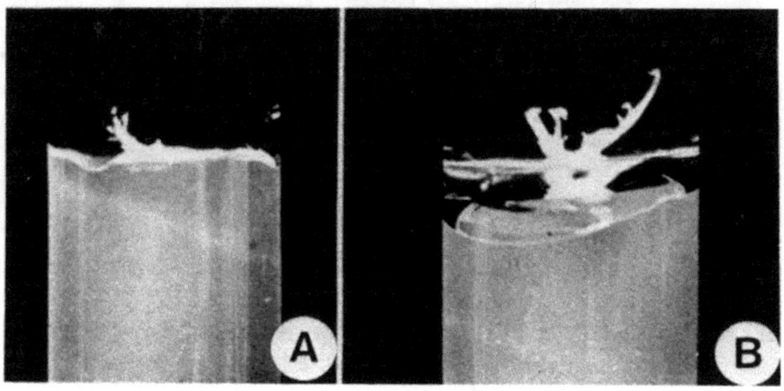

Figure 5 A, B. Production of normal seeds in chickpea plants regenerated from freeze-preserved meristems. Fig. A: A retrieved meristem five weeks after culture on MS + IAA (2 mg/l) + BAP (0.2 mg/l). Fig. B: Same, showing multiple shoot formation.

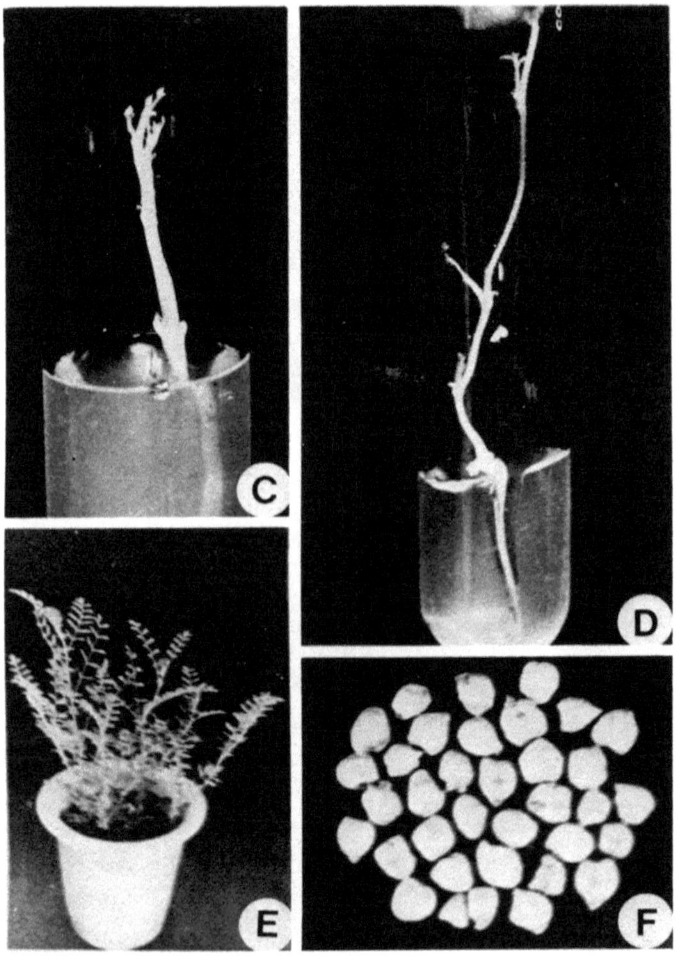

Figure 5 C-F. Figs. C,D: 10 and 12 week old cultures regenerating into plants. Fig. E: Plant regenerated from freeze-preserved meristem, and transferred to soil. Fig. F: Normal chickpea seeds obtained from such a plant. Total time from culture of retrieved meristem to seed production is 21 weeks (20).

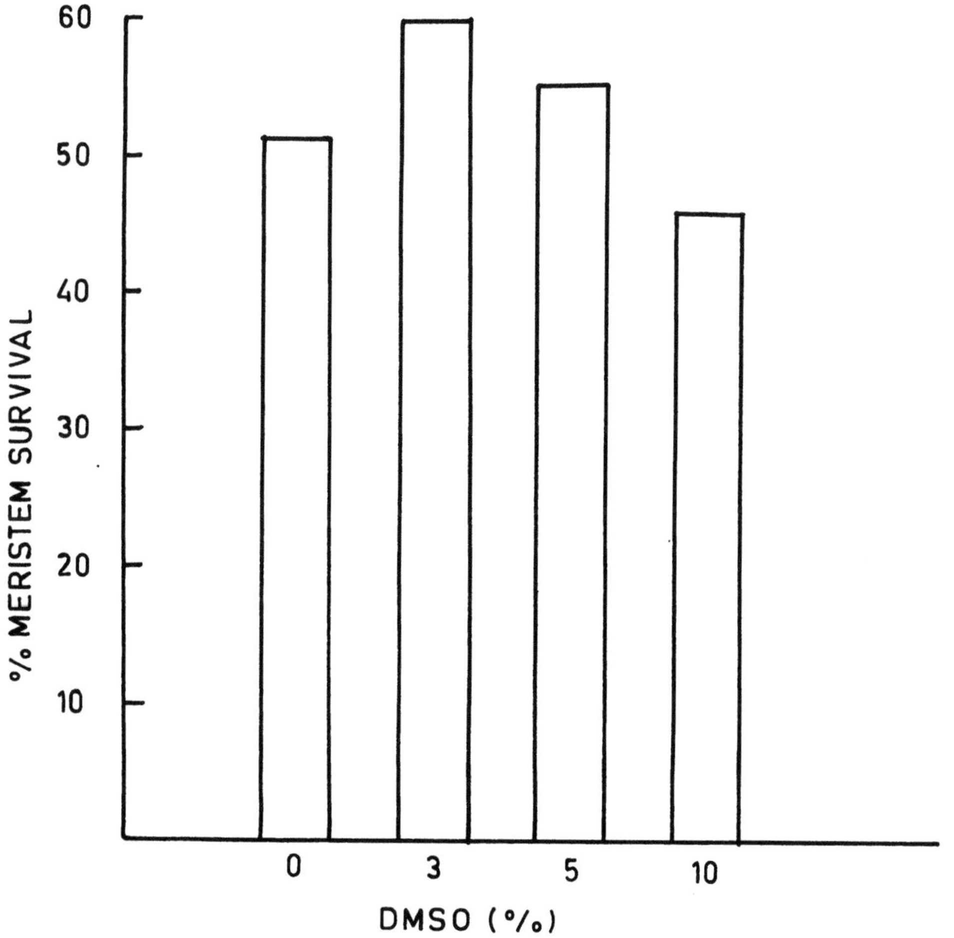

Figure 6. Effect of DMSO in the (pre-) culture medium on survival of meristems of *Cicer arietinum* frozen in liquid nitrogen.

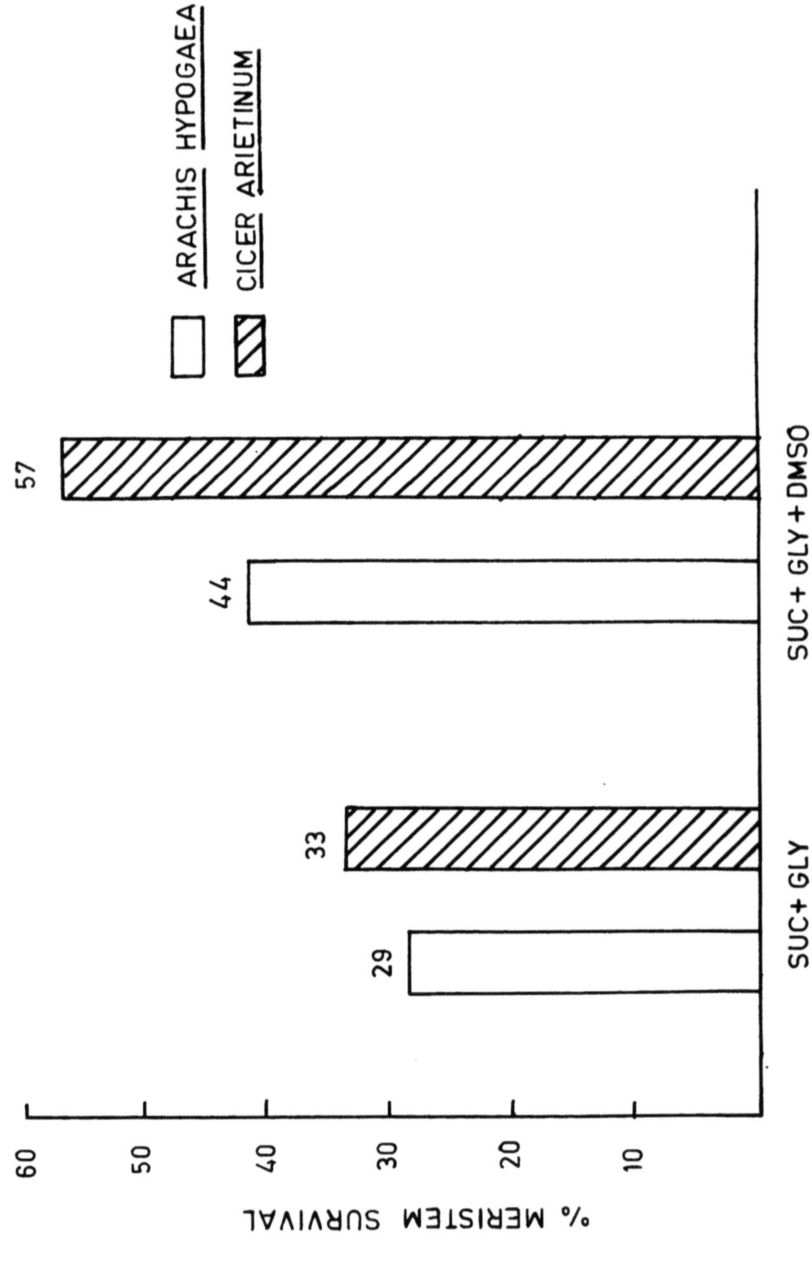

Figure 7. Effect of combination of cryoprotectants on survival of meristems of *Arachis* and *Cicer* frozen in liquid nitrogen.

and for that purpose basic studies need to be carried out on the selection of the material, effect of various cryoprotectants, method and rate of cooling and thawing.

General experience has been that age, nature and physiological state of the cells is of utmost importance. The actively growing and periodically transferred, highly cytoplasmic, thin-walled, non-vacuolated small cells and aggregates of cells are able to withstand freezing, and give high viability. This would also avoid selection pressure.

Before freezing the cultures may be pregrown for a couple of days on a medium containing low concentrations of the cryoprotectant (for instance, 50% DMSO).

A mixture of various cryoprotectants at low concentrations, rather than a single cryoprotectant at high concentration yield better results (Figure 7).

The results on the method of freezing are rather contradictory. Cell survival has been obtained in cultures subjected to sudden freezing, pre-frozen material, as well as those frozen at a slow and regulated rate of cooling. However, efforts should be made to concentrate studies on the regulated rate of cooling in order to reproduce reuslts.

Sudden and brief thawing at 35°C-40°C has proved better than the cultures thawed at room temperature.

The cultures stored at -20°C or -70°C have been observed to undergo deterioration with the passage of time. Thus, for the long-term preservation, the materials should be maintained in liquid nitrogen.

Any results on cell survival based on the staining methods alone can be misleading; thus, resumption of growth should be the criterion for determining viability.

Although no cytological studies have so far been conducted on the frozen plant cell cultures, no changes are presumed to occur if the technique is not faulty. These conclusions are made from the already available information on microorganisms (12) and the animal cells (Registry 1964) freeze-preserved in liquid nitrogen. However, extensive work needs to be carried out on the cytology of the freeze-preserved material to determine if any changes occur during storage.

Freeze-preservation of Meristems

From the practical point of view, meristem cultures are the ideal material for the cryopreservation as well as for the international exchange of germplasm. They are genetically more stable and relatively easy to grow. Plants have been regenrated from freeze-preserved meristems of crops like potato (13 - 15), cassava (16), strawbery (17), pea (18), tomato (19), peanut and chickpea (20). The storage period varied up to 2 years (15), and the plants obtained from such retrieved meristems of potato and chickpea underwent normal tuberization and seed formation respectively (Figures 4 and 5).

Induction of Cold Hardiness and Frost Resistance

In addition to many potentials of the cryopreservation of plant materials, it is pertinent to emphasize that there is the possibility of the induction of hardiness through cold acclimation of cell and tissue culture, and subsequently the selection of "frost-resistant plants". The work on these lines has already yielded some encouraging results. The conditions that favor hardening of the tissues are the ones that retard growth. It is observed that gardenia and cabbage leaves (21) into which sugar was introduced showed a rise in freezing resistance. Likewise, callus cells of *Prunus* (22) have been observed to show increase in hardiness when grown on a high sugar media (10-12%). These cultures exhibited cold resistance when gradually hardened at 2°C and subsequently subjected to sub-freezing temperature of -30°C. The callus of *Chyrsanthemum morifolium* acclimated for 6 weeks at 4.5°C showed complete survival at -16°C, whereas an unacclimated callus mass could withstand freezing to 6.6°C only (23, 24). Similarly, cell suspension of *Nicotiana sylvestris* and *Capsicum annuum* (25) acclimated at -3°C for 21 days showed enhanced chilling resistance. Recently Chen et al. (26) observed the hardening of callus of *Solanum acule* to -9°C after 15 days at 3°C. It is pertinent to point out that cultivated potato (*Solanum tuberosum*) possesses little or no frost tolerance, and normally cannot survive at 3°C. However, by use of ABA frost hardiness is increased in the leaf tissue (26); thus, in the near future there exists the possibility of obtaining frost-resistant potato and other crops through tissue cultures.

"Germplasm Banks" and the International Exchange of Germplasm

The germplasm is traditionally exchanged through the seeds, spores, whole plant or tis cuttings, and underground stems, i.e., bulb, rhizome, corm, etc. Although *in vitro* techniques have been employed for a variety of purposes, both basic and applied (27), exchange of plant tissue cultures (especially of the meristem cultures) holds a great promise for the national and international dissemination of germplasm. The germplasm of disease-free potato and cassava stocks (28, 29) have been successfully transferred through meristem cultures and transplanted in far off places. This has become an accepted method for the international exchange, without having to undergo quarantine. Germplasm of a large number of plant species can be packed in a small box and transported to international locations. For this purpose disposable plastic tubes should be preferable to glass test tubes, as in the latter one runs the risk of breakage and, thus, contamination.

For long-term storage of those germplasms which are rare, but are not expected to be needed in the near future, cryopreservation is perhaps the answer. Such cryopreserved plant material can be stored in small liquid nitrogen Dewers, and then can be flown from one country to the other. This would be just like the exchange of frozen semen.

It is highly desirable that this approach for the conservation as well as for the international exchange of germplasm whould be seriously considered, and "germplasm banks" of tissue cultures (3, 4, 30 - 32, Figure 8) should be established at a few selected laboratories. Such banks would mostly be crop oriented and would cater to the needs for the desired germplasm. International institutes like the Potato Research Institute (Peru), CIAT (Colombia), IRRI (Los Banos), CIMMYT (Mexico), IPRI (California), etc. could be entrusted with the job of storage, maintenance, distribution and exchange of disease-free stocks of their respective crops.

REFERENCES

1. M.W. King and E.H. Roberts, Storage of recalcitrant seeds - achievements and possible approaches, IBPGR, Rome (1979).
2. Y P.S. Bajaj, Gene preservation through freeze-storage of plant cell, tissue and organ culture, Acta Horticulturae 63:75-84 (1976a).
3. Y.P.S. Bajaj, Technology and prospects of cryopreservation of germplasm, Euphytica 28:267-285 (1979a).
4. Y.P.S. Bajaj, Establishment of germplasm banks through freeze-storage of plant tissue culture and their implication in agriculture, in: "Plant Cell and Tissue Culture - Principles and Applications," W.R. Sharp, ed., Ohio State University Press, Columbus (1979b).
5. Y.P.S. Bajaj, Prospects of freeze preservation of plant tissue cultures, Phytomorphology 29:255-259 (1979c).
6. Y.P.S. Bajaj and J. Reinert, Cryobiology of plant cell cultures and establishment of gene-banks, in: "Applied and Fundamental Aspects of Plant Cell, Tissue and Organ Culture,"J. Reinert and Y.P.S. Bajaj, eds., Springer-Verlag, Berlin-Heidelberg-New York (1977).
7. R. Galzy, Recherches sur la croissance de Vitis rupestris Scheele Sain et Court nove cultive in vitro a differentes temperature, Annal Phytopathol. 1:149-166 (1969).
8. G. Morel, Meristem culture techniques for the long term storage of cultivated plants, in: "Crop Genetic Resources for

Today and Tomorrow," O.H. Frankel and J.G. Hawkes, eds., Cambridge University Press, Cambridge (1975).
9. R.H. Mullin and D.E. Schlegel, Cold storage maintenance of strawberry meristem plantlets, Hort. Sci. 11:100-101 (1976).
10. P.J. Dale, A method for in vitro storage of Lolium multiflorum Lam., Ann. Bot. 45:497-502 (1980).
11. Y P.S. Bajaj, Growth and morphogenesis in frozen (-196°C) endosperm and embryos of rice, Curr. Sci. 50:947-948 (1981a).
12. M.J. Ashwood-Smith and J. Farrant, "Low Temperature Preservation in Medicine and Biology," Pitman Medicine, London (1980).
13. Y.P.S. Bajaj, Initiation of shoots and callus from potato tuber sprouts and axillary buds frozen at -196°C, Crop Improv. 4:48-53 (1977b).
14. Y.P.S. Bajaj, Tuberization in potato plants regenerated from freeze-preserved meristems, Crop Improv. 5:137-141 (1978c).
15. Y.P.S. Bajaj, Regeneration of plants from potato meristems freeze-preserved for 24 months, Euphytica 30:141-146 (1981c).
16. Y.P.S. Bajaj, Clonal multiplication and cryopreservation of cassava through tissue culture, Crop Improv. 4:198-204 (1977c).
17. A. Sakai, M. Yamakawa, D. Sakata, T. Harada and T. Yakuwa, Development of a whole plant from an excised strawberry runner apex frozen to -196°C, Low Temp. Sci. Ser. B 36:31-38 (1978).
18. K.K. Kartha, N.L. Leung and O.L. Gamborg, Freeze-preservation of pea meristems in liquid nitrogen and subsequent plant regeneration, Plant Sci. Lett. 15:7-15 (1979).
19. B.W.W. Grout, R.J. Westcott and G.G. Henshaw, Survival of shoot meristems of tomato seedlings frozen in liquid nitrogen, Cryobiology 15:478-483 (1978).
20. Y.P.S. Bajaj, Freeze-preservation of meristems of Arachis hypogaea and Cicer arietinum, Indian J. Exp. Biol. 17:1405-1407 (1979d).
21. A. Sakai and S. Yoshida, The role of sugars and related components in variation of freezing resistance, Cryobiology 5:160-175 (1968).
22. I.I. Tumanov, R.G. Butenko and I.V. Ogolevets, Application of isolated tissue culture technique for studying hardening process in plant cells, Fiziol. Rast. 15:749-756 (1968).
23. P.L. Steponkus and L. Bannier, Cold acclimation of plant tissue cultures, Cryobiology 8:386-387 (1971).
24. L.J. Bannier and P.L. Steponkus, Cold acclimation of Chrysanthemum callus cultures, J. Amer. Soc. Hort. Sci. 101:409-412 (1976).
25. P.J. Dix and H.E. Street, Selection of plant cell lines with enhanced chilling resistance, Ann. Bot. 40:903-910 (1976).
26. H.H. Chen, P. Gavinlertvana and P.H. Li, Cold acclimation of stem cultured explants and leaf callus of Solanum sp., Bot. Gaz. 140:142-147 (1979).

27. J. Reinert and Y.P.S. Bajaj, "Applied and Fundamental Aspects of Plant Cell, Tissue and Organ Culture," Springer-Verlag, Berlin-Heidelberg-New York (1977).
28. W.M. Roca, tissue culture methods for the international exchange and conservation of cassava germplasm, Cassava Newslett. CIAT 6:3-5 (1979).
29. W.M. Roca, J.E. Bryan and M.R. Roca, Tissue culture for the international transfer of potato genetic resources, Amer. Potato J. 55:691-701 (1979).
30. Y.P.S. Bajaj, Freeze-preservation of plant cell and tissue cultures - progress and prospects, in: "Plant Tissue Culture, Genetic Manipulation and Somatic Hybridization of Plant Cells," P.S. Rao et al., eds., BARC, Bombay (1980b).
31. Y.P.S. Bajaj, Plant genetic conservation through tissue culture in: "Proc. Internatl. Tissue Culture Workshop," Dacca, Bangladesh (1981d).
32. Y.P.S. Bajaj, Prospects of the in vitro induction and cryopreservation of genetic variability in crop improvement, in: "Proc. Internatl. Symp. New genetical Approaches to Crop Improvement," Karachi, Pakistan (1982b).
33. Y. Sugawara and A. Sakai, Survival of suspension cultured sycamore cells cooled to the temperature of liquid nitrogen, Plant Physiol. 54:722-724 (1974).
34. G.G. Henshaw, Technical aspects of tissue culture storage for genetic conservation, in: "Crop Genetic Resources for Today and Tomorrow," O.H. Frankel and J.G. Hawkes, eds., Cambridge University Press, Cambridge (1975).
35. L.A. Withers, The freeze-preservation of synchronously dividing cultured cells of Acer pseudoplatanus L., Cryobiology 15:87-92 (1978).
36. L E. Towill and P. Mazur, Effect of cooling and warming rates on survival of frozen Acer saccharum tissue cultures, Plant Physiol. 54Suppl.:10 (1974).
37. K.K. Nag and H.E. Street, Freeze preservation of cultured plant cells I. The pretreatment phase, Physiol. Plant. 34:254-260 (1975).
38. Y.P.S. Bajaj, Survival of Atropa and Nicotiana pollen-embryos at -196°C, Curr. Sci. 46:75-84 (1977a).
39. Y.P.S. Bajaj, Effect of super-low temperature on excised anthers and pollen-embryos of Atropa, Nicotiana and Petunia, Phytomorphology 28:171-176 (1978a).
40. Y.P.S. Bajaj, Regeneration of plants from pollen-embryos frozen at ultra-low temperatures - a method for the preservation of haploids, IV Intl. Palynol. Conf., Lucknow, pp. 343-346 (1978b).
41. R.A. Mazur and J. Hartman, Freezing of plant protoplasts, in: "Plant Cell and Tissue Culture: Principles and Applications," W.R. Sharp et al., eds., Ohio State University Press, Columbus (1979).
42. L.A. Withers and H.E. Street, The freeze-preservation of cul-

tured plant cells: III. The pregrowth phase, Physiol. Plant. 39:171-178 (1977).
43. L.J. Bannier and P.L. Steponkus, Freeze-preservation of callus cultures of Chrysanthemum morifolium Ramat., Hort. Sci. 7:194 (1972).
44. Y.P.S. Bajaj, Regeneration of plants from cell suspensions frozen at -20, -70 and -196°C, Physiol. Plant. 37:263-268 (1976b).
45. R. Latta, Preservation of suspension cultures of plant cells by freezing, Can. J. Bot. 49:1253-1254 (1971).
46. D.L. Dougall and D.F. Wetherell, Storage of wild carrot cultures in the frozen state, Cryobiology 11:410-415 (1974).
47. M. Seibert, Shoot initiation from carnation shoot species frozen to -196°C, Science 191:1178-1179 (1976).
48. M. Seibert and P.J. Wetherbee, Increased survival and differentiation of frozen herbaceous plant organ culture through cold treatment, Plant Physiol. 59:1043-1046 (1977).
49. M. Uemura and A. Sakai, Survival of carnation (Dianthus caryophyllus L.) shoot apices frozen to the temperature of liquid nitrogen, Plant Cell Physiol. 21:85-94 (1980).
50. K.K. Kartha, N.L. Leung and K. Pahl, Cryopreservation of strawberry meristems and mass propagation of plantlets, J. Amer. Soc. Hort. Sci. 105:481-484 (1980).
51. Y.P.S. Bajaj, Survival of anther- and ovule-derived cotton callus frozen in liquid nitrogen, Curr. Sci. (1982a).
52. L.B. Hollen and L.M. Blakely, Effects of freezing on cell suspensions of Haplopappus ravenic, Plant Physiol. 56Suppl.:39 (1975).
53. L.A. Withers and P.J. King, A simple freezing-unit and routine cryopreservation method for plant cell cultures, Cryo-letters 1:213-220 (1980).
54. R.S. Quatrano, Freeze-preservation of cultured flax cells utilizing dimethyl sulfoxide, Plant Physiol. 43:2057-2061 (1968).
55. G.G. Henshaw, J.A. Stamp and R.J. Westcott, Tissue culture and germplasm storage, in: "Plant Cell Culture: Results and Prospectives," F. Sala et al., eds., Elsevier/North Holland Biomedical Press (1979).
56. B.J. Finkle, Y. Sugawara and A. Sakai, Freezing of carrot and tobacco suspension cultures, Plant Physiol. 56Suppl.:80 (1975).
57. Y.P.S. Bajaj and J. Reinert, Regeneration of plants from cells subjected to super-low temperatures, XII Intl. Botanical Congress, Leningrad (1975).
58. F. Sala, R. Cella and F. Rollo, Freeze-preservation of rice cells grown in suspension culture, Physiol. Plant. 45:170-176 (1979).
59. Y.P.S. Bajaj, Induction of androgenesis in rice anthers frozen at -196°C, Cereal Res. Commun. 8:365-369 (1980a).
60. B. Tisserat, J.M. Ulrich and B.J. Finkle, Cryogenic preservation and regeneration of date palm tissue, Hort. Sci. 16:47-

48 (1981).
61. A. Sakai and Y. Sugawara, Survival of poplar callus at super low temperatures after cold acclimation, Plant and Cell Physiol. 14:1202-1204 (1973).
62. Y.P.S. Bajaj, Regeneration of plants from ultra-low frozen anthers of Primula obconica, Sci. Hort. 14:93-95 (1981b).
63. B.J. Finkle and J.M. Ulrich, Effects of cryoprotectants in combination on the survival of frozen sugarcane cells, Plant Physiol. 63:598-604 (1979).
64. W.H. Chen, W. Cockburn and H.E. Street, Preliminary experiments on the freeze preservation of sugarcane cells, Taiwania 24:70-74 (1979).
65. L.E. Towill, Solanum tuberosum: a model for studying the cryobiology of shoot tips in the tuber-bearing Solanum species, Plant Sci. Lett. 20:315-324 (1981).
66. B.W.W. Grout and G.G. Henshaw, Freeze preservation of potato shoot-tip cultures, Ann. Bot. 42:1227-1229 (1978).
67. Y.P.S. Bajaj, Freeze-preservation of plant cells -- a novel approach, in: "Natl. Symp. Wheat Genetics," Haryana Agric. Univ., Hissar (1980c).
68. Registry of Animal Cell Lines - certified by the Cell Culture Collection Committee, U.S. Govt. Printing Press, Washington, D.C. (1964).

EFFECT OF SQUARIC ACID AND ISOASCORBATE ON GLYOXALASE-I CELL DIVISION AND DNA SYNTHESIS IN DATURA CALLUS

O. Ramaswamy, S. Pal, S.K. Sopory and S.G. Mukherjee

School of Life Sciences
Jawaharlal Nehru University
New Delhi 110067
INDIA

INTRODUCTION

The presence of the enzyme glyoxalase was reported by Neuberg (1) and Dakin and Dudley in 1913 (2). Since then, it has been isolated from many animal systems and from yeast. However, except for one casual report, its existence has not been reported in higher plants.

Glyoxalase-I is the first enzyme in which glutathione (GSH) was shown to act as a co-enzyme. It converts and α-ketoaldehyde into the thioester of GSH and corresponding α-hydroxy carboxylic acid. The glyoxalase system involves two enzymic reactions. The first enzyme to act is glyoxalase-I which catalyzes the conversion of α-ketoaldehyde into thioester, with the help of glutathione and the second is catalyzed by glyoxalase-II which involves the hydrolysis of the thioester to regenerate GSH. The biological significance of this enzyme is still unclear, although a number of roles have been assigned to it. It has been suggested to protect the cells against α-ketoaldehyde toxicity.

Lately, the interest in this enzyme has spread to many laboratories following the suggestion of Szent Gyorgyi and his co-workers regarding its role in cell division and regulation (3, 4). According to them, when methyl glyoxal (a dicarbonyl) concentration increases, it inhibits cell division. This leads to the production of more glyoxalase enzyme in the cell which destroys methyl glyoxal or glyoxal by converting them to their less reactive hydroxy acids. Thus, the balance of concentration of the inhibitor and the enzyme controls cell division. It has also been shown that this enzyme acts differently in normal and cancer cells. Recently, this enzyme has been shown to affect microtubule free assembly (5). In this communication, we report not only the presence of this enzyme in plant systems, but also its probable role in controlling cell division in higher plants.

MATERIALS AND METHODS

Datura innoxia callus was obtained by inoculating anthers, after sterilization with 0.1% mercuric chloride, on B_5 + NAA (0.2 ppm) medium as reported earlier (5). The callus obtained was transferred to the same medium and maintained on it.

For enzyme extraction, weighed amount of tissue (250 mg) was homogenized in 0.1 M sodium phosphate buffer, pH 7.0, containing 20 mg $MgSO_4.7H_2O$ and 20 ml of glycerol per 100 ml.

The extract was centrifuged at 50,000 g for 20 min. in a Sorval RC5 centrifuge. The supernatant was fractionated with 40-80% ammonium sulphate and the precipitate was collected, redissolved in the extraction buffer and employed for enzyme assay. The enzyme glyoxalase-I (S-lactoyl, GSH glyoxalase, EC 4.4.1.5) was assayed according to the procedure of Racker (5). The assay mixture contained in a final volume of 1.0 ml, 10 mM sodium phosphate buffer, pH 7.5, 3.5 mM methyl glyoxal, 1.7 mM glutathione; 16.0 mM $MgSO_4.7H_2O$. This mixture was incubated for 7 minutes. Immediately after the addition of the enzyme (0.02 ml) the formation of the thioester, which shows maximum absorption at 240 nm, was measured in a Carl Zeiss PMQ-II spectrophotometer. The molar absorption co-efficient of the thioester (S-lactoyl glutathione) at 240 nm is 3.36 $mM^{-1}cm^{-1}$. The enzyme unit (IU) is defined as the amount of enzyme catalyzing the formation of 1 mole of S-lactoyl glutathione per minute per mg protein in the standard assay system. Protein was determined by the method of Lowry et al. (6).

RESULTS AND DISCUSSION

It was found that the enzyme increased with increase in fresh weight and cell number from 2 to 10 days after inoculation of callus (250 mg). Enzyme increase could also be correlated with increase in DNA synthesis. In order to understand if the enzyme was involved in cell division and regulation, inhibitors of cell division, vinblastine, colchicine and methyl glyoxal were employed which arrested growth and cell division of the callus. It was found that simultaneously the enzyme activity also declined. Besides these general cell division inhibitors, specific inhibitors of glyoxalase-I, squaric acid and isoascorbate were employed. As shown in Figure 1, there was a decrease in the enzyme activity with increasing concentrations

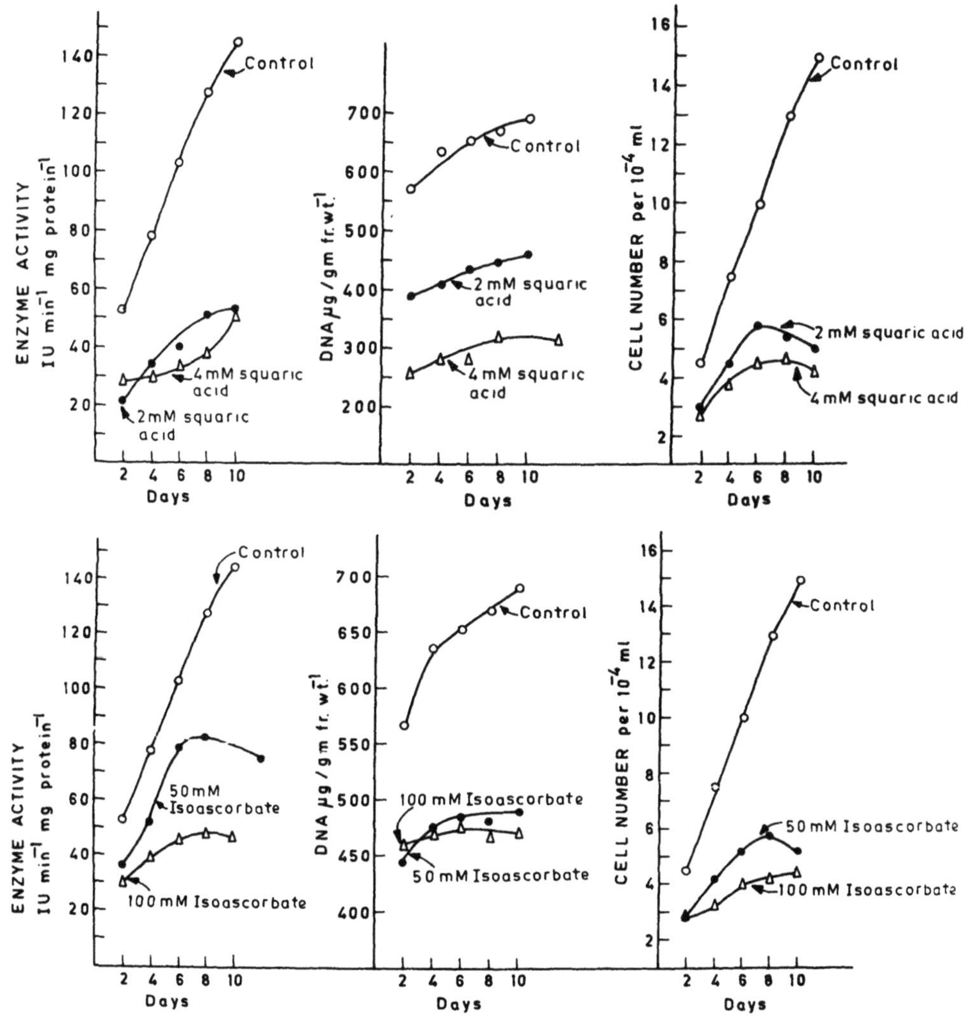

Figure 1. Effect of different concentrations of squaric acid and isoascorbate on glyoxalase-I activity in relation to changes in fresh weight, cell number and DNA content in <u>Datura</u> callus cells.

of these inhibitors which also caused a decrease in cell number and DNA content.

ACKNOWLEDGEMENTS

Two of us (OR and SP) are thankful to the University Grants Commission for the award of Senior Research Fellowships.

REFERENCES

1. C. Neuberg, The destruction of lactic aldehyde and methyl glyoxal by animal organs, Biochem. Z. 49:502-506 (1913).
2. H.D. Dakin and H.W. Dudley, An enzyme concerned in the formation of hydroxy acids from ketonic aldehydes, J. Biol. Chem. 14:155-162 (1913).
3. A. Szent Gyorgyi, Bioelectronics, Science 161:988-990 (1968).
4. A. Szent Gyorgyi, L.G. Egyud and J.A. McLaughlin, Ketoaldehydes and cell division, Science 155:539-541 (1967).
5. E. Racker, The mechanism of action of glyoxalases, J. Biol. Chem. 190:685-696 (1951).
6. O.H. Lowry, N.J. Rosebrough, A.L. Farr and R.J. Randall, Protein measurement with the Folin Phenol reagent, J. Biol. Chem. 193:265-275 (1951).

CELL DENSITY MEDIATED CHANGES IN CYCLIC AMP PHOSPHODIESTERASE ACTIVITY DURING CELL DIFFERENTIATION IN MOSS

S. Sharma and M.M. Johri

Molecular Biology Unit
Tata Institute of Fundamental Research
Homi Bhabha Road,
Bombay 400005,
INDIA

INTRODUCTION

Cyclic 3',5'-adenosine monophosphate (cAMP) enhances the differentiation of chloronema cell type in the protonema cultures of the moss Funaria hygrometrica (1, 2). The endogenous level of cAMP is 4- to 7-fold higher in chloronema cells than that in caulonema cells (1). Thus there is a reduction in the cAMP level during caulonema differentiation. This could involve a modulation of adenylate cyclase and/or cAMP phosphodiesterase activity. Since the initiation of caulonema differentiation (3), and the activity of cAMP phosphodiesterase (4) depend on the amount of inoculum, we have investigated cell-density-mediated changes in cAMP phosphodiesterase activity in relation to caulonema differentiation.

MATERIALS AND METHODS

Cultures of Moss Protonema

The experiments were carried out with the wild type cell line J-2 of the moss Funaria hygrometrica Hedw. Isolation of this cell line, its maintenance and the media used have been described earlier (5). Chloronema cells grown in low calcium medium (LCM) were used as inoculum.

Caulonema cells were obtained by culturing chloronema cells at an initial density of 0.1 mg/ml in minimal medium containing 1% glucose (MMG) and 5 μM IAA (25 ml medium/1000 ml Erlenmeyer flask).

For cell dilution experiments, three to four day old exponentially growing cells were transferred either into sterile distilled water or fresh LCM or spent LCM. The latter was collected by decantation. All steps were carried out under aseptic conditions. Cells were harvested by filtration, homogenized using a French Press at 220 ± 20 kg/cm^2, and the total cell homogenate was used for phosphodiesterase assays.

Assay of cAMP Phosphodiesterase

The assay mixture contained 40 mM soidum acetate buffer, pH 5.4, 10 μM [^{3}H] cAMP (10 nmol = 643,000 cpm) and enzyme preparation in a total volume of 0.5 ml. After incubation for 2 h at 37°C, the reaction was stopped by adding trichloroacetic acid, and the products were determined by TLC (6). One unit of enzyme activity is defined as 1.0 nmol of cAMP hydrolyzed per h of incubation. Protein was determined by the method of Lowry et al. (7) using bovine serum albumin as the standard.

Inhibitor Experiments

The stock inhibitor solutions (1 mg/ml) were sterilized by membrane filtration and added to the cell cultures at the time of inoculation. The cells were harvested 10 h after subculture and the cAMP phosphodiesterase activity was assayed.

RESULTS AND DISCUSSION

cAMP Phosphodiesterase Activity During Caulonema Differentiation

When cultured at a low cell density in the presence of IAA, chloronema cells differentiate and cultures containing 65-70% caulonema filaments are obtained (3). The control cultures, i.e., cells cultured in the absence of IAA, on the other hand show only 5-10% caulonema filaments (3). Changes in the cAMP phosphodiesterase in differentiating and control cultures are shown in Figure 1. In both these cultures, the phosphodiesterase activity increased about 3-fold within 26 h after inoculation. This increase is, therefore, independent of the presence of IAA and seems to be due to a dilution of cells from a high to low density. After 26 h, whereas the phosphodiesterase activity remained at a high level in the differentiating cultures, a gradual decline in activity was observed in control cultures (Figure 1). Increased caulonema differentiation concomitant with the maintenance of enhanced phosphodiesterase activity indicates a relationship between the two. We would like to suggest that the elevated phosphodiesterase activity plays a role in caulonema differentiation, by maintaining low levels of cAMP.

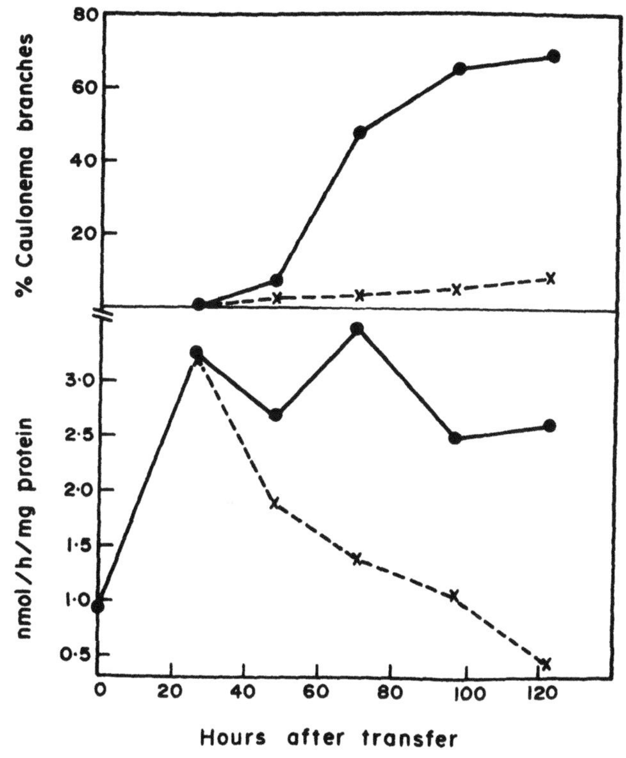

Figure 1. Changes in phosphodiesterase (PDE) activity during caulonema differentiation. Chloronema cells were incubated at an initial cell density of 0.1 mg/ml in MMG alone (x) or MMG containing 5 μM IAA (●). The cAMP PDE activity and the percentage of caulonema filaments were determined at appropriate intervals.

Effect of Lowering Cell Density on Phophodiesterase Activity

As mentioned earlier, the initial increase in the phosphodiesterase activity during differentiation seems to be due to a dilution of cells from a high to low cell density. The relationship between phosphodiesterase activity and cell density has been reported earlier (4). The phosphodiesterase activity is highest at a low cell density (< 2 mg/ml), and decreases about 5-fold at a density of 4-6 mg/ml (4). In all these experiments the phosphodiesterase activity was assayed approximately 24 h after subculturing the cells at various cell densities. A kinetic analysis shows that upon dilution the phosphodiesterase activity increased steadily and the peak was reached after about 16 hours (Figure 2); no lag-period was observed in any of the experiments.

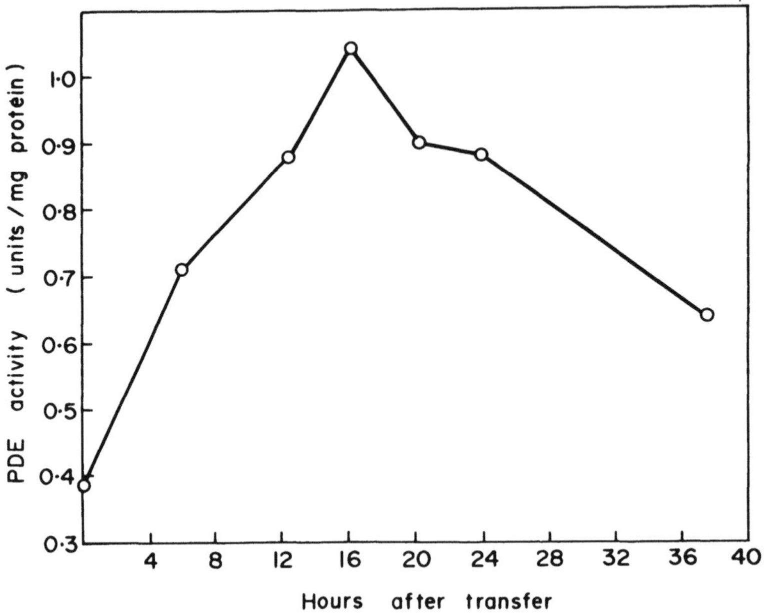

Figure 2. Increase in PDE activity upon dilution. Chloronema cells at the exponential growth phase (density = 13.7 mg/ml) were diluted about 17-fold with fresh LCM medium (final density = 0.8 mg/ml) and the cAMP PDE activity was measured at the indicated intervals.

Effect of Inhibitors of Protein Synthesis on the Increase in Phosphodiesterase Activity

The increase in phosphodiesterase activity was almost completely inhibited by cycloheximide (10 μg/ml), while chloramphenicol (10 g/ml) did not show any significant effect (Table 1). These results indicate a requirement for de novo protein synthesis in the cytoplasm, during the increase of phosphodiesterase activity. The data with chloramphenicol suggest that protein synthesis in mitochondria or chloroplasts is not required for phosphodiesterase increase. We have not yet investigated whether the phosphodiesterase enzyme itself is synthesized; or a different component is synthesized which prevents the degradation of the enzyme or converts inactive enzyme to an active form.

Effect of Diluting Cell Suspension with Water or Spent Medium

The initial increase in phosphodiesterase activity was similar whether the cell suspension was diluted with water or with fresh medium (Figure 3). On the other hand, the cells diluted into the spent medium showed no increase in phosphodiesterase activity, even

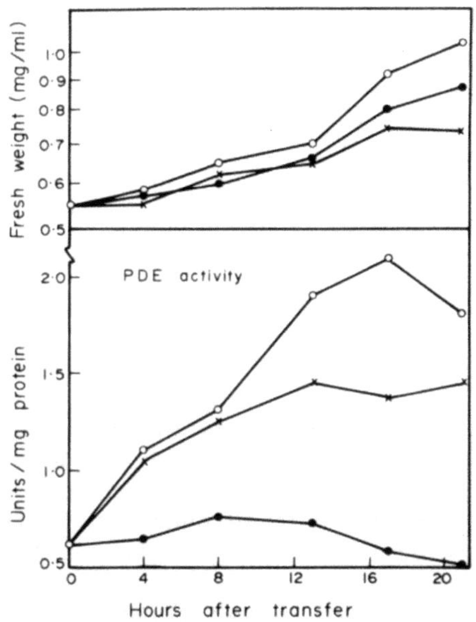

Figure 3. Changes in the fresh weight (top), and PDE activity (bottom) upon dilution of chloronema cells into fresh medium (o), water (x), and spent medium (●). Exponentially growing chloronema cells (density = 11.6 mg/ml) were diluted about 20-fold (final density = 0.55 mg/ml), and the cAMP PDE activity was assayed at the indicated intervals.

Table 1. Effect of Inhibitors of Protein Synthesis on the Increase in Phosphodiesterase Activity

Inhibitor	Concentration µg/ml	% Increase in PDE Activity
None	-	100[a]
Cycloheximide	5	63
	10	9
Chloramphenicol	5	86
	10	93

The inhibitors were added at the time of dilution of chloronema cells. After 10 h of subculture, cells were harvested and phosphodiesterase activity was determined.

[a] The control activity increased from 0.7 units/mg protein to 2.2 units/mg protein.

though their fresh weight increased at the same rate or even at a greater rate than the cells diluted in water. These results indicate that the decrease in the phosphodiesterase activity can be due to a factor (or factors) released by the cells. It seems as though at any growth stage the cells are in equilibrium with the culture medium and upon dilution of cell suspension into fresh medium, or water, the cells experience a reduction in the concentration of effect or components, thereby resulting in an increase in phosphodiesterase activity.

REFERENCES

1. A.K. Handa and M.M. Johri, Cell differentiation by 3',5'-cyclic AMP in a lower plant, Nature 259:480-482 (1976).
2. A.K. Handa and M.M. Johri, Involvement of cyclic adenosine-3',5'-monophosphate in chloronema differentiation in protonema cultures of *Funaria hygrometrica*, Planta 144:317-324 (1979).
3. M.M. Johri and S. Desai, Auxin regulation of caulonema formation in moss protonema, Nature New Biol. 245:223-224 (1973).
4. S. Sharma, R.K. Jayaswal and M.M. Johri, Cell-density-dependent changes in the metabolism of chloronema cell cultures, Plant Physiol. 64:154-158 (1979).
5. M.M. Johri, Differentiation of caulonema cells by auxins in suspension cultures of *Funaria hygrometrica*, in: "Plant Growth Substances," Hirokawa Publishing Co., Tokyo (1974).
6. U. Fischer and N. Amrhein, Cyclic nucleotide phosphodiesterase of *Chlamydomonas reinhardtii*, Biochim. Biophys. Acta 341:412-420 (1974).
7. O.H. Lowry, N.J. Rosebrough, A.L. Farr and R.J. Randall, Protein measurement with the Folin phenol reagent, J. Biol. Chem. 193:265-275 (1951).

DISTRIBUTION PATTERN OF FATTY ACIDS IN CALLUS CULTURES AND PLANT PARTS

T. Halder and V.N. Gadgil

Tissue Culture Section
Bose Institute
93/1 A.P.C. Road
Calcutta 700009
INDIA

Study on lipids and fatty acids in plant callus tissues is of recent origin and a very limited number of reports are available (1). Some of these reports clearly indicate that cultural conditions greatly influence fatty acid composition of callus. In order to have meaningful comparison between the callus derived from six plant species of Cucurbitaceae, we employed identical conditions for initiation of callus and culture of the tissues isolated. Experiments were conducted in horizontal plane using cotyledons and their respective callus tissue of these plants and in vertical plane using root, hypocotyl, cotyledon, stem and leaf and their respective callus cultures of one plant, *Cucumis melo utilissimus*. The plants *C. melo* and *C. melo utilissimus* were used for varietal differences; these and *C. sativus* for differences in the species of a genus; and *Citrullus vulgaris*, *Luffa acutangula* and *Momordica charantia* for generic differences. *M. charantia* was selected as it contains an uncommon fatty acid, α-elaeostearic acid (conjugated octadecatrienoic acid).

Basal medium of Murashige and Skoog (2) solidified with agar, supplemented with NAA (1.0 mg/l) and coconut water (15% v/v) was used for callus initiation and culture. The tissues were grown in dark at 25°C through successive passages of 45 days each.

COTYLEDON AND CALLUS

GLC analysis of methyl esters of fatty acids in total lipid, extracted from tissues showed palmitic, stearic, oleic, linoleic and linolenic acids to be the main components. General pattern of fatty acid composition in cotyledons was similar with obvious difference in individual values. M. charantia differs from the rest in the presence of α-elaeostearic acid (65.9%) which was the major acid in the total pool as well as among the unsaturated fatty acids instead of linoleic (40.2-72.2%) in cotyledons of other species. Further, stearic (24.2%) was the major saturated fatty acid and not the palmitic (12.2-21.3%) as in the rest of the species. In all the cotyledons, the proportion of total unsaturated fatty acids was high (68.6-86.8%). On the other hand, in all the callus cultures, the proportion of total saturated fatty acids was more than that of unsaturated ones. Palmitic and linolenic were the major saturated and unsaturated fatty acids, respectively, in all the cultures except Momordica where oleic acid (15.6%) was the major unsaturated one. In Momordica callus, α-elaeostearic acid of the cotyledon was not detected. The relative distribution pattern of fatty acids in the callus cultures resembled more closely than that in cotyledons. More importantly, the position with respect to proportions of the two types of acids and of linoleic and linolenic acids was reversed in callus cultures.

POINT OF REVERSAL IN FATTY ACID PATTERN

Callus was initiated on cotyledon segments (devoid of growing axis) excised after seed germination. Callus was removed form the segment and cultured through seven successive passages prior to analysis. So, in order to determine the point at which the observed reversal occurred, fatty acid composition was determined during different stages of seedling growth, initiation of callus and culture up to the 20th passage. Fatty acid composition of cotyledon during different stages of seedling growth showed more or less the same pattern as that in the cotyledon of the seed. The excised and enlarged cotyledon segment, prior to visible callus formation showed an increase in palmitic acid content (14.8→21.1%). There was further increase in palmitic acid value (21.1→31.7%) when callusing became visible. But even at this stage total unsaturated acid value (57.1%) was high and linoleic acid (43.6%) was the major component. This progressive increase in palmitic acid indicated increased synthesis of saturated acid in meristematic cells induced. Excised callus, entirely composed of newly formed cells after the 2nd passage of growth exhibited reversal in the proportion of unsaturated fatty acid, drastic reduction in linoleic acid (43.6→5.5%) and an increase in linolenic acid (4.6→17.5%) which became the major unsaturated fatty acid. In subsequent passages, the distribution pattern of fatty acid remained more or less constant. Long chain fatty acids, arachidic (4.9%) and behenic (0.9%), were detected after the 2nd and 20th passages of growth respectively.

PLANT PART AND CALLUS

Different parts of *C. melo utilissimus* showed wide variation in the proportion of unsaturated (56.3-75.1%) and linleic (6.8-46.2%) and linolenic (0.4-43.5%). On the other hand the callus cultures from these parts showed close similarity. Thus, the proportion of unsaturated acid varied only from 30.5 to 41.1%, of linoleic 2.5 to 10.1%, and linolenic 18.6 to 30.0%. The composition of callus differed from that of the respective part of the plant. The difference was exhibited by high proportion of saturated acid; by the decrease in linoleic acid content and by an increase or decrease in linolenic acid value. The major unsaturated fatty acid was linolenic.

CONCLUSION

The following conclusions emerge from the observations of such a study. Distribution pattern of fatty acid in callus culture was influenced by the character of the species. Again, if the species of a family show similarity, the callus derived form these will not exhibit marked differences. The reversal in the fatty acid composition observed in the callus of the 7th passage was not the effect of cultural regime but was the manifestation of the changed pattern of fatty acid synthesis in the newly formed cells under the influence of growth factors. However, the change was not detected in the excised enlarged cotyledon segments and the segments with visible callusing since the majority of cells of the segment analyzed were not involved in the process of proliferation. We propose that under the influence of the same growth factors, the newly formed callus cells derived form different parts of a plant follow a somewhat fixed pattern of fatty acid synthesis, irrespective of the composition of the part concerned. This explains the similarity in fatty acid composition of total lipid in callus of root - an underground organ, cotyledon - a storage part and leaf - a photosynthetic one.

REFERENCES

1. S.S. Radwan and H.K. Mangold, The lipids in plant tissue cultures, *in*: "Advances in Lipid Research," R. Paoletti and D. Kritchevsky, eds., Academic Press, New York (1976).
2. T. Murashige and F. Skoog, A revised medium for rapid growth and bioassays with tobacco tissue cultures, *Physiol. Plant.* 15:473-497 (1962).

STUDIES ON SHOOT ORGAN CULTURES OF *ATROPA BELLADONNA* AND *RAUWOLFIA SERPENTINA*

M.R. Heble, B.D. Benjamin, P.C. Roja and M.S. Chadha

Bio-Organic Division
Bhabah Atomic Research Centre
Trombay
Bombay 400085
INDIA

INTRODUCTION

Studies concerning the regulatory aspects of secondary metabolism in plant tissue and cell cultures have shown that very few systems can produce the characteristic plant products in significant yields (1). Although it has now been possible to isolate high yielding competent cell lines from a heterogeneous cell population (2, 3), the stability of such selected lines is rather poor. Our studies with cultured tissues of *Tylophora indica* and *Atropa belladonna* have shown that the process of organogenesis and cellular differentiation, brought about during the developmental stages of plant growth, have profound influence on the expression of secondary metabolism (4, 5). These observations suggested that *in vitro* culture of isolated plant organs may prove advantageous for the production of secondary products. In this report we present our results concerning the shoot organ cultures of *Atropa belladonna* and *Rauwolfia serpentina*.

MATERIALS AND METHODS

Shoot Cultures

Atropa belladonna shoot cultures, established from the shoot tips (2-3 mm) of the field grown plant, were taken as source material. The cultures were initially maintianed on Wood and Braun's (6) medium supplemented with naphthaleneacetic acid (NAA) and kinetin. Subsequently the cultures were established on MS medium (7) supplemented with LS vitamins (8) and benzyl adenine (BA, 5 ppm).

Rauwolfia serpentina shoot cultures were established from mature nodal segments of the plants maintained in the Experimental Field Station, Trombay, Bombay. Shoot cultures were first initiated on MS agar medium supplemented with NAA (0.1 ppm) and BA (1.0 ppm) and subsequently in MS liquid suspension with the same hormonal levels. The suspension cultures were shaken at 90 rpm under continuous fluorescent light at 25 ± 2°C. The tissues were periodically subcultured after 30 days.

Chemical Analysis

For chemical investigation, the tissues were harvested after 35 days incubation, oven dried at 50°C for 48 hours, powdered and extracted with organic solvents. Atropa belladonna tissues were extracted and analyzed by the methods described previously (9). Rauwolfia tissues (12 g dry weight) were extracted with methanol (cold) and the extracts were separated into three fractions, neutral, moderately basic and strongly basic. The latter two fractions contained similar components and were, therefore, pooled (basic fraction). The extracts were analyzed by TLC and GLC. Characterization of the compounds was done by spectroscopic methods.

RESULTS

Shoot Cultures of Atropa belladonna

The cultures showed uniformly good growth on MS medium supplemented with BA (5 ppm). Fresh shoot buds developed into thin leaves from proliferating mass of unorganized tissues at the base and were generally devoid of roots (Figure 1 A and B). Growth of the tissues was maximum during the third and the fourth week of incubation (Figure 2 A). The shoot cultures, upon transfer to MS medium containing different auxin-cytokinin combinations such as BA + NAA, Kn + IAA and Kn + 2,4-D, showed distinct differences in their growth and morphogenetic characteristics. Cultures grown in the absence of BA, formed roots at the base. BA + NAA in the medium caused broadening of the leaves without rooting. Cultures grown on Kn + IAA medium showed sporadic rooting and Kn + IAA combination suppressed the shoot bud development. 2,4-D in the medium caused proliferation of unorganized tissues resulting in a cell suspension.

TLC analysis of the tissue extracts showed that the alkaloidal profile was not comparable to that of the plant shoot extract. Although the major component of the cultured tissues behaved similar to scopolamine in TLC mobility but GLC analysis indicated that the compound was not scopolamine. The major alkaloid of the plant shoot, atropine, was present in trace amounts in the cultured tis-

Figure 1 A: Shoot buds of Atropa belladonna originating from proliferating mass of unorganized tissues on MS + BA (5 ppm) medium.

B: Established shoot cultures of A. belladonna grown for 35 days on MS + BS (5 ppm) medium.

sues. The total alkaloid content of the shoot cultures grown on BA (5 ppm) medium was 10 mg %. The total alkaloid content varied significantly in tissues grown on media containing different hormonal combinations as follows: BA, 10 mg %; BA + NAA, 16 mg %; Kn + 2,4D, 15 mg %; Kn + NAA, 14 mg %; Kn + IAA, 15 mg %. None of the combinations gave alkaloid content similar to the plant shoot (146 mg %).

Shoot Culture of Rauwolfia serpentina

The cultures grew uniformly on MS medium supplemented with BA (1 ppm) and NAA (0.1 ppm) (Figure 1 D). Multiple shoots developed from axillary meristems of the stem explants. Further development of the stem explant also occurred resulting in the formation of fresh meristematic centers (Figure 1 C). Growth of the tissues was maximum during the fourth week resulting in nearly an eleven-fold increase in dry weight (Figure 2 B).

The basic fraction obtained from the shoot cultures and the plant shoots was examined by TLC. Tissue extracts showed the presence of several UV fluorescent compounds of which three major compon-

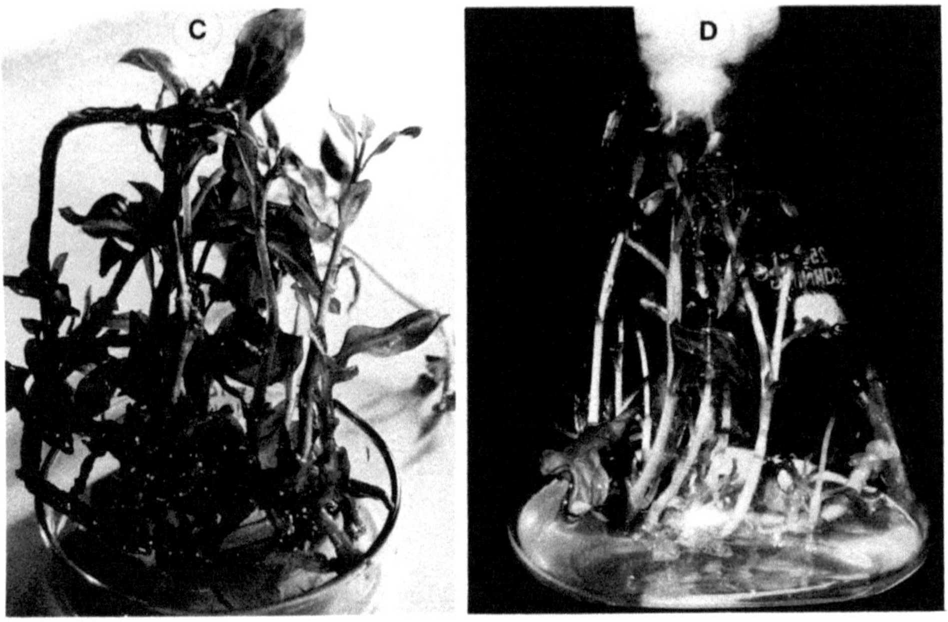

Figure 1 C: Shoots of *Rauwolfia serpentina* developing from axillary meristems of the explanted tissue on MS + BA (1 ppm) + NAA (0.1 ppm). Note the simultaneous development of the stem portion also.

D: Established shoot cultures of *R. serpentina* grown for 35 days.

ents, I - 10 mg, II - 20 mg and III - 16 mg, were separated by column and thin layer chromatography. All the three compounds showed the characteristic UV absorption of indole alkaloids. MeOH max nm: I - 291, 246; II - 287, 281; III - 290, 248. Compound III co-chromatographed with ajmaline on TLC. These components were also present in the plant shoot extracts but to a lesser extent. The alkaloid accumulation is known to be less in the shoots of the plant (10).

DISCUSSION

Studies of *in vitro* culture of isolated plant parts have been extensive, particularly with respect to root cultures and secondary metabolism (11). Although shoot cultures have been successfully utilized for germplasm preservation and clonal multiplications (12),

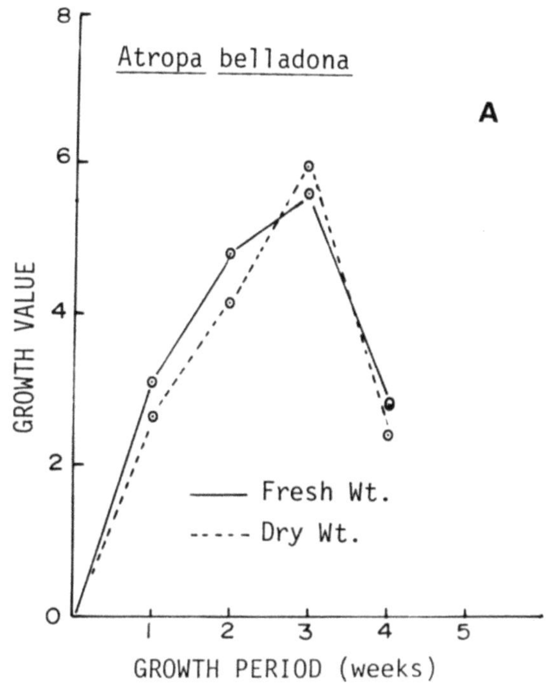

Figure 2 A: Growth characteristics of Atropa belladonna shoot cultures on MS + BA (5 ppm) medium.

very few reports are known in regard to their biosynthetic abilities. Our studies with Atropa belladonna and Rauwolfia serpentina shoot cultures showed that the biomass production, during three to four week incubation is almost similar to the cell culture systems. The different morphogenetic and biosynthetic responses of the Atropa cultures to various hormonal combinations indicated that the cultures are amenable to hormonal regulations like cell cultures. Heble and Staba (13) have reported significant variations in diosgenin production in shoot cultures of Dioscorea composita subjected to different levels of BA. In organ cultures of Digitalis lanata, Lui and Staba (14) have reported the influence of different growth hormones on digoxin production.

The shoot cultures of Atropa belladonna produced very low levels of alkaloids as compared to the shoots of the intact plant

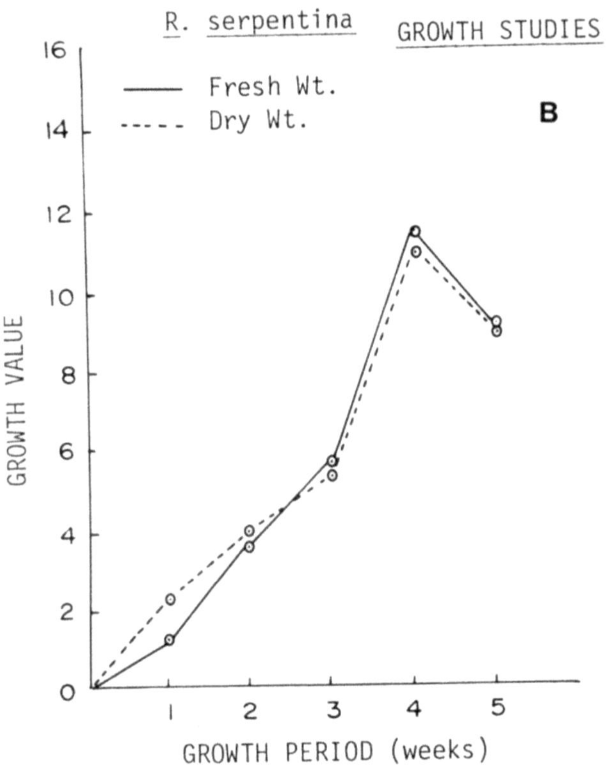

Figure 2 B: Growth characteristics of Rauwolfia serpentina shoot cultures on MS + BA (1 ppm) + NAA (0.1 ppm).

and the alkaloid profile of the cultures was also different from the plant shoots. On the other hand, shoot cultures of Rauwolfia serpentina produced significantly higher levels of indole alkaloids of which the major components were also found in the plant shoots. The striking differences in the metabolic events could possibly be due to the origin of the cultures and their subsequent mode of development. In Atropa the cultures originated from the somatic tissues of the shoot tip, but subsequent development of the shoot was through proliferating unorganized mass of tissues (Figure 1 A). This could possibly be the reason for the variation in the biosynthetic potential of the shoot cultures. However, in shoot cultures of R. serpentina the shoots originated from the axillary meristems through multiple shoot bud formation and further development of the cultures was through stem elongation and multiple shoot bud formation from newly formed axillary buds (Figure 1 C). Those tissues have not only retained the synthetic ability of the parent but were also able to produce significantly high levels of indole alkaloids. Heble and Staba (13) observed that shoot cultures of *Dioscorea composita* obtained form axillary buds contained almost similar levels of diosgenin as in the plant shoot. The data clearly demonstrated that cultures which result from organized development are advantageous in the production of secondary products.

REFERENCES

1. M.S. Chadha and M.R. Heble, Biotechnological applications of plant tissue and cell cultures: problems and prospects, *in*: "Golden Jubilee Commemoration Volume," U.S. Srivastava, ed., National Academy of Sciences, Calcutta, India (1980).
2. M.R. Heble, Plant tissue culture, a source of natural products. Biosynthesis of steroids and naphthoquinones, *in*: "Cultivation and Utilization of Medicinal and Aromatic Plants," C.K. Atal and B.M. Kapur, eds., Regional Research Laboratory, Jammu-Tawi (1977).
3. M.H. Zenk, The impact of plant cell culture on industry, *in*: "Frontiers of Plant Tissue Culture," T.A. Thorpe, ed., International Association for Plant Tissue Culture (1978).
4. B.D. Benjamin, M.R. Heble and M.S. Chadha, Alkaloid synthesis in tissue cultures and regenerated plants of *Tylophora indica* Merr., *Z. Pflanzenphysiol.* 92:77-84 (1979).
5. S. Eapen, T.S. Rangan, M.S. Chadha and M.R. Heble, Biosynthetic and cytological studies in tissue cultures and regenerated plants of haploid *Atropa belladonna*, *Can. J. Bot.* 56:2781-2784 (1978).
6. H.N. Wood and A.C. Braun, Studies on the regulation of certain essential biosynthetic systems in normal and crown-gall tumor, *Proc. Natl. Acad. Sci.* USA 47:1907-1913 (1961).
7. T. Murashige and F. Skoog, A revised medium for rapid growth and bioassay with tobacco tissue culture, *Physiol. Plant.* 15:473-497 (1962).
8. M. Lin and E.G. Staba, Peppermint and spearmint tissue cultures. I. Callus formation and submerged culture, *Lloydia* 24:139-145 (1961).
9. S. Eapen, T.S. Rangan, M.S. Chadha and M.R. Heble, Morphogenetic and biosynthetic studies on tissue cultures of *Atropa belladonna* L., *Plant Sci. Lett.* 13:83-89 (1978).
10. A. Krishnamurthy, "The Wealth of India - Raw Materials, Vol. VIII," Publications and Information Directorate, CSIR, New Delhi (1969).
11. G.C. Mitra, Effect of amino acids on growth of excised *belladonna* root cultures and on their atropine synthesis, *in*: "Plant Tissue Culture, Genetic Manipulation and Somatic Hybridization of Plant Cells," P.S. Rao, M.R. Heble and M.S Chadha, eds., BARC, Bombay (1980).

12. T. Murashige, The impact of plant tissue culture on agriculture, *in*: "Frontiers of Plant Tissue Culture," T.A. Thorpe, ed., International Association for Plant Tissue Culture (1978).
13. M.R. Heble and E.J. Staba, Diosgenin synthesis in shoot cultures of *Dioscorea composita*, *Planta Medica Supp.*:120-123 (1980).
14. J.H.C Lui and E.J. Staba, Effects of age and growth regulators on serially propagated *Digitalis lanata* leaf and root cultures, *Planta Medica* 41:90-95 (1981).

STUDIES OF STEROIDAL SAPOGENINS FROM YUCCA ALOEFOLIA L. IN VIVO AND IN VITRO TISSUE CULTURES

Pushpa Khanna and Paresh V. Purohit

Laboratory of Pl. Physiology and Biochemistry
Department of Botany, University of Rajasthan
Jaipur 302015, INDIA

INTRODUCTION

The genus Yucca is among the promising source with regard to their valuable steroidal sapogenin contents. Variety of Yucca species have been examined by wall M.E. and co-workers (1-5). Tomova and Panova (6) has examined sapogenins from leaves of Yucca aloefolia L. grown in Bulgaria. Recently Bedour and co-workers (7) have reported seasonal variation in sapogenin contents of Y. aloefolia. We have investigated steroidal sapogenin contents of Y. aloefolia L., leaves rhizomes, flowers and its eight month old callus tissue established on revised Murashige and Skoog's (RT) medium from leaves.

MATERIALS AND METHODS:

Unorganized callus tissue of Y. aloefolia L. was raised from leaves on revised (8) Murashige and Skoog's (9) (RT) medium supplemented with 1.ppm of 2,4-D and 1% agar. Tissue grown as static cultures was harvested at the culture age of 2, 4, 6, 8 and 10 weeks and their growth indices (GI) were calculated (Final dry weight tissues - initial dry weight of tissue/initial dry weight of tissue separately. The tissue thus harvested were dried at 100^oc for 15 min. followed by 60^oc at constant weight and powdered separately.

Leaves rhizomes and flowers were collected from three years old plants grown in botanical garden, University of Rahasthan, Jaipur during the month of July 1980 dried similarly and powdered separately.

Each powdered sample was hydrolysed with 15% Hcl in ethanol for 4 hr (10), (12). The filtrate was extracted (4) with ethylacetate. The ethylacetate fraction was washed with distilled water till neutral, brought to dryness and reconstituted in chloroform for further analysis. Each extract so obtained were subjected to TLC (Silica Gel G: Benzene - Ethylacetate 3:2) and purified further by preparative TLC in the same solvent system as well as other solvent system as given in (Table Ia). Each corresponding band was eluted with chloroform and crystallized from methanol/acetone. the mp was obtained on each compound and each was subjected to IR spectral studies. The samples, as well as mixture of authentic reference compounds were also subjected to GC along with co-injections for further confirmation.

The quantitative estimation of identified sapogenin were carried out colorimetrically with a spectrophotometer (Carl Zeiss, Jena DDR, VSU-2p) according to the method of Sanches et al. (11). The results are the average of independent determination of five replicates (SE < 0.05%).

RESULTS AND DISCUSSION

Leaves, rhizomes and flowers collected from three year old plants yielded quite a substantial amount of sapogenins as shown in (Table 2). Sapogenins were identified as smilagenin, sarsasapogenin, tigogenin, hecogenin, gitogenin and chlorogenin in each sample. The characteristic IR spectral peaks in all six sapogenin were found to be similar to those of their respective reference samples. The GC analysis also shows single well defined peaks with no evidence of decomposition. Leaves yielded 1.15% of total sapogenins with maximum of 0.45% tigogenin while rhizomes yielded 0.90% of total sapogenin having 0.34% of tigogenin whereas flowers yielded the total of 0.60% having 0.18% sarsasapogenin as the main sapogenin.

The callus tissue established from leaves showed its maximum growth index (1.7) in 8 week old tissue. The maximum total yield of sapogenin were also observed here 0.82% having 0.30% tigogenin. The production of sapogenins showed its linear relation to the growth of tissue up to maximum growth period of 8 weeks and then a decline was observed at 10 week of culture age.

Khanna et. al., (12 and (13) have reported tigogenin, hecogenin and gitogenin (0.44%, 0.10% and traces) respectively from leaves of *A. wightii*. The production of sapogenins from callus cultures showed a higher content of total as well as individual sapogenins (except tigogenin 0.43%), indicitating that tissue culture system is able to biosynthesize these substances in quite significant amounts. However, the contents from *in vivo* tissue systems in *Yucca aloefolia* in the present study shows a higher concentration

Table I. Rf x 100 values in different solvent system tested for study on sapogenin contents of *Y. aloefolia* L. on silica Gel G. plates.

	Benzene/ Ethylacetate 3:2	Hexene Ethylacetate 1:1	Chloroform/ Methanol/ Water 188:12:1	Water Saturated Dichloro-methane/ Ether.1:1
Smilagenin	82	72	69	80
Sarsas-apogenin	76	51	66	76
Tigogenin	71	46	64	64
Hecogenin	58	21	54	54
Gitogenin	28	09	30	24
Chlorogenin	10	04	12.5	11

Gas Chromatographic analysis of sapogenin contents of *Y. aloefolia* L.

Model: 5712 A: Hewlett Packard

Column st. phase: 10% SE-30 — Length: 2m: ¼ O.D.

Support: Chromosorb W 60-80 — Temp: 225°

Carrier flow: H_2; 60 ml/min — TCD: 300°

Amount injected: 0.4 1. — Atten: 2

Chart speed: 1.00"/min — Time: 7 min.

Retention time relative to cholestane as reference.

Smilagenin:	2.47
Sarsasapogenin:	2.57
Tigogenin:	2.71
Gitogenin:	4.81
Hecogenin:	4.96
Chlorogenin:	5.40

All the compounds gave single well defined peaks with no evidence of decomposition.

possibly due to differentiation in organized plant system.

A significant increase in sarsapogenin 0.18% in total sapogenin of 0.60% from flowers preferably shows the accumulation and further supply during seed formation stage. Since seeds of most yucca species shows very high content of sarsasapogenin. The ratio of hecogenin to tigogenin remains 1:3 in rhizomes as well as leaves but in flowers it becomes 1:1.5 further suggesting interconversion during the flowering process. Thus leaves show the site of active synthesis for sapogenin and its deposition is shown by rhizomes. Further accumulation in flowering reveals the unique role of biosynthesis and distribution of sapogenin in yucca aloefolia L.. The tissue cultures established from leaves also show quite significant potentiality of total sapogenin production 0.66% with 75% of tigogenin as main sapogenin at the culture age of 8 weeks having its maximum growth index as compared to leaves from a three year old plant.

Table 2. Sapogenin content (% on dry weight basis) from static cultures of Yucca aloefolia L. as well as plant parts.

Age of culture (weeks)	GI*	Smila-genin	Sarsa-sapo-genin	Tigo-genin	Heco-genin	Gito-genin	Chloro-genin	Total Sapo-genin
2	0.25	0.08	0.05	0.12	0.04	0.05	0.02	0.36
4	0.80	0.10	0.08	0.18	0.06	0.07	0.03	0.42
6	1.40	0.15	0.13	0.22	0.08	0.08	0.05	0.52
8	1.70	0.18	0.15	0.30	0.09	0.06	0.04	0.82
10	1.20	0.14	0.12	0.21	0.05	0.04	0.03	0.30
Rhizomes		0.16	0.12	0.34	0.11	0.08	0.09	0.90
Leaves		0.26	0.22	0.45	0.14	0.05	0.03	1.15
Flowers		0.10	0.18	0.13	0.08	0.06	0.05	0.60

*GI = Growth index (final dry weight of tissue - initial dry weight of tissue/initial dry weight of tissue).

REFERENCES

1. M.E. Wall, et.al., J. Am. Pharm. Assoc. Sci. Ed., 43:1 (1954).
2. M.E. Wall, et.al., J. Am. Pharm. Assoc. Sci. Ed., 43:503 (1954).
3. M.E. Wall, et.al., J. Am. Pharm. Assoc. Sci. Ed., 44:438 (1955).
4. M.E. Wall, et.al., J. Am. Pharm. Assoc. Sci. Ed., 46:653 (1957).
5. M.E. Wall, et.al., J. Am. Pharm. Assoc. Sci. Ed., 50:1001 (1961).
6. M. Tomova, D. Panova, Arch. Pharm., 296:553-6 (1963).
7. M.S. Bedour, M.H.A. Elgamal, International Research Congress of Natural Products as Medicinal Agents, Strasbourg, France, July 6-11, 24:11-44 (1980).
8. B. Kaul, E.J. Staba, Lloydia, 31:171 (1968).
9. T. Murashige, F. Skoog, Physiol Pl., 15:473 (1962).
10. Y. Tomita, A. Vomori and H. Minato, Phytochemistry, 9:111 (1970).
11. G.L. Sanches, J.C. Medina Achevedo and R.R. Sato, Analyst, 97:973 (1972).
12. P. Khanna, O.O. Sharma and S.C. Jain, Indian J. Exp. Biol., 17:446 (1979).
13. P. Khanna, O.P. Sharma, Jour. of Nat. Prod., 43:4-459 (1980).

ISOLATION AND CHARACTERIZATION OF MUTANT PLANT CELL CULTURES

Jack M. Widholm

Department of Agronomy
University of Illinois
Urbana, IL 61801, USA

INTRODUCTION

Mutant plant cell cultures are useful for biochemical and physiological studies, for markers in various genetic manipulation experiments and for the direct improvement of plants if the desired trait can be selected for and plants then be regenerated which express the trait. Many of the lines selected so far cannot truly be called mutants since genetic analysis, at the whole plant level, has not been carried out.

The field of cell culture mutants has been reviewed recently by Maliga (1, 2), Widholm (3) and Handro (4). This review will emphasize work not covered by my review of 1978, as well as additional selection schemes and mutants which would seem to be useful for improving plants. Also, recent work which has shown the expression of the trait in the regenerated plant and the heritability of the trait by sexual progeny will be emphasized. Several studies which show a correlation between cell culture responses and those of whole plants will also be discussed.

DISEASE RESISTANCE SELECTION

Ingram (5) and Helgeson et al. (6) had shown earlier that callus of susceptible, but not resistant, cultivars of *Solanum tuberosum* and *Nicotiana tabacum*, respectively, would support the growth of spores of pathogenic fungi. More recently Ruyack et al. (7) have demonstrated a similar response using callus from susceptible and resistant *Gossypium hirsutum* varieties and the bacterial pathogen *Xanthomonas malvacearum*.

Earlier selection work by Gengenbach et al. (8) had produced disease resistant male-fertile *Zea mays* plants from Texas male-sterile cytoplasm calli by selection against the toxin from *Helminthosporium maydis* race T. The toxin resistance was inherited maternally. Brettell and Thomas (9) later reported that unselected Texas male-sterile cytoplasm calli could also produce T-toxin resistant plants which were male-fertile. Gengenbach and Green (10) had shown that several isolated mitochondrial activities were affected by the T-toxin if the mitochondria originated from Texas male-sterile derived calli, but not in mitochondria from normal calli or calli selected as toxin resistant. Restriction endonuclease analysis of mitochondrial DNA from three selected resistant plant lines showed that these DNAs were different from each other and from both normal and Texas cytoplasm types (11). This indicates that unique mitochondrial types were obtained during the *in vitro* toxin selection process.

Behnke (12) selected dihaploid calli of *S. tuberosum* which were resistant to the culture filtrate of *Phytophthora infestans*, which causes late blight. Leaves from plants regenerated from the resistant lines formed callus on medium containing the filtrate more readily than did leaves from the control plants.

Subsequently, Behnke (13) selected dihaploid *S. tuberosum* calli resistant to the culture filtrate of *Fusarium oxysporum* an organism which can cause a potato wilt and rot. In this case leaves of the plants regenerated from selected lines were shown to be resistant to bleaching by the culture filtrate, indicating resistance at the plant level as well.

HERBICIDE RESISTANCE SELECTION

Oswald et al. (14) found that suspension cultures of the metribuzin-sensitive *Glycine max* cultivar Coker 102 were much more sensitive to this herbicide than cultures of the more resistant cultivar Bragg. The Bragg cells readily detoxified the herbicide while the Coker 102 cells produced an inhibitor of the detoxification mechanism.

Differential sensitivity to 2,4-dichlorophenoxyacetic acid inhibition during the initial two transfers were seen with callus cultures initiated from field selected tolerant and wild type cultivars of *Lotus corniculatus* (15).

Chaleff and Parsons (16) selected picloram resistant lines by plating *N. tabacum* suspension cultures on plates containing inhibitory concentrations. Plants were regenerated from four lines which transmitted the resistance trait as a single dominant Mendelian allele. The resistance was expressed both by seedlings germinated in picloram and by callus initiated from the plants. The four mutant alleles are distributed on three distinct linkage groups (17), and the resistance is due to a mechanism other than an alteration in the uptake or metabolism of picloram.

Miller and Hughes (18) selected paraquat resistant lines from calli and plated suspensions of N. tabacum. Prior treatment with X-rays increased the resistance frequency. Some of the plants regenerated from the resistant lines were shown to be resistant to paraquat using a leaf bioassay test. Calli reinitiated from most of the regenerated plants retained resistance.

A variation on the tissue culture selection methods was used by Radin and Carlson (19) to select for resistance to Bentazon and Phenmedipham which do not inhibit callus growth, but do kill tobacco leaves. Gamma-irradiated haploid N. tabacum plants were sprayed with the herbicides, and the green areas, which remained following the yellowing of most of the tissue, were removed and placed into culture. Plants were regenerated and diploidized with colchicine. Some of the regenerated plants showed resistance when sprayed with the respective herbicides. Sexual crossing showed that several different recessive gene loci were involved.

SODIUM CHLORIDE RESISTANCE SELECTION

Recently the tolerance to salt (NaCl) of whole plants has been correlated with that of tissue cultures of several species. Tal et al. (20) found that calli initiated from Lycopersicon esculentum, a salt sensitive species, was less tolerant to salt in the culture medium when compared with calli of L. peruvianum and Solanum pennellii, which are salt tolerant relatives. Orton (21) showed that Hordeum vulgare plants were more salt sensitive than H. jubatum plants. In studies with calli initiated from the two species there were indications that H. vulgare was again more sensitive. Smith and McComb (22) showed that Beta vulgaris seedlings were capable of growing in three-times the salt concentration withstood by Phaseolus vulgaris seedlings. The halophytes Atriplex undulata and Suaeda australis were even more tolerant and required salt for optimal growth. When calli of these species were tested, an almost perfect correlation was found between the plant and culture sensitivity with B. vulgaris and P. vulgaris. However, the halophytes were as sensitive as P. vulgaris in culture probably because the plant resistance is caused by the presence of salt glands which are not present in the cultured cells.

In 1980 Nabors et al. (23) reported the regeneration of salt tolerant N. tabacum plants from suspension cultures which were selected to grow in a medium containing 6.4 g/l NaCl. The F_1 seedlings obtained from these regenerated plants showed a somewhat higher survival rate than the wild type seedlings when grown in soil watered with salt solutions (26.2 g/l). The F_2 seedlings grown from seed obtained from unstressed F_1 plants showed greater resistance than that of the F_1 generation seedlings. The explanation of how the resistance can increase in the F_2 without selection in the F_1, and how this resistance, which appears not to fit a dominant expression model, can be selected for with an amphidiploid like _N. tabacum_, remains to be answered. The data does indicate, however, that heritable resistance to salt has been recovered.

Tyagi et al. (24) selected a _Datura innoxia_ line from calli placed on 1% NaCl medium which was resistant to concentrations about three-fold those tolerated by the wild type. Plants could also be regenerated from this line on 1% NaCl where the wild type would not regenerate. This resistance trait was also carried by calli initiated from the resistant plants.

Thus, it appears that salt tolerance expressed at the plant level can also be expressed in cultured cells, and that selection for salt tolerance with cultured cells can successfully produce plants which carry this trait.

An interesting approach has been proposed by Wong and Sussex (25, 26) for the selection of osmotic stress tolerant cell lines. They found that mannitol-stressed _Nicotiana sylvestris_ suspension cultured cells contained increased levels of abscisic acid. Cell growth could also be inhibited by exogenous abscisic acid, but when this occurs the endogenous levels of abscisic acid were much higher than those found in mannitol-stressed cells. Lines were selected which could tolerate ca. ten-times the normal levels of exogenous abscisic acid, but, unfortunately, these lines showed no cross resistance to mannitol-induced water stress. Such hormone resistant variants, which might be altered in the hormonal response, could also be valuable for various studies.

AMINO ACID ANALOG RESISTANCE SELECTION

Many studies have shown that 5-methyltryptophan resistant cell lines can be selected, most of which contain 30-fold or more increased levels of free tryptophan, due to the presence of an altered form of the feedback control enzyme anthranilate synthase (3). In one case, plants were regenerated from 5-methyltryptophan resistant carrot cells which when recultured gave lines again resistant to the analog. Here the 5-methyltryptophan resistance was due to altered uptake.

Plants were regenerated from one resistant tobacco line which produced resistant cultures again from leaf pieces (27). However, the altered form of anthranilate synthase found in the cultures was not detected in leaf extracts and shoots from the resistant line did not form roots in the presence of 5-methyltryptophan. Here apparently the feedback resistant form of anthranilate synthase is expressed at the cell culture level, but not in the plants.

Recently (Ranch and Widholm, unpublished) we have selected 5-methyltryptophan resistant D. innoxia cell lines and have regenerated plants from several of these. The resistance trait is carried to cultures reinitiated from these plants. The regenerated shoots are able to form roots in the presence of normally inhibitory levels of 5-methyltryptophan and both cultures and leaves of the regenerated plants contain a feedback altered form of anthranilate synthase and increased levels of free tryptophan. Thus, in this case, 5-methyltryptophan resistance is expressed at the whole plant level. Seeds have been produced by some of the resistant diploid plants so studies of the inheritance of the trait are forthcoming.

Suspension cultured diploid Medicago sativa cells were plated on ethionine (methionine analog) containing medium to obtain some stably resistant lines (28). Some of the lines showed resistance to growth inhibition by lysine plus threonine and some also contained increased levels of free methionine. However, leaves of regenerated plants did not contain increased free methionine (E.T. Bingham, personal communication).

Bright et al. (29) screened azide-mutagenized M_2 excised H. vulgare embryos on sterile medium containing aminoethyl-cysteine, a lysine analog. One embryo which showed resistance was grown to a complete plant. Crossing experiments showed that the resistance was inherited by progeny as a single recesssive nuclear gene. A subsequent study (30) showed that the amino acid metabolism of this mutant was not altered, but root uptake of both aminoethyl-cysteine and lysine were decreased by two-thirds.

Calli of Z. Mays were placed on medium containing inhibitory levels of lysine plus threonine to select resistant lines (31). The one line studied was stably resistant as callus, both before and after passage through a regenerated plant. Increases of from two- to nine-fold in the aspartate derived amino acids, lysine, threonine, isoleucine and methionine were observed in the resistant line. These increases were apparently caused by the presence of an altered form of aspartokinase which was less sensitive to feedback inhibition by lysine. Unfortunately measurements of free amino acids and aspartokinase activity were not reported for the regenerated plants.

Several cell lines have been selected which are resistant to the proline analogs, hydroxyproline and azetidine-2-carboxylate, including a carrot line which accumulates from 15- to 30-times the normal level of free proline (32). So far biochemical characterization beyond the free amino acid levels have not been carried out, and plant regeneration has not been reported. Since increases in free proline have been shown to be associated in some manner with water and saline stresses, studies were carried out with the carrot line containing the high free proline levels to determine if this line is more or less resistant to such stresses applied at the cellular level. These studies showed that these cells grew no better than the wild type in the presence of inhibitory concentrations of NaCl, mannitol or polyethyleneglycol (Widholm, unpublished).

Kueh and Bright (33) selected for hydroxyproline resistance using excised, mature H. vulgare embryos in sterile culture. Four resistant plants were recovered from 20,000 embryos. One of the plants carried the resistance as a single partially dominant nuclear gene. In progeny homozygous for this trait, the leaf free proline levels were increased by up to six-fold. Preliminary tests of water stress tolerance did not indicate that this mutant was more tolerant than the wild type.

Early reports concerning P-fluorophenylalanine resistance beginning with Palmer and Widholm in 1975 (34) and ending with Berlin in 1980 (35) described the selection and characterization of resistant lines of D. carota, Acer pseudoplatanus, N. tabacum and Nicotiana glauca. Three mechanisms of resistance were found: 1) overproduction of phenylalanine to compete with the analog; 2) decreased uptake of the analog; and 3) increased phenylalanine ammonia lyase activity which metabolizes the analog and also results in higher phenolic levels.

Evans and Gamborg (36) reported the selection of a P-fluorophenylalanine resistant D. innoxia line from plated, haploid derived suspension cultures. The line was resistant to about 20-fold higher concentrations of P-fluorophenylalanine and could form shoots in normally inhibitory levels of the analog.

Flick et al. (37) isolated N. tabacum lines which were about seven-times more resistant to P-fluorophenylalanine than wild type by plating unmutagenized suspensions on agar plates. Plants could be regenerated from some, but not all of the lines, in the presence of inhibitory levels of the analog. Protoplasts isolated from resistant lines and callus initiated from at least one shoot showed resistance. Unfortunately, the Su albino mutant was used in these studies so plants could not be grown to maturity to obtain seeds.

Since some P-fluorophenylalanine resistant lines accumulate increased levels of phenolics, which have been linked to insect and disease resistance in some cases, it will be interesting to see if such a selection system might be used to obtain plants which are insect or disease resistant.

Thus far none of the selected lines with increased free amino acids have been shown to improve the amino acid composition of the intact plant appreciably. Recent work by Thompson et al. (38) indicates that the composition of the free amino acids available to the seed can affect the amino acid composition of the seed protein. In this study with cultured immature G. max cotyledons, the addition of methionine to the medium stimulated growth and increased the protein methionine content by 22%.

METAL RESISTANCE SELECTION

Wu and Antonovics (39) initiated callus from zinc and copper-tolerant and sensitive clones of the grass Agrostis stolonifera. This differential sensitivity was maintained by the calli and also by plants regenerated from the calli. The cultures and plants showed similar metal uptake patterns, with the tolerant ones taking up more of the metal than the sensitive ones.

Lines of L. esculentum which tolerated elevated levels of aluminum were selected by Meredith (40) by plating suspensions and calli on inhibitory concentrations. The frequency of occurrence was near that expected for spontaneous mutations, and most of the variants were stably resistant. Both the selected resistant cells and the sensitive parental cells took up aluminum at similar rates. No plants were regenerated.

The selection of Petunia hybrida cells resistant to $HgCl_2$ was reported by Colijn et al. (41). Mutagen treatment increased the resistance frequency. The resistance was retained by lines grown away from the selection agent for at least 35 generations and was not due to lessened uptake in the two lines tested.

Since tungstate is known to inactivate nitrate reductase by binding to the enzyme in place of molybdate, the normal cofactor (42), one might be able to select for cells with elevated levels of nitrate reductase by selecting for tungstate resistance. Cells with higher than normal levels of the enzyme may require higher tungstate concentrations for complete enzyme inactivation. Since nitrate reductase is a key inducible enzyme in nitrogen metabolism, increased levels could be valuable at the whole plant level.

If this selection scheme is a valid one, the tungstate inhibition of cell growth should be reversible by molybdate just as the tungstate inactivation of nitrate reductase activity is reversible by molybdate (42). We were also able to show with the same tobacco system that molybdate did prevent nitrate reductase inactivation by tungstate over a 24 hr period, but by 72 hr the tungstate treated cells were dead even in the presence of molybdate (43). Accordingly, in 10 day growth studies, 300 μM molybdate did not reverse the growth inhibition caused by a range of tungstate concentrations.

If tungstate acts only by inhibiting nitrate reductase activity then it should have little effect on the growth of tobacco cells growing with amino acids as the nitrogen source. However, tungstate inhibited the cell growth of both wild type and a nitrate reductase lacking mutant in a way similar to that of wild type cells on nitrate medium.

Thus, tungstate apparently inhibits tobacco cell growth in some way other than by affecting nitrate reductase, so the selection of cells resistant to tungstate will probably not select for variants with increased levels of nitrate reductase.

SELECTION FOR DECREASED PHOTORESPIRATION

There are several tissue culture selection approaches which could possibly be used to attempt to produce plants with lower levels of photorespiration. Berlyn (44) selected ultraviolet light-irradiated N. tabacum suspension cultured cells on agar plates containing inhibitory levels of isonicotinic acid hydrazide. This compound is known to inhibit glycine decarboxylase, an enzyme of the photorespiratory pathway. Twenty resistant lines were recovered, some of which produced plants which gave resistant cultures again. The resistance trait was inherited dominantly when tested as callus since seed germination tests with isonicotinic acid hydrazide did not differentiate between sensitive and resistant seedlings.

A later study (45) showed that mitochondrial NAD-dependent glycine decarboxylase from a resistant line and plants regenerated from this line was less sensitive to inhibition by isonicotinic acid hydrazide, pointing to this site as the cause of resistance.

Lines of haploid derived N. tabacum were also selected as resistant to another glycine decarboxylase inhibitor, glycine hydroxamate by Lawyer et al. (46). The resistant lines contained elevated levels of the 12 free amino acids analyzed, including glycine and serine. The increased level of glycine could account for the resistance by competition with glycine hydroxamate for the presumed inhibitory site, glycine decarboxylase. Some of the regenerated plants did give calli which retained the resistance, but the plants themselves did not contain elevated levels of the free amino acids.

No information has been presented about any effect of the resistance to isonicotinic acid hydrazide or glycine hydroxamate on photorespiration, however. Information from other sources indicates that the photorespiration pathway must remain intact for C_3 plant survival (47), and that the active site on the enzyme ribulose-biphosphate carboxylase/oxygenase is probalby the site which needs to be altered. The desired alteration would either increase the affinity of the active site for CO_2 or lower the affinity for O_2 or be a combination of both. While screening for whole plants with these traits has not been successful (3, 48), it still might be possible to use a cell culture selection system. Photoautotrophic cultures would be grown under low CO_2 conditions to select cells which survive. Since billions of cells could be screened for such an ability, the search should continue even though only a few reports of photoautotrophic growth have appeared and in all cases much higher than air levels of CO_2 are needed to support growth.

CHILLING TOLERANCE SELECTION

Chilling tolerance, which has been found to be stable at the cellular level, has been selected with N. sylvestris, Capsicum annuum and D. carota cells (49, 50). However, following plant regeneration and seed formation, the trait was lost with N. sylvestris (51). Likewise the trait was not expressed once embryos were induced from the chilling tolerant D. carota cultures (50).

AMINOPTERIN RESISTANCE SELECTION

Haploid-derived suspension cultured D. innoxia cells were selected for aminopterin resistance by plating in inhibitory levels (52). Aminopterin inhibits the enzyme dihydrofolate reductase. Following testing at increased aminopterin concentrations, a resistance frequency of 10^{-7} was realized. The resistance trait was retained by cells grown away from the inhibitor and was expressed by two lines which formed shoots in normally inhibitory aminopterin concentrations.

CARBOHYDRATE UTILIZATION SELECTION

A stable sugarcane cell line was selected by Maretzki and Thom (53) which could grow at a rapid rate with the normally toxic sugar galactose as the sole carbon and energy source. The adapted cells contained higher levels of the enzyme UDP-galactose 4-epimerase which apparently allows these cells to convert UDP-galactose, which accumulates in the unadapted cells, to UDP-glucose which can be readily metabolized.

A N. tabacum mutant capable of utilizing glycerol as the sole carbon and energy source was isolated at low frequency by plating cells in a medium containing glycerol by Chaleff and Parsons (54). Calli initiated from leaves of progeny from one of the regenerated plants, obtained by selfing and reciprocal crosses with wild type, were tested for the ability to grow on glycerol. These tests indicated that the trait was carried as a dominant nuclear gene.

Soybean suspension cultures which could grow rapidly with maltose as the sole carbon and energy source were selected by Limberg et al. (55). The selected lines apparently grow because of a more rapid maltose uptake. The ability to grow on maltose was not lost following growth on sucrose for 100 generations.

REGENERATION OF MORPHOLOGICALLY ALTERED PLANTS

Reisch et al. (28) selected 91 ethionine resistant cell lines from a diploid M. sativa suspension culture. Plants from 23 of these lines showed morphological abnormalities while few abnormal plants were regenerated from lines treated only with the mutagen, ethylmethane sulfonate. The altered traits included increased yield, multifoliolate leaves, elongated petioles, compact growth habit, plant height and branch number (56). This increased variation was largely attributed to the ethionine treatments since this compound has been shown to be a mutagen.

Similar useful variations were noted in S. tuberosum plants regenerated from leaf protoplasts by Shepard et al. (57). In this case, however, no mutagen treatments were included so the cause of this variability is not clear. However, the production of useful variation might be very valuable especially with crops where such variation cannot be obtained easily in other ways.

AUXOTROPH SELECTION

Auxotroph selection started early when Carlson in 1970 (58) reported the selection of six lines of N. tabacum which showed a partial requirement for hypoxanthine, biotin, P-aminobenzoic acid, arginine, lysine or proline. The selection procedure utilized a 5-bromodeoxyuridine "suicide" approach with haploid cells, whereby auxotrophic cells, which cannot grow on minimal medium, do not incorporate the toxic nucleotide analog into their DNA. The growing prototrophic cells will incorporate the analog and, thus, will be killed. The auxotrophs are rescued by the addition of various compounds which may fulfill the growth requirement. The specific requirement is found by testing the compounds individually for growth stimulation.

The general 5-bromodeoxyuridine "suicide" method was not used successfully again until Malmberg in 1979 (50) used the method to obtain N. tabacum lines which would not grow at the restrictive temperature of 33°C (temperature-sensitive), but would grow at 26°C. Growth at 33°C could not be stimulated by the addition of coconut water or casamino acids. The temperature-sensitive lesion in one line may be two enzymes in the polyamine biosynthetic pathway which are very low in activity (60).

Polacco (61) devised another general "suicide" selection scheme which utilized the phosphate analog arsenate to kill growing Glycine max cells. One line was selected from 2 x 10^7 cells which required a mixture of amino acids for growth. This trait was lost after nine months, however.

The specific "suicide" agent chlorate was first used with plant cells by Muller and Grafe (61) to select N. tabacum cells which lack nitrate reductase. Chlorate itself is not toxic to cells, but it can be reduced to the toxic compound chlorite by the enzyme nitrate reductase. Thus, cells lacking nitrate reductase should survive the chlorate treatment while those with the enzyme will kill themselves. Muller and Grafe (62) plated mutagenized cells in a medium containing chlorate for killing and amino acids to supply reduced nitrogen to the cells lacking nitrate reductase. Cells were selected which lacked either the nitrate reductase apoenzyme or the molybdenum-containing cofactor.

King and Khanna (63) also used chlorate to select a D. innoxia cell line lacking nitrate reductase activity while Murphy and Imbrie (64) selected Rosa damascena lines which either lacked nitrate reductase activity or could somehow tolerate chlorate and its reduction products.

Another directed suicide method has been proposed by Widholm (65) to select cells which lack the enzyme tryptophan synthase. Several indole analogs can be converted by this enzyme into toxic tryptophan analogs. Even though this selection system would seem to be useful for selecting cells lacking tryptophan synthase activity, we have not been successful in doing so thus far.

Recently several auxotrophs have been selected using "brute force" methods whereby mutagenized haploid cells or protoplasts are plated in enriched medium. The individual colonies which form are removed by hand and plated on minimal and complete medium for growth. Savage et al. (66) and King et al. (67) have described D. innoxia lines which require pantothenate and adenine, respectively. Sidorov et al. (68) selected two N. plumbaginifolia lines which required either uracil or isoleucine for growth. Plants could be regenerated only from the isoleucine-requiring line. These plants also required isoleucine apparently due to the absence of threonine deaminase activity. Strauss et al. (69) and Gebhardt et al. (70) have described several mutants selected from Hyoscyamus muticus protoplasts. These include three which require a group of amino acids or certain single amino acids for growth due to the lack of nitrate reductase. Two other clones required histidine, one required tryptophan and three others required nicotinamide. This H. muticus selection also included testing of the clones at 32°C for temperature-sensitives and two were recovered which either stopped growing or exhibited chlorosis at this restrictive temperature.

The results using "brute-force" methods show that such methods can be utilized to obtain auxotrophic and temperature-sensitive mutants. It is also evident that a wide variety of deficiencies can be selected for.

CONCLUSION

The many mutants described in this review have already provided much information concerning basic biochemical and genetic processes. The lines have been and will be used also in many genetic manipulation experiments as markers. There are also several examples where in vitro selection can result in the regeneration of plants which are disease, herbicide, NaCl, amino acid analog and enzyme inhibitor (isonicotinic acid hydrazide and aminopterin) resistant. Thus, when plant regeneration techniques are developed for more crop species, these selection methods for useful traits can be applied with every reason to expect success in improving crop plants.

ACKNOWLEDGEMENTS

I wish to thank the Organizing Committee for inviting me to this symposium and for paying part of the travel expenses. The unpublished results reported here were obtained using funds from the Illinois Agricultural Expereiment Station, the National Science Foundation and the United States Department of Agriculture, Science and Education Administration.

REFERENCES

1. P. Maliga, Resistance mutants and their use in genetic manipulation, in: "Frontiers of plant Tissue Culture," T. Thorpe, ed., University of Calgary, Calgary (1978).
2. P. Maliga, The need and the search for genetic markers in plant cell cultures, in: "Plant Cell Cultures: Results and Perspectives," F. Sala, B. Parisi, R. Cella and O. Ciferri, eds., Elsevier/North-Holland Biomedical Press, Amsterdam (1980).
3. J. M. Widholm, The selection of agriculturally desirable traits with cultured plant cells, in: "Propagation of Higher Plants Through Tissue Culture - A Bridge Between Research and Application," K.W. Hughes, R. Henke and M. Constantin, eds., Technical Info. Center, U.S. Dept. of Energy, Washington, DC (1978).
4. W. Handro, Mutagenesis and in vitro selection, in: "Plant Tissue Culture Methods and Applications in Agriculture," T.A. Thorpe, ed., Academic Press, New York (1981).
5. D.S. Ingram, The expression of R-gene resistance to *Phytophthora infestans* in tissue cultures of *Solanum tuberosum*, *J.*

Gen. Micro. 49:99-108 (1967).
6. J.P. Helgeson, J.D. Kemp, G.T. Haberlach and D.P. Maxwell, A tissue culture system for studying disease resistance: the black shank disease in tobacco callus cultures, Phytopathol. 62:1439-1443 (1972).
7. J. Ruyack, M.R. Downing, J.S. Chang and E.D. Mitchell, Growth of callus and suspension culture cells from cotton varieties (Gossypium hirsutum L.) resistant and susceptible to Xanthomonas malvacearum (E.F.SM.) Dows., In Vitro 15:368-373 (1979).
8. B.G. Gengenbach, C.E. Green and C.M. Donovan, Inheritance of selected pathotoxin resistance in maize plants regenerated form cell cultures, Proc. Natl. Acad. Sci. USA 74:5113-5117 (1977).
9. R.I.S. Brettell, E. Thomas and D.S. Ingram, Reversion of Texas male-sterile cytoplasm maize in culture to give fertile, T-toxin resistant plants, Theor. Appl. Genet. 58:55-58 (1980).
10. B.G. Gengebach and C.E. Green, Selection of T-cytoplasm maize callus cultures resistant to Helminthosporium maydis race T pathotoxin, Crop Sci. 15:645-649 (1975).
11. B.G. Gengenbach, J.A. Connelly, D.R. Pring and M.F. Conde, Mitochondrial DNA variation in maize plants regenerated during tissue culture selection, Theor. Appl. Genet. 59:161-167 (1981).
12. M. Behnke, Selection of potato callus for resistance to culture filtrates of Phytophthora infestans and regeneration of resistant plants, Theor. Appl. Genet. 55:69-71 (1979).
13. M. Behnke, Selection of dihaploid potato callus for resistance to the culture filtrate of Fusarium oxysporum, Z. Pflanzenzuchtg. 85:254-258 (1980).
14. T.H. Oswald, A.E. Smith and D.V. Phillips, Phytotoxicity and detoxification of metribuzin in dark grown suspension cultures of soybean, Pesticide Biochem. Physiol. 8:73-83 (1977).
15. E.B. Swanson and D.T. Tomes, In vitro responses of tolerant and susceptible lines of Lotus corniculatus L. to 2,4-D, Crop Sci. 20:792-795 (1980).
16. R.S. Chaleff and M.F. Parsons, Direct selection in vitro for herbicide-resistant mutants of Nicotiana tabacum, Proc. Natl. Acad. Sci. USA 75:5104-5107 (1978).
17. R.S. Chaleff, Further characterization of picloram-tolerant mutants, Theor. Appl. Genet. 58:91-95 (1980).
18. D.K. Miller and K.W. Hughes, Selection of paraquat-resistant variants of tobacco from cell cultures, In Vitro 16:1085-1091 (1980).
19. D.N. Radin and P.S. Carlson, Herbicide-tolerant tobacco mutants selected in situ and recovered via regeneration from cell culture, Genetical Res. 32:85-89 (1978).
20. M. Tal, H. Heikin and K. Dehan, Salt tolerance in the wild relatives of the cultivated tomato: responses of callus tissue of Lycopersicon esculentum, L. peruvianum and Solanum

pennelli to high salinity, Z. Pflanzenphysiol. 86:231-240 (1978).
21. T.J. Orton, Comparison of salt tolerance between Hordeum vulgare and H. jubatum in whole plants and callus cultures, Z. Pflanzenphysiol. 98:105-118 (1980).
22. M.K. Smith and J.A. McComb, Effect of NaCl on the growth of whole plants and their corresponding callus cultues, Aust. J. Plant Physiol. 8:267-275 (1981).
23. M.W. Nabor, S.E. Gibbs, C.S. Bernstein and M.E. Meis, NaCl-tolerant tobacco plants from cultured cells, Z. Pflanzenphysiol. 97:13-17 (1980).
24. A.K. Tyagi, A. Rashid and S.C. Maheshwari, Sodium chloride resistant cell line from haploid Datura innoxia Mill. A resistance trait carried from cell to plantlet and vice versa in vitro, Protoplasma 105:327-332 (1981).
25. J.R. Wong and I.M. Sussex, Isolation of abscisic acid-resistant variants from tobacco cell cultures. I. Physiological bases for selection, Planta 148:97-102 (1980).
26. J.R. Wong and I.M. Sussex, Isolation of abscisic acid-resistant variants from tobacco cell cultures. II. Selection and characterization of variants, Planta 148:103-107 (1980).
27. J M. Widholm, Differential expression of amino acid biosynthetic control isoenzymes in plants and cultured cells, in: "Plant Cell Cultures: Results and Perspectives," F. Sala, B. Parisi, R. Cella and O. Ciferri, eds., Elsevier/North-Holland Biomedical Press, Amsterdam (1980).
28. B. Reisch, S.H. Duke and E.T. Bingham, Selection and characterization of ethionine-resistant alfalfa (Medicago sativa L.) cell lines, Theor. Appl. Genet. 59:89-94 (1981).
29. S.W.J. Bright, P.B. Norbury and B.J. Miflin, Isolation of a recessive barley mutant resistant to S(2-aminoethyl)L-cysteine, Theor. Appl. Genet. 55:1-4 (1979).
30. S.W.J. Bright, L.C. Featherstone and G.J. Miflin, Lysine metabolism in a barley mutant resistant to S(2-aminoethyl)cysteine, Planta 146:629-633 (1979).
31. K.A. Hibberd, T. Walter, C.E. Green and B.G. Gengenbach, Selection and characterization of a feedback-insensitive tissue culture of maize, Planta 148:183-187 (1980).
32. J.M. Widholm, Selection and characterization of cultured carrot and tobacco cells resistant to lysine, methionine and proline analogs, Can. J. Bot. 54:1523-1529 (1976).
33. J.S.H. Kueh and S.W.J. Bright, Proline accumulation in a barley mutant resistant to trans-4-hydroxy-L-proline, Planta 153:166-171 (1981).
34. J.E. Palmer and J.M. Widholm, Characterization of carrot and tobacco cells resistant to P-fluorophenylalanine, Plant Physiol. 56:233-238 (1975).
35. J. Berlin, Para-fluorophenylalanine resistant cell lines of tobacco, Z. Pflanzenphysiol. 97:317-324 (1980).

36. D.A. Evans and O.L. Gamborg, Effects of para-fluorophenylalanine on ploidy levels of cell suspension cultures of Datura innoxia, Environ. Expt. Bot. 19:269-275 (1979).
37. C.E. Flick, R.A. Jensen and D.A. Evans, Isolation , protoplast culture, and plant regeneration of PFP-resistant variants of Nicotiana tabacum Su/Su, Z. Pflanzenphysiol. 103:239-245 (1981).
38. J.F. Thompson, J.T. Madison, M.A. Waterman and A.E. Muenster, Effect of methionine on growth and protein composition of cultured soybean cotyledons, Phytochem. 20:941-945 (1981).
39. L. Wu and J. Antonovics, Zinc and copper tolerance of Agrostis stolonifera L. in tissue culture, Amer. J. Bot. 65:268-271 (1978).
40. C.P. Meredith, Selection and characterization of aluminum-resistant variants from tomato cell cultures, Plant Sci. Lett. 12:25-34 (1978).
41. C.M. Colijn, A.J. Kool and J.J.J. Nijkamp, An effective chemical mutagenesis procedure for Petunia hybrida cell suspension cultures, Theor. Appl. Genet. 55:101-106 (1979).
42. Y.M. Heimer, J.L. Wray and P. Filner, The effect of tungstate on nitrate assimilation in higher plant tissues, Plant Physiol. 44:1197-1199 (1969).
43. J.M. Widholm, J.P. Ranch and K. Wakasa, Tungstate as a selective agent for high nitrate reductase activity in plant cells, In Vitro 17:211 (1981).
44. M.B. Berlyn, Isolation and characterization of isonicotinic acid hydrazide-resistant mutants of Nicotiana tabacum, Theor. Appl. Genet. 58:19-26 (1980).
45. I. Zelitch and M.B. Berlyn, Altered glycine decarboxylation inhibition in isonicotinic acid hydrazide-resistant mutant callus lines and in regenerated plants and seed progeny, Plant Physiol. 69:198-204 (1982).
46. A.L. Lawyer, M.B. Berlyn and I. Zelitch, Isolation and characterization of glycine hydroxamate-resistant cell lines of Nicotiana tabacum, Plant Physiol. 66:334-341 (1980).
47. C.R. Somerville and W.L. Ogren, A phosphoglycolate phosphatase deficient mutant of Arabidopsis, Nature 280:833-836 (1979).
48. C.R. Somerville and W.L. Ogren, Photorespiration mutants of Arabidopsis thaliana deficient in serine-glyoxylate aminotransferase activity, Proc. Natl. Acad. Sci. USA 77:2684-2687 (1980).
49. P.J. Dix and H.E. Street, Selection of plant cell lines with enhanced chilling resistance, Ann. Bot. 40:903-910 (1976).
50. K.M. Templeton-Somers, W.R. Sharp and R.M. Pfister, Selection of cold-resistant cell lines of carrot, Z. Pflanzenphysiol. 103:139-148 (1981).
51. P.J. Dix, Chilling tolerance is not transmitted sexually in plants regenerated from Nicotiana sylvestris cell lines, Z. Pflanzenphysiol. 84:223-226 (1977).

52. J.M. Mastrangelo and H.H. Smith, Selection and differentiation of aminopterin resistant cells of Datura innoxia, Plant Sci. Lett. 10:171-179 (1977).
53. A. Maretzki and M. Thom, Characteristics of a galactose-adapted sugarcane cell line grown in suspension culture, Plant Physiol. 61:544-548 (1978).
54. R.S. Chaleff and M.F. Parsons, Isolation of a glycerol-utilizing mutant of Nicotiana tabacum, Genetics 89:723-728 (1978).
55. M. Limberg, D. Cress and K.G. Lark, Variants of soybean cells which can grow in suspension with maltose as a carbon-energy source, Plant Physiol. 63:718-721 (1979).
56. B. Reisch and E.T. Bingham, Plants from ethionine-resistant alfalfa tissue cultures: variation in growth and morphological characteristics, Crop Sci. 21:783-788 (1981).
57. J.F. Shepard, D. Bidney and E. Shahin, Potato protoplasts in crop improvement, Science 208:17-24 (1980).
58. P.S. Carlson, Induction and isolation of auxotrophic mutants in somatic cell cultures of Nicotiana tabacum, Science 168:487-489 (1970).
59. R.L. Malmberg, Temperature-sensitive variants of Nicotiana tabacum isolated from somatic cell culture, Genetics 92:215-221 (1979).
60. R.L. Malmberg, Biochemical, cellular and developmental characterization of a temperature-sensitive mutant of Nicotiana tabacum and its second site revertant, Cell 22:603-609 (1980).
61. J.C. Polacco, Arsenate as a potential negative selection agent for deficiency variants in cultured plant cells, Planta 146:155-160 (1979).
62. A.J. Muller and R. Grafe, Isolation and characterization of cell lines of Nicotiana tabacum lacking nitrate reductase, Molec. Gen. Genet. 161:67-76 (1978).
63. J. King and V. Khanna, A nitrate reductase-less variant isolated from suspension cultures of Datura innoxia (Mill.), Plant Physiol. 66:632-636 (1980).
64. T.M. Murphy and C.W. Imbrie, Induction and characterization of chlorate-resistant strains of Rosa damascena cultured cells, Plant Physiol. 67:910-916 (1981).
65. J.M. Widholm, Utilization of indole analogs by carrot and tobacco cell tryptophan synthase in vivo and in vitro, Plant Physiol. 67:1101-1104 (1981).
66. A.D. Savage, J. King and O.L. Gamborg, Recovery of a pantothenate auxotroph from a cell suspension culture of Datura innoxia Mill., Plant Sci. Lett. 16:367-376 (1979).
67. J. King, R.B. Horsch and A.D. Savage, Partial characterization of two stable auxotrophic cell strains of Datura innoxia Mill., Planta 149:480-484 (1980).
68. V. Sidorov, L. Menczel and P. Maliga, Isoleucine-requiring Nicotiana plant deficient in threonine deaminase, Nature

194:87-88 (1981).

69. A. Strauss, F. Bucher and P.J. King, Isolation of biochemical mutants using haploid mesophyll protoplasts of Hyoscyamus muticus. I. A NO_3^- non-utilizing clone, Planta 153:75-80 (1981).

70. C. Gebhardt, V. Schnebli and P.J. King, Isolation of biochemical mutants using haploid mesophyll protoplasts of Hyoscyamus muticus. II. Auxotrophic and temperature-sensitive clones, Planta 153:81-89 (1981).

DEVELOPMENT OF ZYGOTIC BARLEY EMBRYOS *IN VITRO*

J.M. Dunwell

John Innes Institute
Colney Lane
Norwich,
UNITED KINGDOM

INTRODUCTION

Both mature and immature embryos have frequently been used as sources of material for studies of callus initiation and regeneration in cereals (1). Although they have proved to be the explants most amenable to culture, the substantial inter- and intraspecific differences in response remain an obstacle to any application of cell culture methods to variety improvement. As part of a continuing investigation into this problem a previous study conducted on three contrasting barley genotypes examined the influence of incubation temperature and medium composition on the growth of mature embryos *in vitro* (2). These results were discussed in relation to the known dormancy characteristics of some of these genotypes (1). It was found that there was a significant difference in the response of these genotypes to particular combinations of 2,4-D and sucrose present in the culture medium.

The present study examines in detail the variation of growth under a range of temperature regimes and involves the transfer of cultures between different temperatures at a number of intervals during the culture period. A similar type of analysis of temperature effects had previously been conducted on dormant oat seeds with a view to studying the control of dormancy (3).

MATERIALS AND METHODS

The two lines of barley used in this study were R231 and 4082. Their source and the methods of embryo isolation and culture have been described previously (2). In the first experiment equal numbers of embryos were cultured on each of nine media comprising all possible combinations of 2,4-D (2, 4 and 8 mgl^{-1}) and sucrose (3, 6 and 9%). Embryos were incubated in the dark at 5°C and equal numbers were then transferred after 7 d, 14 d or 21 d to one of the following temperatures: 5, 10, 15, 20 or 25°C. After a total of 49 d incubation, embryos were scored for shoot length as well as fresh and dry weight. In order to facilitate analysis of the large amount of data, mean figures for each parameter were reduced to zero (2).

In the second experiment the media combinations were reduced to four (2,4-D, 4 and 8 mgl^{-1}; sucrose, 3 and 6%). Embryos were incubated at one of the five temperatures before equal numbers were transferred after either 7 d or 14 d to each of the other four temperatures. As a control, one sample of embryos was kept at each temperature throughout the entire culture period. Embryos were scored as above after 49 d incubation. Data were then averaged for each parameter to give values for all cultures which had been incubated at each temperature initially, and also the value for all cultures transferred into each temperature. Analysis of these two groups of values thus enabled the relative importance of the two culture periods to be estimated.

RESULTS

The influence of temperature on shoot length, dry weight and water content is given in Figures 1, 2 and 3 respectively. Considering first the shoot length data (Figure 1), there appears to be a gradual increase as the temperature increases for R231, whereas 20°C is the optimum temperature for 4082. With R231 embryos there is a decline in length as both the sucrose and 2,4-D concentration increases, whereas 4082 shows the similar decline for 2,4-D but has a maximum value at the intermediate sucrose level.

Dry weight data (Figure 2) for R231 show an overall increase with temperature, whereas results with 4082 are much more variable; 20°C seems to give maximum values although 7 days (10°C) also gives a high value. In all cases there is a steady increase with increasing 2,4-D/sucrose concentration.

For water content (Figure 3) there is little variation with temperature; for R231 20°C has the highest value and widest range. For 4082 the variation is greatest at the low temperatures and as the temperature increases, variation decreases. In all cases the values decline with increasing sucrose concentration, and there is little variation with increasing 2,4-D concentration.

When the data are plotted using 100 as the value at the lowest concentration, then it is simpler to compare the influence of the

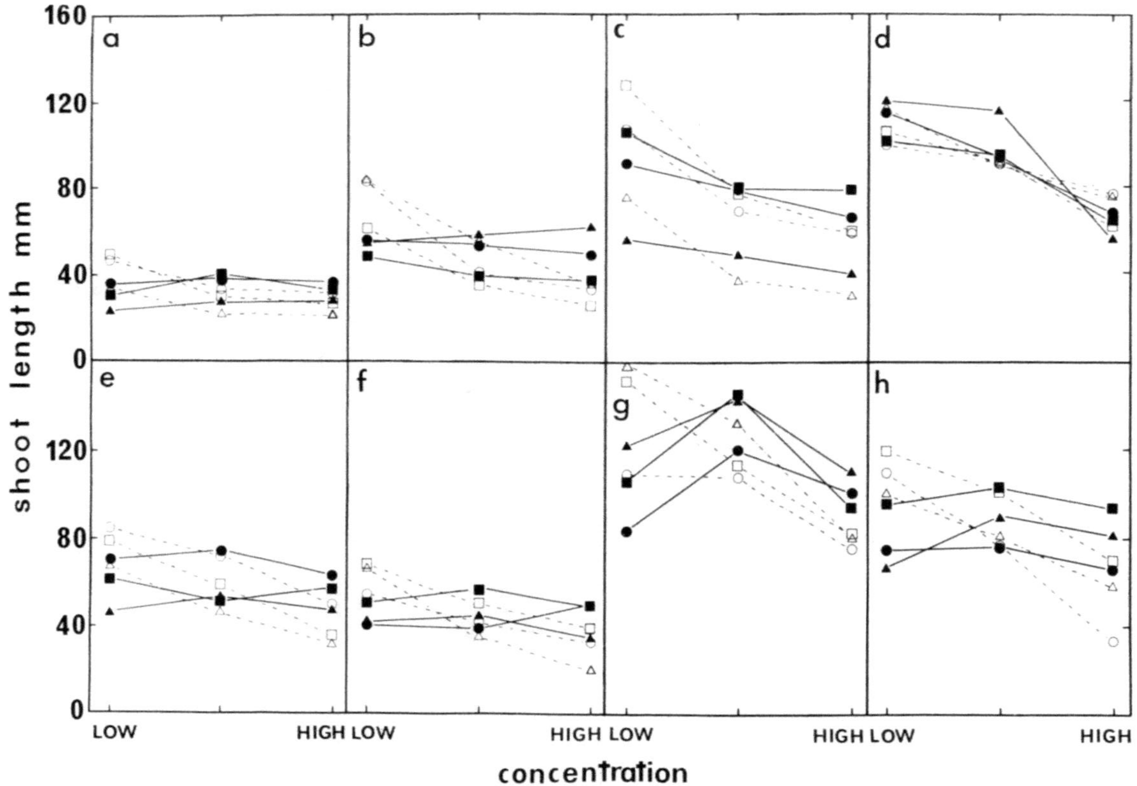

Figure 1. Graphs showing the effect of various concentrations of sucrose (3, 6, 9%) (———) and 2,4-D (2, 4, 8 mgl^{-1}) (-----) on the shoot length of barley embryos of two genotypes, R231 (a, b, c, d) and 4082 (d, e, f, g) incubated at 5°C for 7 d (□ ■), 14 d (△ ▲) or 21 d (○ ●) before transfer to 10 (a, e), 15 (b, f), 20 (c, g) or 25°C (d, h). Cultures were scored after a total of 49 d culture.

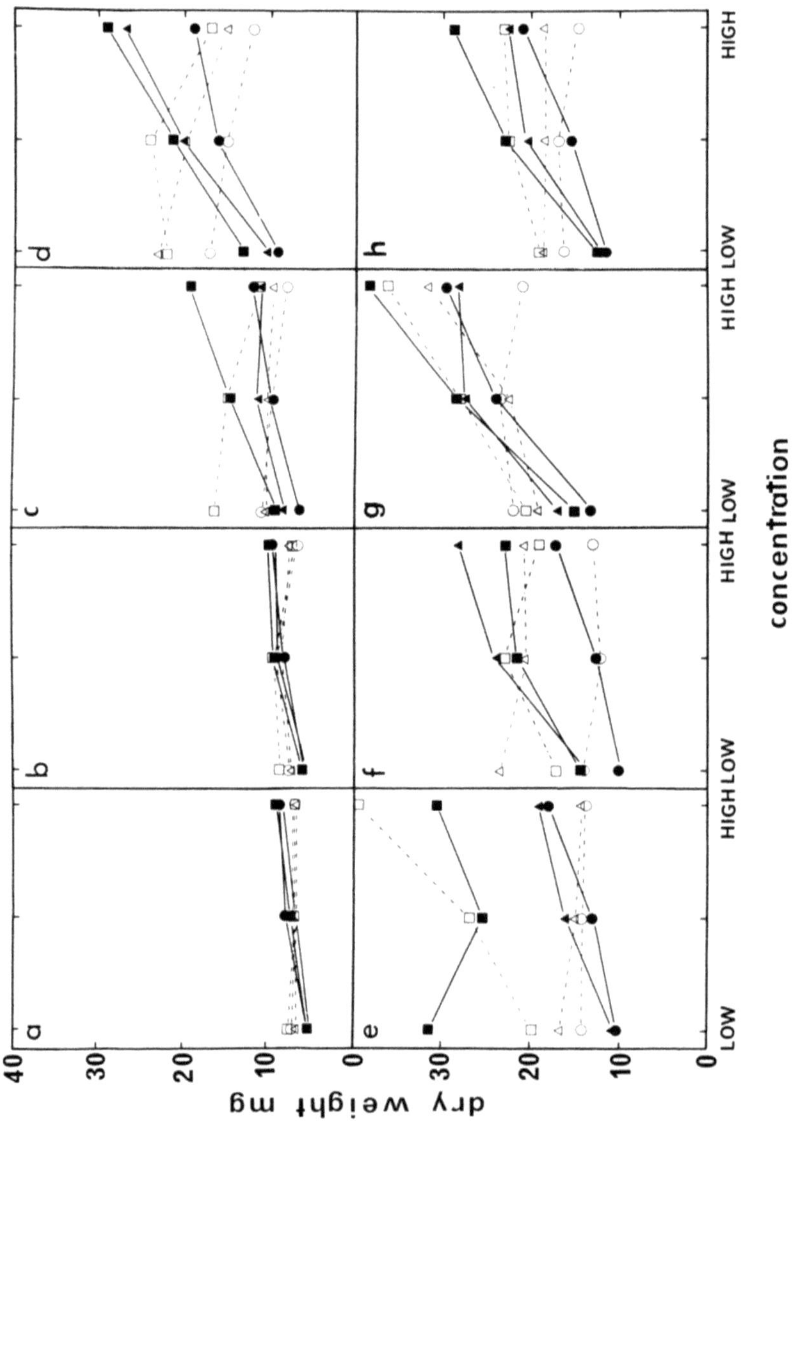

Figure 2. Graphs showing the variation in dry weight of embryos. Details as in Figure 1.

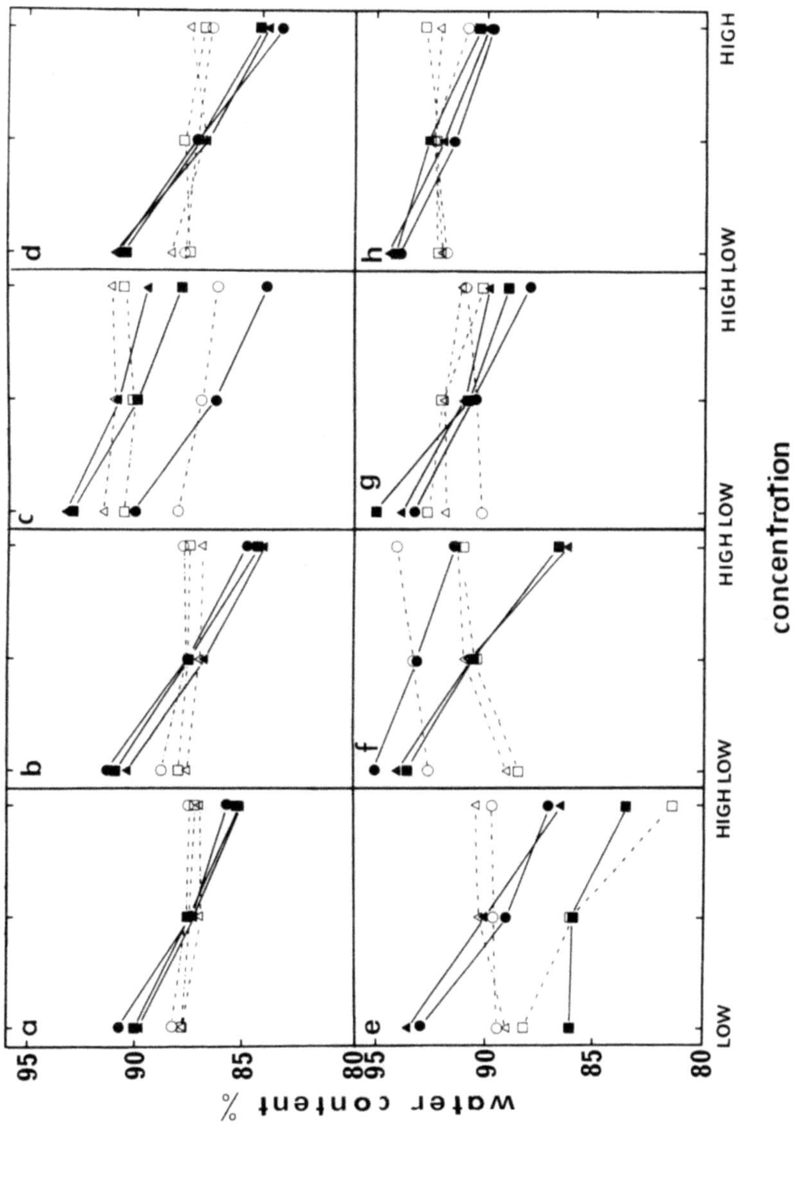

Figure 3. Graphs showing the variation in water content of embryos. Details as in Figure 1.

different periods at 5°C before transfer (Figures 4, 5 and 6). Shoot length (Figure 4) is the parameter which varies most with temperature, and it appears to be relatively unaffected by the length of low temperature treatment. On the other hand, dry weight data (Figure 5) do show a clear effect of pretreatment time, at least with the 4082 genotype. Whereas after the shortest time (7 days) before transfer, temperature has a large influence on response, after 21 days the temperature effect is reduced and the reactions to increasing concentrations of the two media components are more consistent. Water contents (Figure 6) show a similar pattern of response with R231 embryos being relatively unaffected by the length of time spent at the lower temperature before transfer and 4082 embryos showing a decreasing effect of temperature as the pretreatment period is increased.

Data from temperature interchanges are given for R231 in Figure 7 and 4082 in Figure 8. Considering first the data for R231, it seems that the temperature of the first culture period has little effect on shoot length and only a slight effect on dry weight, while these parameters show a more pronounced increase with temperature for the second culture period. Water content data show little effect of temperature during either culture period, except for 5°C during the second period. The only parameters to show a distinct difference between the 7 and 14 day transfer time are the shoot length and more especially the water content data for this 5°C second culture period. The effects of 2,4-D and sucrose, particularly on dry weight, are as described previously and increase in magnitude as the temperature increases.

The 4082 data for shoot length demonstrate a similar increase with temperature for both the first and second culture period with 2,4-D and sucrose having a greater effect than that shown with R231. Dry weight and water content values, however, seem to be affected by temperature only at the higher 2,4-D and sucrose concentrations. None of the three parameters show any consistent difference between the 7 and 14 day transfer.

DISCUSSION

Considered in conjunction with the results of the previous studies (2, 4), the present data extend the information available on the response of these two barley genotypes to particular culture media and incubation temperature combinations. Although certain other studies (5, 6) have investigated the effect of various 2,4-D concentrations on isolated barley embryos, they did not include an examination of the influence of temperature. It can now be seen that different growth parameters have different optimal temperature regimes, and that such regimes depend upon the genotype employed. It has long been known that temperature is of utmost importance in

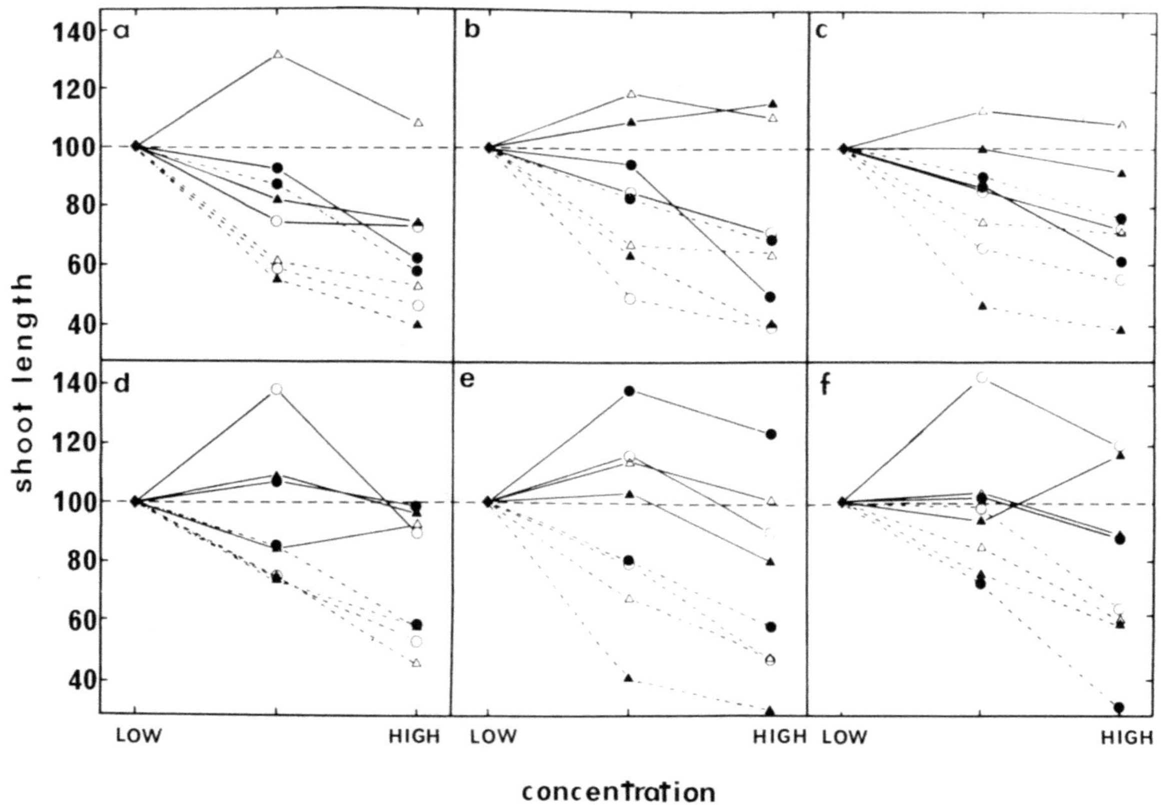

Figure 4. Graphs showing the effect of various concentrations of sucrose (———) and 2,4-D (– – –) on the shoot length of barley embryos of two genotypes, R231 (a, b, c) and 4082 (d, e, f) after adjusting the value at the lowest concentration to equal 100. Embryos were cultured at 5°C for 7 (a, d), 14 (b, e) or 21 d (c, f) before transfer to 10 (Δ), 15 (▲), 20 (○) or 25°C (●).

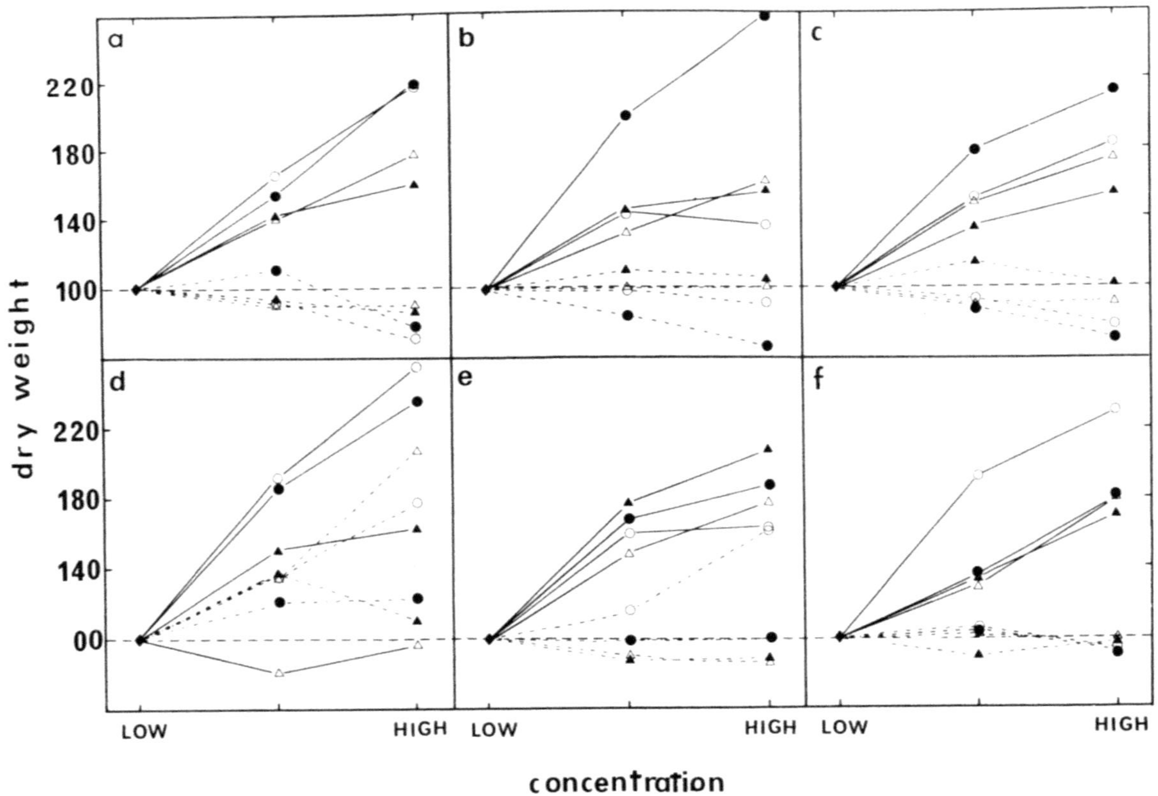

Figure 5. Graphs showing the variation in dry weight of embryos. Details as in Figure 4.

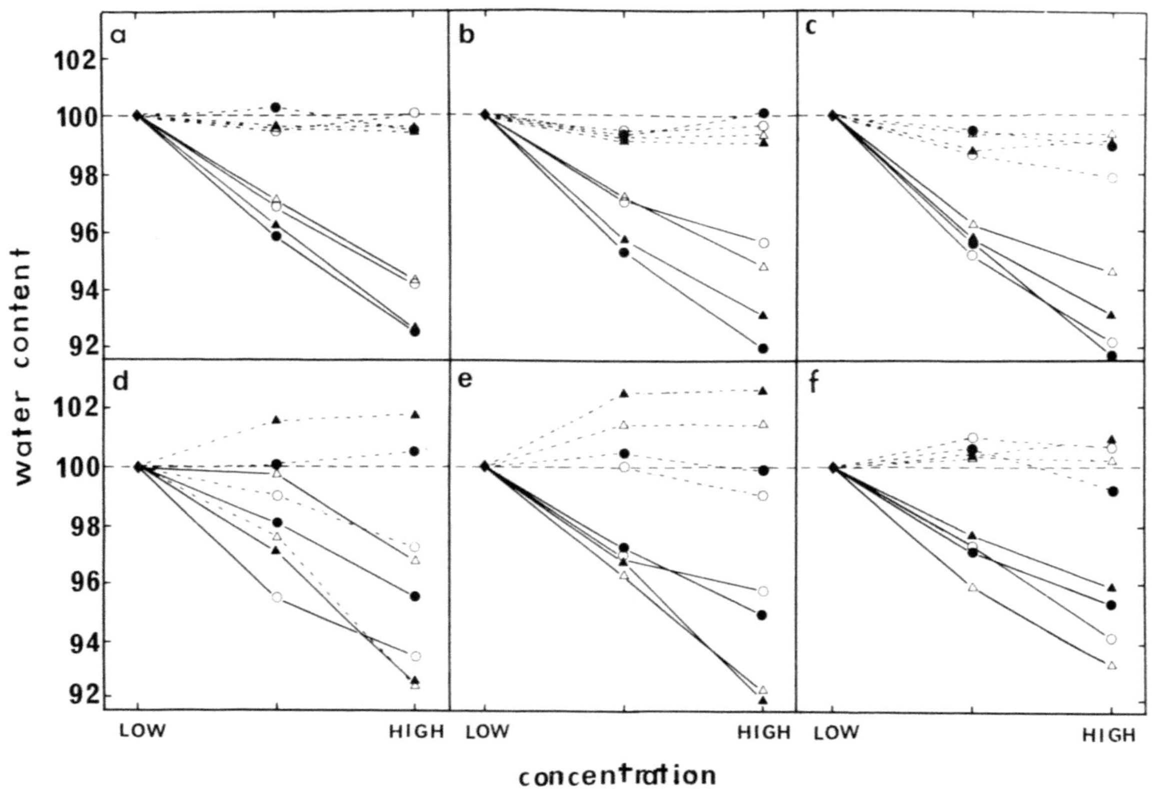

Figure 6. Graphs showing the variation in water content of embryos. Details as in Figure 4.

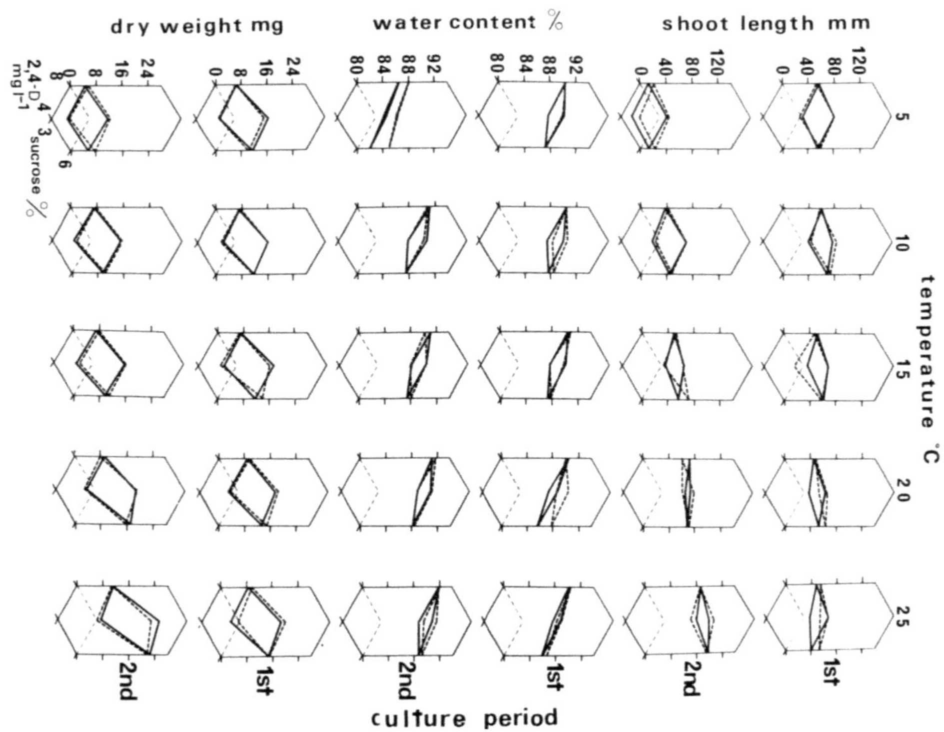

Figure 7. Graphs showing the effect of sucrose and 2,4-D on shoot length, water content and dry weight of barley embryos of R231 genotype incubated at 5, 10, 15, 20 or 25°C for 7 d (——) or 14 d (----). After this initial period the cultures from each temperature were distributed amongst each of the five temperatures for the second period of incubation. Cultures were scored after a total of 49 d. Data were combined and are plotted to show the values for all cultures incubated at each temperature for either the first or second period of culture.

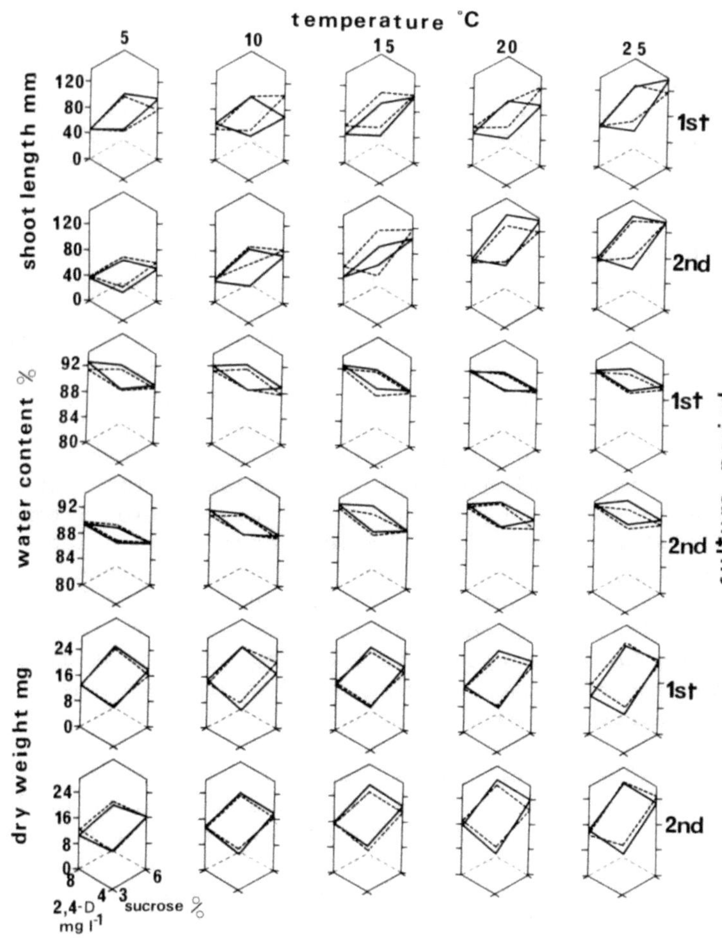

Figure 8. Graphs showing the effect of sucrose and 2,4-D on 4082 embryos. Details as in Figure 7.

determining germination rates (3, 7, 8), and it is, therefore, not unexpected that there are comparable effects on growth of isolated embryos.

The dormancy expressed by a mature caryopsis develops during its maturation and is dependent upon the physical environment during that maturation (9). It is also known that callus initiation on mature embryos depends not only upon the genotype but also upon the batch of seed, that is, upon the environment during the development of that seed (5). It has been concluded specifically that low temperatures during the drying of immature caryopses modify the subsequent behavior of the aleurone layer, delaying the onset and reducing the rate of gibberellic acid-induced processes which culminate in the production of α-amylase (10). Since it is already known from other species that the behavior of a range of explants in culture is dependent upon the environment under which the parent plant is grown (11), it may be possible to increase the responsiveness of embryos in culture by such environmental manipulation.

One of the important distinctions to be drawn between the study described here and other investigations of barley embryos in culture is the difference in development age. The source of callus and sites of cell division appear to differ between immature embryos where the scutellum is the responsive tissue (6), and more mature embryos where mesocotyl proliferation occurs (5, 12). Callus from these two sources has a different regenerative potential with the compact, nodular scutellar callus being the more amenable. As in other members of the Gramineae (13), this regeneration probably proceeds via somatic embryogenesis (14). Differentiation from the soft, friable callus formed from the mesocotyl is much less efficient and apparently proceeds via an organogenetic process. It has been suggested that the former process is much more stable genetically and is less liable to produce aberrant regenerants. Therefore, it will probably be advantageous in the long term to study methods for the production of slower growing nodular callus rather than the larger celled watery callus.

As a means of altering callus morphology it may be useful to investigate culture media of higher osmotica than often used (15). Previous studies on regeneration from barley callus have concentrated on low (3-4%) sucrose concentrations (6, 16 - 20) with the exception of Granatek and Cockerline (15), who examined the influence of higher sucrose concentrations (9%). However, levels of sucrose between 1 and 5% have been tested on Triticale embryo callus where 2% was found to produce the highest fresh weight (21) (cf. Figures 2, 5). During the maturation of an embryo, its water content is continuously declining, to rise again on germination (22). However, when an immature embryo is removed from its normal environment to the artificial one provided by culture, the process of dehydration is halted and precocious germination is often induced (23, 24). In order to maintain the continuing development of an embryo and suppress precocious germination, the usual culture medium can be adjusted by the addition of abscisic acid (ABA) and extra ammonium ions (25). It is not clear whether further addition of 2,4-D as supplied in the present study would stimulate the proliferation of a scutellum whose normal growth is promoted by ABA.

Although, in comparisons of genotypes, the varieties Golden Promise (26) and the related Maythorpe (14) have been found to be especially easy to regenerate from immature scutella, it is not known whether this characteristic is controlled by any particular gene(s). It is hoped that such a control will soon be identified by the use of marked genetic stocks such as those employed in this study.

ACKNOWLEDGEMENT

The author is most grateful for the careful technical assistance of Margaret Cornish.

REFERENCES

1. J.M. Dunwell, Dormancy and germination in embryos of Hordeum vulgare L. Effect of dissection, incubation temperature and hormone application, Ann. Bot., 48:203-213 (1981a).
2. J.M. Dunwell, Influence of genotype and environment on growth of barley embryos in vitro, Ann. Bot. 48:535-542 (1981b).
3. C.J. Andrews and V.D. Burrows, Germination response of dormant seeds to low temperature and gibberellin, Can. J. Pl. Sci. 52:295-303 (1972).
4. J.M. Dunwell, Genotypic variation in the response of barley embryos to growth in vitro, Proc. 4th Intl. Barley Genetics Symp. (in press) (1982).
5. M.W. Bayliss and S.D.M. Dunn, Factors affecting callus formation from embryos of barley (Hordeum vulgare), Pl. Sci. Lett. 14:311-316 (1979).
6. P.J. Dale and E. Deambrogio, A comparison of callus induction and plant regeneration from different explants of Hordeum vulgare, Z. Pflanzenphysiol. 94:65-77 (1979).
7. I. Nishiyama, A break on the Arrhenius plot of germination activity in rice seeds, Pl. Cell Physiol. 16:533-536 (1975).
8. E.H. Roberts and R.D. Smith, Dormancy and the pentose phosphate pathway, in: "The Physiology and Biochemistry of Seed Dormancy and Germination," A.A. Khan, ed., Elsevier/North Holland Biomedical Press, Amsterdam (1977).
9. T. Buraas and K. Aastveit, Investigations on seed dormancy in barley, Proc. 4th Intl. Barley Genetics Symp. (in press) (1982).

10. P.B. Nicholls, Development of responsiveness to gibberellic acid in the aleurone layer of immature wheat and barley caryopses: effect of temperature, Austr. J. Pl. Phys. 7:645-653 (1980).
11. J.M. Dunwell, A comparative study of environmental and developmental factors which influence embryo induction and growth in cultured anthers of Nicotiana tabacum, Env. Exp. Bot. 16:109-118 (1976).
12. C.H. Granatek and A.W. Cockerline, Callus formation of cultured early differentiating barley embryos, Bull. Torrey Bot. Club 106:85-96 (1979).
13. P.J. Dale, Embryoids from cultured immature embryos of Lolium multiflorum, Z. Pflanzenphysiol. 100:73-77 (1980).
14. C.J. Jensen, Regenerating monoploid barley plants from haploid cells, tissues and organs of various origin, Proc. 4th Intl. Barley Genetics Symp. (in press) (1982).
15. C.H. Granatek and A.W. Cockerline, Callus formation versus differentiation of cultured barley embryos: hormonal and osmotic interactions, In Vitro 14:212-217 (1978).
16. T.Y. Cheng and H.H. Smith, Organogenesis from callus culture of Hordeum vulgare, Planta 123:307-310 (1975).
17. H-J. Jacobsen, Genotypic and environmental influences on germination and callus induction in Hordeum distichum, Biochem. Physiol. Pflanzen 169:453-460 (1976).
18. T.J. Orton, A quantitative analysis of growth and regeneration from tissue cultures of Hordeum vulgare, H. jubatum and their interspecific hybrid, Env. Exp. Bot. 19:319-335 (1979).
19. T.J. Orton, Haploid barley regenerated form callus cultures of Hordeum vulgare x H. jubatum, J. Hered. 71:280-282 (1980a).
20. T.J. Orton, Chromosomal variability in tissue cultures and regenerated plants of Hordeum, Theor. Appl. Genet. 56:101-112 (1980b).
21. G.C. Sharma, L.L. Bello, V.T. Sapra and C.M. Peterson, Callus initiation and plant regeneration from Triticale embryos, Crop Sci. 21:113-118 (1981).
22. C.A. Adams and R.W. Rinne, Moisture content as a controlling factor in seed development and germination, Intl. Rev. Cytol. 68:1-8 (1980).
23. K. Norstog, Factors relating to precocious germination in cultured barley embryos, Phytomorph. 22:134-139 (1975).
24. K. Norstog and R.M. Klein, Development of cultured barley embryos. II. Precocious germination and dormancy, Can. J. Bot. 50:1887-1894 (1972).
25. P.F. Umbeck and K. Norstog, Effects of abscisic acid and ammonium ions on morphogenesis of cultured barley embryos, Bull. Torrey Bot. Club 106:110-116 (1979).
26. B.J. Miflin, S.J. Bright and E. Thomas, Towards the genetic manipulation of barley, Proc. 4th Intl. Barley Genetics Symp. (in press) (1982).

27. T. Murashige and F. Skoog, A revised medium for rapid growth and bioassays with tobacco tissue culture, Physiologia Plantarum 15:473-497 (1962).

SIGNIFICANCE OF EMBRYO DIFFERENTIATION IN BARLEY INTERSPECIFIC HYBRIDIZATION AND HAPLOID FORMATION

N.C. Subrahmanyam

School of Life Sciences
Central University of Hyderabad
Hyderabad - 500134
INDIA

Reciprocal crosses between cultivated barley *Hordeum vulgare* (2x) and *H. bulbosum* (2x) results in haploids of *H. vulgare* (1) through selective elimination of *bulbosum* chromosomes from hybrid embryos (2, 3). In the hybrid plants that do not undergo elimination, chromosome instability exists but the extent of instability varies with the genome balance (ratio) of the original cross (4 - 7). The genome balance phenomenon is under the control of genetic factors located on chromosomes 2 and 3 of *vulgare* (8 - 10) whose dose ratio to *bulbosum* chromosomes is critical to the elimination process (2). The phenomenon of selective chromosome elimination leading to haploid formation is widespread among *Hordeum* interspecific hybrids (11 - 15) and intergeneric crosses involving *H. bulbosum* or *H. vulgare* (16 - 19). The evidence indicates that *Hordeum* species differ in their ability to eliminate, and the elimination is dependent on the ratio of the parental genomes in any cross combination. These findings in *Hordeum* interspecific hybrids led to the proposal of a hierarchy of species in dominance in selective chromosomal elimination (20).

In the present study, interspecific cross combinations involving *H. bulbosum* and/or *H. vulgare* with other *Hordeum* species (13 - 15) were compared with respect to embryo morphology and the nature of progeny following the standard technique (21).

The parentage of seed set in each cross (Tables 1 - 3) was variable depending on the vigor of the parents and the environmental conditions at the time of pollination. Therefore, the difference in seed setting ability among different crosses are not comparable. At about two weeks after pollination the caryopses are small and begin

Table 1. Crosses Involving Diploid *Hordeum bulbosum*.

Parental species and ploidy (x = 7)	No. of florets pollinated	Seed set %	Embryos cultured Ud	Embryos cultured Vd	Plants obtained Ud	Plants obtained Vd	n	2n
H. VULGARE (2x)	4086	61	1596	722	147	362	506	3
H. BRACHYANTHERUM (4x)	558	18	15	58	1	20	19	2
H. DEPRESSUM (4x)	285	40	20	94		41	8	33
H. PARODII (6x)	1164	8	18	52		35	35	
H. PROCERUM (6x)	447	12	-	26		15	15	
H. LECHLERI (6x)	796	18	1	77		16	16	
H. ARIZONICUM (6x)	215	47	13	66		34	34	

Table 2. Crosses Involving Tetraploid *Hordeum bulbosum*.

Parental species and ploidy (x = 7)	No. of florets pollinated	Seed set %	Embryos cultured Ud	Embryos cultured Vd	Plants obtained Ud	Plants obtained Vd	n	2n
H. BRACHYANTHERUM (4x)	214	25.7	-	33		3		3
H. DEPRESSUM (4x)	429	27.0	35	70		10		10
H. PARODII (6x)	422	6.9	0	14		13		13
H. PROCERUM (6x)	784	11.2	-	69		23	5	18
H. LECHLERI (6x)	267	26.6	-	49		-		
H. ARIZONICUM (6x)	258	25.9	4	48		30	18	3

Table 3. Crosses Involving *Hordeum vulgare*.

Parental species and ploidy (x = 7)	No. of florets pollinated	Seed Set %	Embryos cultured Ud	Embryos cultured Vd	Plants obtained Ud	Plants obtained Vd	n	2n
H. PARODII (6x)	172	11.1	-	11		10		10
H. PROCERUM (6x)	893	6.9	2	50		23	8	11
H. LECHLERI (6x)	119	22.7	-	14		5	5	-
H. ARIZONICUM (6x)	277	15.5	5	30	2	17	13	4

to turn yellow. Usually endosperm becomes exhausted and embryo culture is essential to recover the progeny. The embryos from interspecific hybridizations were always smaller as compared to the embryos obtained from the same spike following self-pollination. At the time of dissection, prior to culturing, embryos were scored as underdeveloped (ud) or visibly differentiated (vd). In caryopses from crosses producing only hybrid progeny or only haploid progeny, all embryos were differentiated. However, the proportion of underdeveloped embryos varied considerably among different crosses, progeny of each of which consisted of haploids and hybrids with varying proportions. The embryos that gave rise to hybrid progeny were always found to be differentiated prior to culturing, whereas haploid progeny originated from both types of embryos (Tables 1 - 3) irrespective of the cross combination. As reported earlier (13 - 15, 20, 21), haploid progeny in any of the crosses resulted from the selective elimination of either bulbosum or vulgare chromosomes during early development of hybrid embryos depending upon the cross combination, yet, a majority of the haploid progeny from different crosses developed from differentiated embryos.

Thus, it is likely that chromosomally balanced proembryos, cells of which have completed selective chromosome elimination (haploids) or with no elimination (hybrids), undergo differentiation within two weeks after pollination, while underdeveloped embryos originate from chromosomally imbalanced proembryos with incomplete and/or ongoing chromosome elimination. A small proportion of the underdeveloped embryos develop into haploid progeny through completion of elimination process at least in cell lines which undergo organogenesis.

There are differences in the proportions of haploids (among the progenies obtained) from hybridizations (Tables 1 - 3) involving bulbosum (2x or 4x) or vulgare (2x). For instance, the crosses involving diploid bulbosum invariably resulted in haploids of the other parent but with varying frequencies ranging from 20% with depressum (4x) to 100% with any of the hexaploids (procerum, parodii, arizonicum or lechleri). Hybridizations involving tetraploids bulbosum resulted in 22% procerum haploids and 85% arizonicum haploids. Similarly, the progenies from the crosses involving vulgare give varying frequencies of haploids ranging from 42% with procerum to 76% with arizonicum via selective elimination of vulgare chromosomes. If these differences in the proportions of the haploid progenies are taken as a reflection of the extent of embryos exhibiting complete elimination of either bulbosum or vulgare chromosomes (depending upon the cross combination), there are differences in the "genetic strength" in eliminating different chromosomes. Species like arizonicum, lechleri or procerum are capable of eliminating up to two bulbosum genomes or one vulgare genome, whereas other species listed are capable of eliminating a single bulbosum genome only.

REFERENCES

1. K.J. Kasha and K.N. Kao, High frequency haploid production in barley (*Hordeum vulgare* L.), *Nature* 225:874-876 (1970).
2. N.C. Subrahmanyam and K.J. Kasha, Selective chromosomal elimination during haploid formation in barley following interspecific hybridization, *Chromosoma* 42:111-125 (1973).
3. M.D. Bennett, R.A. Finch and I.R. Barclay, The time rate and mechanism of chromosome elimination in *Hordeum* hybrids, *Chromosoma* 54:175-200 (1976).
4. W. Lange, Crosses between *Hordeum vulgare* L. and *Hordeum bulbosum* L. 1. Production, morphology and meiosis of hybrids, haploids and dihaploids, *Euphytica* 20:14-29 (1971a).
5. W. Lange, Crosses between *Hordeum vulgare* L. and *H. bulbosum* L. 2. Elimination of chromosomes in hybrid tissues, *Euphytica* 20:181-194 (1971b).
6. K.J. Kasha and R.S. Sadasivaiah, Genome relationships between *Hordeum vulgare* L. and *H. bulbosum* L., *Chromosoma* 35:264-287 (1971).
7. K. Noda and K.J. Kasha, Chromosome elimination in different meristematic regions of hybrids between *H. vulgare* L. and *H. bulbosum* L., *Jap. J. Genet.* 56:193-204 (1981).
8. I.R. Barclay, K.W. Sheperd and D.H. Sparrow, Chromosome elimination in *Hordeum vulgare* - *Hordeum bulbosum* hybrids, *in*: "Biennial Report, Waite Agric. Res. Inst.," University of Adelaide, S. Australia (1970-71).
9. K.J. Kasha, E. Reinbergs, W.A. Johns, N.C. Subrahmanyam and K.M. Ho, Barley haploid studies, *Barley Genet. Newsl.* 2:36-41 (1972).
10. K.M. Ho and K.J. Kasha, Genetic control of chromosome elimination during haploid formation in barley, *Genetics* 81:263-275 (1975).
11. T. Rajhathy and S. Symko, High frequency of haploids from crosses of *Hordeum lechleri* (6x) x *H. vulgare* (2x) and *H. jubatum* (4x) x *H. bulbosum* (2x), *Can. J. Genet. Cytol.* 16:468-472 (1974).
12. R. Islam and D.H.B. Sparrow, Chromosome elimination in interspecific hybrids involving *Hordeum arizonicum* (6x) x *H. vulgare* (2x), *Barley Genet. Newsl.* 4:36-37 (1974).
13. N.C. Subrahmanyam, Haploidy from *Hordeum* interspecific crosses, 1. Polyhaploids of *H. parodii* and *H. procerum*, *Theor. Appl. Genet.* 49:209-217 (1977).
14. N.C. Subrahmanyam, Haploidy from *Hordeum* interspecific crosses, 2. Dihaploids of *H. brachyantherum* and *H. depressum*, *Theor. Appl. Genet.* 55:139-144 (1979).
15. N.C. Subrahmanyam, Haploidy from *Hordeum* interspecific crosses, 3. Trihaploids of *H. arizonicum* and *H. lechleri*, *Theor. Appl. Genet.* 56:257-263 (1980).
16. I.R. Barclay, High frequencies of haploid production in wheat (*Triticum aestivum*) by chromosome elimination, *Nature*

256:410-411 (1975).
17. T. Shigenobu and S. Sakamoto, Production of a polyhaploid plant of _Aegilops crassa_ (6x) pollinated by _Hordeum bulbosum_, _Jap. J. Genet._ 52:397-401 (1977).
18. A. Kruse, Intergeneric hybrids between _Hordeum vulgare_ L. sp. _distichum_ (V. Pallas, 2n = 14) and _Secale cereale_ L. (V. Petkus, 2n = 14), _Kgl. Vetlandbohojsk Arsskr._ 82-92 (1967).
19. G. Fedak, Haploids from barley x rye crosses, _Can. J. Genet. Cytol._ 19:15-19 (1977).
20. N.C. Subrahmanyam, Species dominance in chromosome elimination in barley hybrids, _Curr. Sci._ (in press) (1981).
21. K.J. Kasha, N.C. Subrahmanyam and A. Ali, Effect of gibberellic acid treatment, and nutrient supply through detached tillers, upon haploid frequency in barley, _Theor. Appl. Genet._ 51:169-175 (1978).

CALLUS INITIATION AND PLANT REGENERATION FROM HAPLOID INTERNODES IN RICE

G.M. Reddy

Department of Genetics
Osmania University
Hyderabad - 500007
INDIA

INTRODUCTION

Ever since the successful demonstration of haploid production through pollen cultures by Guha and Maheswari (1), haploids have been induced in many crop plants, although not much utilization has been made in crop improvement. Haploid cultures play a significant role in the rapid development of homozygous lines and help in genetic manipulation especially in isolation of desirable mutants in the very first generation. In rice, anther cultures have been exploited in the production of haploids (2, 3, 4).

Although there are numerous reports of regenerating potential of anther culture in different crops, there are few studies dealing with the comparative morphogenetic potential of different ploidy levels particularly in cereals. One of the reasons could be that cereals are not easily amenable to culture studies. In the present study, plant regeneration from matured plant explants of rice haploid internodes, beside roots of certain genotypes have been exploited for rapid multiplication, cloning, etc.

MATERIALS AND METHODS

Spontaneous haploids which were selected and being maintained in the greehnouse by Dr. J.N. Rutger, Rice Geneticist, U.S.D.A., University of California, Davis, are used in the present investigation. Internodes from fresh tillers of SD7 and also regenerated haploids were cut in sodium hypochlorite (2.6%) and sterilized with the same for about 15 minutes in individual beakers.

After three rinses with sterile deionized water, internodes were cut horizontally into two halves and cut surfaces were placed on the medium in petri dishes. Similarly, roots from several diploid and tetraploid rice genotypes were plated on the medium in petri dishes. Linsmaier and Skoog's (5) media was used for callus initiation with 2,4-dichlorophenoxy acetic acid (2,4-D), and the plant regeneration was obtained with or without IAA and Kinetin. After plating of internodes onto the media, the petri dishes were sealed with two strips of parafilm and kept in the dark at 27°C ± 1°C. The healthy actively growing callus of different genotypes were transferred to regeneration media after several subcultures. Five types of media, one for callus initiation, three for regeneration, one for test tube "wick culture" were used. For regeneration, LS medium 2 mg/l of IAA and 4 mg/l of kinetin; LS medium with 1 mg/l of IAA and 6 mg/l of kinetin, or LS medium without auxins were used. The wick culture medium consisted of LS medium without agar, yeast extract and auxins. After six to eight weeks, the healthy active calli, ranging from small clumps of cells to about 1 gm of tissue, were inoculated onto the above three types of regeneration media, and simulatneously onto the original callus initiation medium for subculture maintenance, and the subcultures were carried out monthly. The callus for regeneration was kept under continuous white fluorescent light (60 $UE/m^2/sec$) at 27°C ± 1°C.

RESULTS AND DISCUSSION

The results of the experiments are presented in Table I and illustrated in Figures 1 and 2. The internodal haploid stem explants started callusing at the internodal portions by initiating roots within one week and a conspicuous callus mass could be seen within two to three weeks. The transfer of five different haploid internode calli onto the regeneration medium exhibited different percentages of totipotency as evidenced by plant regeneration. The RR-9 haploid callus was found to be superior, and the regenerating ability was 50% even after 228 days which is a significant finding. Earlier studies have shown that the regenerating ability was lost after 60 days only (6). The haploid callus was also cultured on LS liquid media without agar, and plantlets were regenerated. Besides internodes, the rice seedlings were regenerated from root callus of other diploid and tetraploid genotypes (7).

A comparative study of the morphogenetic potential of the diploid and tetraploid calli suggested that in general all the calli regenerated within 105 days or alternatively within 160 days with the exception of the SD7 tetraploid. The regenerating ability generally decreased with the increase in age of calli. The regeneration capacity varied among different genotypes with the diploids generally being better than tetraploids. Among the diploids, M-101 showed the best regenerating ability, with 8 out of 11 petri dishes exhibit-

Table I. Totipotency of Calli Derived from Haploid Internodes, Diploid and Tetraploid Seedling Roots

Genotypes	No. of Petri Dishes	No. of Days when the Regenerating Ability was Studied	Regenerating Ability in Percentage
SD7 Haploid Internode (RR-9)	14	228	50
SD7 Diploid	12	105	75
SD7 Tetraploid	11	105	66
Calrose 76 Diploid	22	160	59
Calrose 76 Tetraploid	12	160	25
CS-M3 Diploid	6	160	50
M-101 Diploid	11	160	72

ing plantlet regeneration at 188 days. Many regenerated plantlets were observed in each petri dish. Several actively growing diploid plantlets were transferred to test tube wick culture, then to vermiculite pots and finally to the greenhouse. Three of the regenerated plants reached maturity and had normal seed set and phenotypic appearance similar to diploid controls.

In these studies diploids were superior to tetraploids in frequency of callus initiation, amount of callus growth and frequency of plantlet regeneration. Since the most successfully cultured and regenerated plant tobacco is a natural polyploidy, we had speculated that polyploidy might have some positive effect on callus initiation and plant regeneration. However, our preliminary studies on rice suggest that tetraploids are not so favorable in comparison to diploids (8). In field studies it has been seen that the rice diploid lines grow faster than the tetraploids. Thus the only generalization that seems valid is that tissue culture and regenerative ability, like plant growth, are best at the naturally adapted ploidy level of the crop.

The ease with which homozygosity can be achieved by diploidization in a single step offers a great advantage with haploid calli. The exposure of cells to specific mutagens may lead to the isolation

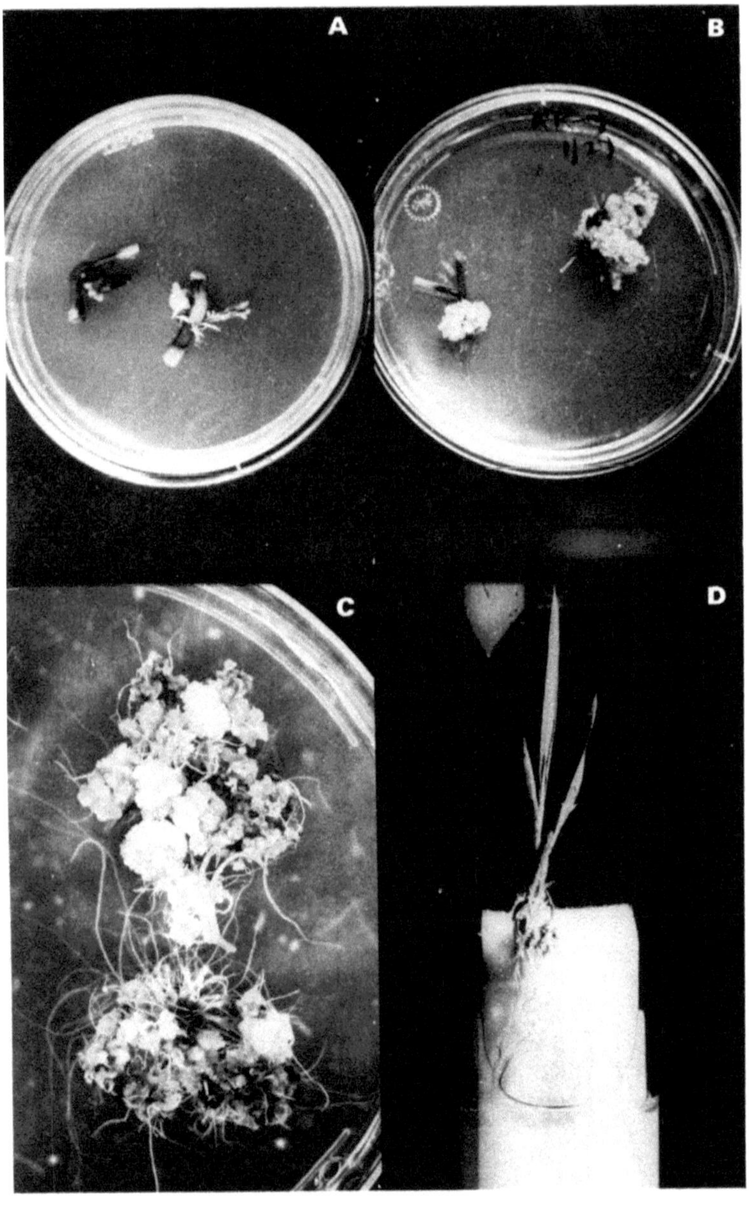

Figure 1. Induction of callus and regeneration of plantlets from haploid internodes.
a, b) Root initiation, callus formation and proliferation from internodes;
c) Regeneration of plantlets;
d) Seedling in wick culture.

Figure 2. a, b) Regenerated haploid (sterile) and regenerated diploid (fertile) plants;
c, d) Suspension cultures of haploid calli: root initiation and recovered plantlet.

of cell lines resistant to pathological toxins, herbicides and salt, etc., which on regeneration may lead to the isolation of plants carrying the altered trait. The successful initiation of suspension cultures from haploid calli and their regeneration into plants provides a simpler mechanism analogous to microorganisms in screening large population of cells for the selection of desirable plant types in breeding programs aimed at the improvement of rice crops.

ACKNOWLEDGEMENTS

My grateful thanks to Drs. J.N. Rutger and D.W. Raines of the Plant Growth Laboratory, University of California, Davis, for providing the facilities during my visit as a Fulbright Scholar.

REFERENCES

1. S. Guha and S.C. Maheswari, In vitro production of embryos from anthers of Datura, Nature 204:497 (1964).
2. H. Niizeki and K. Oono, Induction of haploid rice plant from anther culture, Proc. Jap. Acad. 44:554-557 (1968).
3. T. Nishi and S. Mitsuoka, Occurence of various ploidy plants from anther and ovary culture of rice plant, Jap. J. Genet. 44:341-346 (1969).
4. S. Guha, R.D. Iyer, N. Gupta and M.S. Swaminathan, Totipotency of gametic cells and the production of haploids in rice, Curr Sci. 39:174-176 (1979).
5. E.M. Linsmaier and F. Skoog, Organic growth factor requirements of tobacco tissue cultures, Physiol. Plant. 18:100-127 (1965).
6. A.D. Genovesi, "Advances in the Development of Somatic Cell Genetic System for Oryza sativa (Rice) via Anther Culture," Ph.D. Thesis, Texas A & M University, USA (1978).
7. G.M. Reddy and J.N. Rutger, Callus initiation and plant regeneration from diploid and tetraploid seedling roots and rice, Proc. 18th Rice Technical Working Grp. (1980).
8. G.M. Reddy, Tissue and mutational studies in rice improvement, Proc. Symp. Tissue Cult. Economically Important Plants, National University of Singapore (1981).

PROPAGATION OF SANDALWOOD (*SANTALUM ALBUM* LINN) USING TISSUE AND ORGAN CULTURE TECHNIQUE

P.S. Rao and N.V. Raghava Ram

Plant Morphogenesis and Tissue Culture Section
Bio-Organic Division
Bhabha Atomic Research Centre
Trombay, Bombay 400085
INDIA

INTRODUCTION

A majority of the woody plants are difficult to propagate by vegetative means. Traditional methods such as rooting of the excised branches, and grafting, although successful in some instances, still has limitations and cannot be applied for all commercially valuable species. The technique of tissue and organ culture, especially the culture of organs and buds and the *in vitro* regeneration of whole plants through them, could be attempted as an alternate method in the propagation in woody plants (1).

In this laboratory a program has been initiated on tissue culture of woody species with a view to developing techniques for clonal multiplication of elite species and the following is a brief report on the successful attempts made on Indian sandalwood tree (*Santalum album* Linn).

MATERIALS AND METHODS

The experimental material was obtained from the Forest Research Laboratory campus, Bangalore, and consisted of (a) juvenile shoots from mature trees, (b) green fruits, and (c) mature seeds. The cut ends of shoot segments were sealed with molten wax. Shoot segments as well as green fruits were washed thoroughly with detergent soap and tap water. The washed material was then sterilized in 0.1% aqueous solution of mercuric chloride for 5 - 8 minutes and rinsed several times with autoclaved distilled water. In addition, green fruits

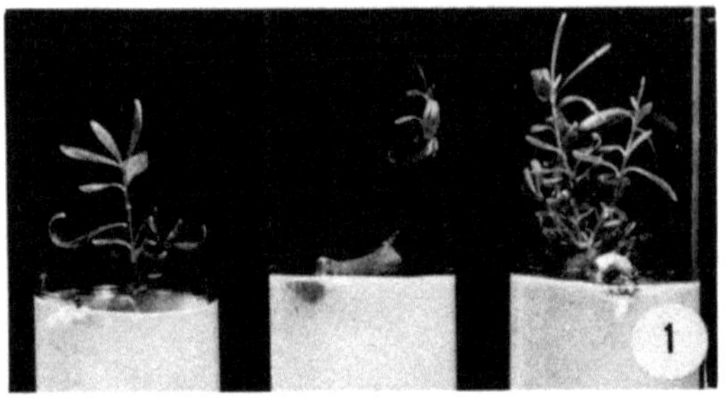

Figure 1. (Left to right) Shoot bud regeneration on hypocotyl segments cultured on MS + NAA (1 mg/l), MS + NAA (1 mg/l) + Kn (1 mg/l), and MS + NAA (1 mg/l) + BA (1 mg/l), respectively.

yielded a vigorously growing callus mass. The growth of callus tissue could be maintained through periodic subcultures. Callus maintained on MS medium fortified with 2,4-D was more friable as compared to tissues grown on MS + 2,4-D + Kn.

Callus Initiation in Endosperm Tissues

Endosperm tissues excised from mature fruits were cultured on MS medium fortified with various growth substances. Best results were obtained on MS + 2,4-D (1 mg/l) alone or in combination with Kn (0.2 mg/l). Within 3 weeks of culture callus initiation occurred all over the surface of endosperm (Figure 2), and, during the course of 6 weeks, a good proliferating callus tissue was obtained. The tissue was friable and could be repeatedly subcultured by transferring the tissue to fresh medium every four weeks.

Embryogenesis in Callus Tissues Derived from Stem and Endosperm

In order to induce differentiation the undifferentiated callus tissues resulting from stem segments and endosperm tissues were grown on MS medium supplemented with various auxins and cytokinins individually and in combinations. Various factorial experiments were conducted using several concentrations of these substances. The sucrose level of the medium was raised to 5% and agar was used at 0.6%. In callus tissues transferred to MS + IAA (1 mg/l) + BA (1 mg/l) a large number of embryos differentiated within a period of 4 weeks. Morphologically and functionally the embryos showed similarities to embryos obtained in seeds. All the developmental stages from globular to mature embryos were discernible among the population (Figures 3, 4). If the embryos were allowed to grow on the same medium, they developed into fasciated structures. Such fasciated structures developed shoots without roots. However, fasciation and irregular growth of embryos could be considerably reduced if the

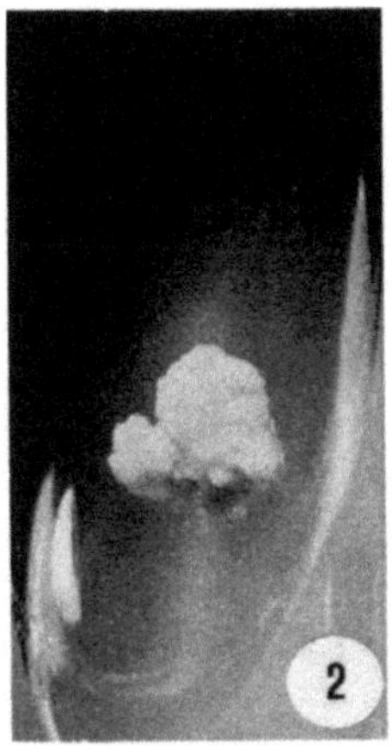

Figure 2. Callus tissue developed from endosperm on MS + 2,4-D (5 mg/l).

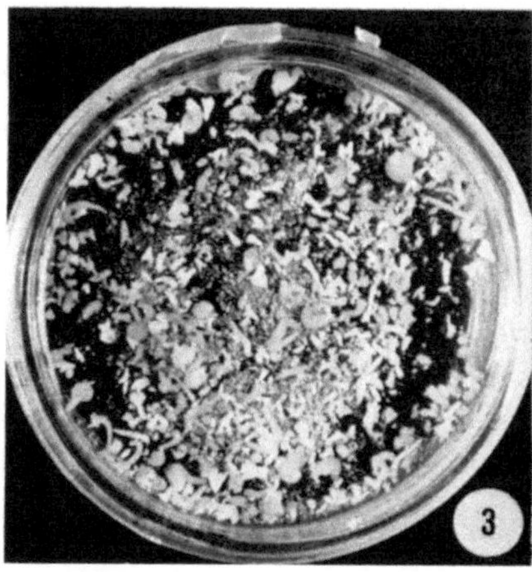

Figure 3. (Left) Embryos differentiated in stem callus tissues on MS + IAA (1 mg/l) + BAP (1 mg/l).

Figure 4. (Right) Differentiation of embryos in endosperm callus tissue on MS + IAA (1 mg/l) + BAP (1 mg/l).

major elements in the MS medium were used at half the normal strength.

Development of Plantlets

Well developed embryos were separated from callus tissues and individually planted on MS medium supplemented with IAA (1 mg/l) + IBA (0.5 mg/l) + GA_3 (0.5 mg/l). Sucrose concentration was maintained at 5% and agar at 0.6%. Initially the cultures were grown in dark for a period of 72 hrs, during which root growth was initiated, after which the cultures were exposed to continuous light. Small plantlets possessing two or three pairs of leaves and roots developed. At this stage the young plantlets were floated on filter paper bridges in large size tubes having greater volume of medium. After maintaining the plantlets for 4-6 weeks under these conditions, they were transferred to paper cups containing vermiculite. Plantlets were allowed to harden gradually in sunlight and were eventually transferred to earthen pots with soil (Figures 5, 6).

DISCUSSION AND CONCLUSIONS

The results presented here and those reported earlier by us (3, 4, 5) on the development of techniques for obtaining plant regeneration in the cultured parts of hypocotyl, stem segments and endosperm tissues of Santalum album offer the promise of clonal propagation in this species in which vegetative propagation by cuttings has not been successful.

Figure 5. Complete plantlet with well-developed root.

Figure 6. *In vitro* obtained sandalwood plants successfully established in soil.

were dipped in ethyl alcohol and flamed. 1.5 cm long segments were prepared from sterilized shoots. In case of green fruits, they were cut longitudinally, and well-developed embryos were completely removed leaving only the endosperm tissue. For culturing hypocotyl explants seeds were aseptically germinated, and from 4 week old seedlings 5 mm hypocotyl segments were isolated. Stem segments and the endosperm tissues as well as hypocotyl explants were placed on Murashige and Skoog's nutrient medium (2) containing 2% sucrose and jelled with 0.7% agar. The pH of the medium was adjusted to 5.6 before autoclaving. The final medium was sterilized by autoclaving at 1.06 kg/cm^2 pressure for fifteen minutes. The basal medium was supplemented with plant growth hormones as per requirements. All cultures were exposed to continuous light (1000 lux) and maintained at 25°C $\pm$ 2°C at a relative humidity of 55%.

RESULTS

Responses of Hypocotyl Segments

Shoot Bud Regeneration. Hypocotyl explants were cultured on MS medium supplemented with various concentrations of auxins, cytokinins and combinations of auxins and cytokinins. Shoot bud regeneration was observed generally on all auxin media within 4 weeks. Bud formation was moderate on IAA, IBA and NOA, whereas it was intense on NAA. Amongst the cytokinins tested, shoot bud formation was optimum on BA. The number of shoot buds per explant also varied. 8-10 Shoot buds per explant were produced on MS medium supplemented with kn, Z, SD 8339 and Ad, whereas on BA medium 15-20 buds regenerated per explant (Figure 1). Rooting in *in vitro* obtained shoot buds was achieved by growing the excised shoot buds on MS + NAA (0.5 mg/l) + IBA (5 mg/l). Complete plants were recovered by this process.

Embryogenesis. When the hypocotyl explants in which the regenerated shoot buds had been removed were transferred to fresh medium of the same composition, they initiated fresh shoot buds. In some instances they proliferated into a callus tissue. After 4 weeks, numerous globular embryos differentiated from the callus. Such embryos were isolated and transferred to MS liquid medium supplemented with various growth substances to stimulate root and shoot morphogenesis. Vigorous shoot development followed in isolated embryos cultures on MS + BA (1 mg/l). Fully organized embryos derived in liquid cultures were grown on MS agar medium supplemented with IAA (1 mg/l) and plantlets were obtained.

Callus Initiation in Stem Explants

Isolated stem segments cultured on MS medium supplemented with 2,4-D (1 mg/l) alone or in combination with Kn (0.2 mg/l) showed proliferation at the excised regions and within 8 weeks of culture

The salient features which have emerged from this study relate to the diverse morphogenetic pathways obtained in organ cultures of Santalum. Hypocotyl explants, stem segments and the endosperm tissues show immense morphogenetic potential. Of special interest is the fact that embryogenesis can be induced in endosperm tissues and the resultant plants are triploid. The importance of triploidy in plant improvement has been well emphasized (6), and in forest trees triploids are useful because wood quality and yield are of prime importance (7).

REFERENCES

1. D.F. Karnosky, Potential for forest tree improvement via tissue culture, Bioscience 31:114-120 (1981).
2. T. Murashige and F. Skoog, A revised medium for rapid growth and bioassay with tobacco tissue cultures, Physiol. Plant. 15:473-497 (1962).
3. P.S. Rao and V.A. Bapat, Vegetative propagation of sandalwood plants through tissue culture, Can. J. Bot. 56:1153-1156 (1978).
4. V.A. Bapat and P.S. Rao, Somatic embryogenesis and plantlet formation in tissue cultures of sandalwood (Santalum album L.), Ann. Bot. 44:629-630 (1979).
5. P.S. Rao and V.A. Bapat, Morphogenetic investigations on tissue and organ cultures of sandalwood tree, in: "Plant Tissue Culture, Genetic Manipulation and Somatic Hybridization of Plant Cells," P.S. Rao, M.R. Heble, M.S. Chadha, eds., Proc. Natl. Symp. BARC, Bombay (1980).
6. B.M. Johri and S.S. Bhojwani, Triploid plants through endosperm culture, in: "Plant Cell, Tissue and Organ Culture," J. Reinert and Y.P.S. Bajaj, eds., Springer-Verlag, Berlin (1977).
7. L. Winton, Morphogenesis in clonal propagation of woody plants, in: "Frontiers of Plant Tissue Culture," T.A. Thorpe, ed., University of Calgary, Calgary, Canada (1978).

THE CONCEPT OF MORPHOGENIC COMPETENCE WITH REFERENCE TO ANTHER AND POLLEN CULTURE

Norman Sunderland

John Innes Institute
Colney Lane
Norwich, United Kingdom

INTRODUCTION

A morphogenically competent cell may be defined as one which has the potential or acquires the potential *in vitro* to develop as embryo, callus or other structure given appropriate cultural conditions. Most tissue culturists probably regard competence mainly as a property of culture and one that is acquired and regulated by the particular combination and concentration of hormones included in the medium. In this paper I propose to query this assumption and to put the case that (a) competence may be more readily induced before culture, and (b) expression of that competence is dependent as much upon endogenous as upon exogenous factors. Pollen will be used to illustrate this theme because of the ease with which competent cells can be recognized in suitably stained squash preparations of anthers. I shall refer mainly to the peonies and barley, since these are the two groups of plants in which I first encountered competent cells *in vivo*, a phenomenon subsequently referred to as pollen dimorphism (1).

POLLEN DIMORPHISM

The formation of multicellular embryo-like structures in anthers of potato hybrids (2) and of proembryo-like structures in experimental cultivars of the anemone (3) is good evidence of competence *in vivo*. A single plant growing from a *Dianthus* flower (4) has also been interpreted as originating from the pollen. In general, however, anthesis cuts short expression of such competent pollens.

Variant forms of pollen, throwing light on pollen ontogeny, have been described from a wide range of plants (5). Their role as morphogenically competent cells first became evident from their similarity to embryogenic pollen grains (E grains) formed in anther cultures of *Nicotiana tabacum* White Burley (6) and *Datura innoxia* (7). Dimorphism could not be seen in these plants *in vivo*. The morphogenic competence of the variant forms was formally demonstrated in the peonies. With cultured anthers already containing two distinct fractions, division was seen to begin in the variant forms -- the true pollen grains degenerated (8). The products of culture were thus assumed to have arisen from the variants, and the variety of structures produced (bipolar or multipolar embryos, embryo clusters, shoots without roots, roots without shoots, callus and other indefinable entities) to be a reflection of natural heterogeneity within the variant population. Dale (9) working with barley anthers and Horner and Street with a Burley tobacco (10), both of which showed pollen dimorphism *in vivo*, likewise concluded that the calluses and embryos produced respectively were derived from the variant pollens.

There thus emerged the concept of a fraction of competent pollens (the E grain fraction) formed *in vivo* in some species and *in vitro* in others (11). Culture yields were seen to be determined initially by the magnitude of the E grain fraction, and the varied cell lineages established by differences in polarization and disposition of nuclei at the onset of division (5, 11, 12).

GENOTYPIC DIFFERENCES

Potential E grains *in vivo* are distinguishable from true pollen grains because they are generally smaller, have a poor affinity for cytoplasmic stains and often show extra divisions of the gametophytic cells (1, 3, 11). They vary in frequency between species, between varieties of the same species and between anthers of the same plant. The peonies show the widest known genotypic variation, some of the garden hybrids having as many as one quarter to half of the population as variants (Table 1). The peonies show a similar diversity in anther and pollen culture. Table 1 summarizes results collected over almost a decade. Progress is slow because buds are available for only a few days each year: some species flower only intermittently. Moreover, the rate of growth in culture is slow: the first embryos appear after about 4 months but others may still be emerging after more than 12 months (13, 14). The pollen cultures referred to in Table 1 mostly date from 1976-77 and in many instances are *ab initio* in the true sense of the term, i.e., the pollen was isolated immediately after removal of the anthers from the plant according to the Nitsch (15) procedure. Successful pollen cultures, but not anther cultures, have been obtained in *P. delavayi* and *P. mlokosewitschii* x *tenuifolia*. To my knowledge, these are the only instances in which pollen culture has so far proved more effectual than anther culture.

Table 1. Species and Cultivars of Paeonia in the John Innes Collection with Reference to Pollen Dimorphism, Anther and Pollen Culture.

Species or cultivar	Percentage of variant pollen in mature anthers	Successful in anther culture	Successful in pollen culture
P. delavayi	1	No	Yes
P. delavayi (seedling)	NT	Yes	NT
P. potanini trollioides	4	Yes	Yes
P. peregrina	9	No	NT
P. triternata	9	Yes	NT
P. emodi	14	No	NT
P. officinalis	15	No	NT
P. mlokosewitschii x. tenuifolia	16	No	Yes
P. emodi Lady Beatrice	2	Yes	Yes
P. lactiflora Elwes hybrid	2	No	No
Chas. Leveque	2	Yes	Yes
Victoria	8	Yes	Yes
Bride	18	Yes	Yes
Dreadnought	23	Yes	Yes
Cleopatra	25	Yes	Yes
Darius	45	Yes	Yes
Othello	NT	Yes	Yes

NT: Not Yet Tested

ENVIRONMENTAL DIFFERENCES

The frequency of potential E grains *in vivo* also varies with the way in which plants are cultivated. The spring barley, Sabarlis, for instance, shows pollen dimorphism in glasshouse plants, particularly those grown out of season, whereas little dimorphism occurs in plants grown under natural conditions in the field. Plant growth temperature greatly influences the frequency of variant pollens, a drop in temperature from 15°C to 10°C with Sabarlis barley increasing the frequency several fold (11). More recently, Heberle-Bors and Reinert (16) have exploited both temperature and photoperiod to increase dimorphism in Badischer Burley tobacco and obtained *ab initio* pollen cultures by first isolating the pollen by the Nitsch (15) procedure from relatively mature anthers (in which two distinct populations of pollen could be distinguished) and second separating the two by the Wernicke (17) density gradient fractionation procedure.

INDUCTIVE PRETREATMENTS

Competence can easily be induced before culture by temperature stress of excised floral parts. The technique has been developed from the original cold-shock treatment of Nitsch and Norreel (18). The excised tissues are kept in darkness in sealed containers for a period determined by temperature, developmental stage of anthers and the flowering age of the plants (5, 19, 20). The method of pretreatment is exceedingly important (24).

The role of the stress pretreatment in the induction of competent pollen populations was first demonstrated on excised flower buds of White Burley tobacco stored at 7°C in polythene bags (11, 21). After about 2 weeks, typical E grains were formed when buds were excised from the plants just after completion of the first pollen division. The process was one of cytoplasmic regression in the presumptive vegetative cell like that observed and characterized earlier in cultured anthers of the same cultivar (6, 22, 23). It was at once evident that all the inductive changes previously ascribed to culture could proceed simply as a result of excision. Maintenance of the excised buds in the cold preserves pollen viability long enough to ensure a high frequency of induction. At normal incubation temperatures, only small competent fractions are formed in cultures initiated direct from the plant, because most of the pollen degenerates before induction is assured.

Table 2. Differences Between the Peonies and Barleys with Respect to Pretreatments, Anther and Pollen Culture.

	Peony	Barley
Length of anthers	up to 10 mm depending on genotype	1.5 - 2.0 mm
Number of anthers	from 200 - 300 per bud	from 60 - 90 per spike
Method of pretreatment	whole buds in Petri dishes sealed with parafilm	whole spikes or spikelets in sealed Petri dishes
Temperature of pretreatment	14 - 20 °C	4 - 7°C
Duration of pretreatment	7 - 14 days depending on genotype	21 - 28 days for maximal callus production; 35 - 42 days for maximal regeneration
Preincubation of anthers	as effective as bud pretreatment	not known
Media for anther culture	MS (36) basal or N_6 (37) basal with charcoal 0.5 - 1.0%, agar 0.8%	not applicable
Incubation	25°C for 4 - 12 months	not applicable
Media for float culture	not applicable	potato (38) or N_6 (37) plus sucrose 9%, kinetin 0.5 mg/l, 2,4-D 1.5 mg/l
Incubation	not applicable	28°C for 28 days
Isolation of pollen	mechanical	natural shedding
Media for pollen culture	as above with/without charcoal and without agar	as above
Pollen density	pollen from 1 - 2 anthers/ml	pollen from 60 - 120 anthers/ml

The pretreatment has thus proved to be a means of inducing competence in those species, like Datura innoxia, Hyoscyamus niger and White Burley tobacco, which do not show pollen dimorphism in vivo, and of increasing the frequency in those species which do.

Subsequent studies have shown that better culture yields may be attained in some species, and more quickly, with pretreatment at mild rather than cold temperatures (5). Details for the peonies and barley are given in Table 2. Optimal pretreatment times may differ for anther and pollen culture as in P. Dreadnought (Table 3), and for maximal callus production and plant regeneration in Sabarlis barley (24). Pretreatment may also alter the optimal developmental stage as in H. niger (20).

Pollen degeneration in anthers cultured direct from the plant can be reduced by preincubation at cold temperatures, but in general such preincubation has been less productive than bud pretreatment (see e.g. 20). However, for P. Bride, preincubation of anthers at 14°C for 14 days, particularly in water, is as productive as bud pretreatment (Table 4). The high productivity of both anther and pollen cultures after water preincubation reinforces the argument that endogenous factors meet all the requirements for induction of competence.

Preincubation of anthers has found its most important application in Brassica anther culture (25) for which an elevated temperature (35°C) is beneficial.

DEVELOPMENTAL STAGE

For the stress pretreatment to exert its maximal influence, the floral parts need to be excised before, during or just after the first pollen division (1, 12). With P. Bride (Table 5), and also P. Dreadnought, premitotic anthers give the most productive cultures, whereas in White Burley tobacco and D. innoxia mitotic and postmitotic anthers are best (11, 19). Barley also belongs to the premitotic class as do other important cereals and forage grasses. Removal of barley spikes during the G1 interphase of the spore cycle is essential (26).

With buds of P. Dreadnought held at 14°C (Table 3), it takes up to 7 days for the microspores to enter the first pollen division and a further 7 days before competent E grains can be distinguished. At what appears to be the best pretreatment time for pollen culture, the competent cells cannot be distinguished and with the best time for anther culture, the true magnitude of the induced fraction is not manifest. With Sabarlis barley held at 4°C, the spores move exceedingly slowly through the cell cycle and even after the optimal time may still not have divided. A reasonably accurate assessment

Table 3. Influence of Bud Pretreatment at 14°C in Anther (AC) and Pollen (PC) Cultures of Paeonia Dreadnought.

Duration of pre-treatment days	Percent responding anthers in AC	No. of* embryos from PC	Cytology E/D	U	M	EBi	M-LBi	E grains
0	3.4 (59)+	0 (10)++	36	64	0	0	0	0
7	3.3 (60)	25 (10)	47	41	1	11	0	0
10	8.2 (61)	38 (10)	45	22	3	25	5	0
14	68.3 (60)	8 (10)	44	14	0	16	14	12 (3)+++

E/D: empty or degenerating U: unicellular M: mitosis
EBi: early bicellular M-LBi: mid- to late-bicellular.
* These values represent the largest structures transferred at the first scoring. Total numbers not recorded.
\+ Number of anthers cultured. ++ Number of anthers used.
+++ % dividing.

of the E grain fraction can be made if the chilled spikes are transferred for a day or so to a higher temperature before the anthers are dissected out. The increased temperature causes the spores to divide and the resulting presumptive pollen grains go immediately into morphogenic division (see Table 6). In some anthers, virtually all the spores may be induced and thus all the pollen grains rapidly become multicellular within the excised spikes (5).

It cannot be too strongly emphasized that with the best pretreatment conditions two distinct size classes of pollen may not be produced as can be seen in mature anthers in vivo. Moreover, morphogenic division no longer appears to be restricted to variant pollens; it may occur in starch-filled, dense staining pollen grains which to all intents and purposes are fulfilling their normal gametophytic program (11, 26).

ISOLATION OF POLLEN

With many species, anthers dehisce rapidly in culture following an optimal inductive pretreatment. If such anthers are floated on liquid media, pollen is liberated into it. This natural dehiscence has been exploited to produce high yielding pollen cultures with White Burley tobacco (19, 27). The technique has since been applied to rice (28) and barley (29).

Table 4. Influence of Anther Preincubation and/or Bud Pretreatment in Anther (AC) and Pollen Culture (PC) of *Paeonia* Bride. (Incubation temperature for pollen cultures 25°C throughout.)

Treatment	AC	% anthers with embryos	Time of pollen isolation	No. of * embryos from PC	E/D	U	M	EBi	M-LBi	E grains
None	Nutrient agar throughout, 25°C	15.1 (33)+	day 0 (10)++	2	16	84	0	0	0	0
Anthers on nutrient agar for 14d/25°C			day 14 (16)	62	31	5	0	0	0	64 (27)+++
Anthers on nutrient agar for 14d/14°C	to 25°C at 14d	8.3 (12)	day 14 (16)	15	38	5	0	13	10	34 (14)
Anthers on water agar for 14d/25°C	to nutrient agar at 14d	0 (13)	day 14 (16)	15	60	2	0	0	0	38 (7)
Anthers on water agar for 14d/14°C	to nutrient agar at 14d, at 25°C	38.5 (13)	day 14 (16)	63	47	11	0	0	0	42 (7)
Anthers *in situ* in bud kept in P.D. for 14d	to nutrient agar at 14d, at 25°C	21.4 (14)	day 14 (16)	67	40	6	4	32	4	14 (9)

* Numbers transferred at the first scoring
Symbols as in Table 2

\+ number of anthers cultured
++ number of anthers used
+++ % dividing

Table 5. Anther (AC) and Ab Initio Pollen Culture (PC) in Paeonia Bride at Different Developmental Stages.

Bud Stage	No. of Buds	% Anthers Responding	No. of* embryos from PC	Cytology E/D	U	M	EMi	M-LBi	E grains
Pre-mitotic	5	5.4 (56)+	21 (75)++	19	81	0	0	0	0
Mitotic	4	4.8 (42)	15 (60)	22	53	8	17	0	0
Post-mitotic	2	0 (30)	4 (30)	14	3	1	20	48	14+++

* Numbers transferred at the first scoring.
\+ Number of anthers cultured.
++ Number of anthers used.
+++ % dividing.
Symbols as in Table 2.

Table 6. Percentage of Potential E Grains in Anthers of Two Barley Cultivars after a Long Cold Pretreatment (21 Days at 4°C) of Excised Tillers Followed by a Short Warm Pretreatment (25°C).

Cultivar	No. of Tillers Examined	Time at 25° Days	Empty or Degenerating	Normal and Unclassifiable Pollens	Potential E grains
var abyssinicum	3	1	17	27	56 (43)+
	1	2	40	0	60 (34)
	2	3	37	3	60 (39)
	2	4	73	7	20 (20)
var Nepal	2	1	34	31	35 (19)
	3	2	19	37	44 (11)
	5	3	46	22	32 (24)
	4	4	51	21	28 (13)

\+ % dividing.

Culture of shed pollen is simpler in operation than that of isolated pollen. Elaborate preparative procedures are avoided, the pollen does not suffer isolation damage (30), and in principle the technique is applicable to any species regardless of the size of anthers. Many economically important species for which pollen plants would be an asset have small anthers relative to those of the tobaccos, peonies and daturas. Small anthers are tedious to dissect out, more difficult to process and the pollen cannot be isolated without severe damage. Unfortunately, all recent work on isolated pollen culture (16, 31) has been carried out yet again on tobacco so that the relevance of the technique to other economically important plants is still unknown. This work has done nothing more than to demonstrate that so-called *ab initio* culture is feasible with Badischer Burley tobacco.

The major drawback to isolated pollen culture in tobacco [and this is pinpointed by the recent inclusion of the Wernicke (17) fractionation] is that it is less efficient, on a per anther basis, than the shed pollen procedure (21, 27). This is also the case with barley. Anthers of the Sabarlis cultivar, dehisce within 24 hours in float culture. Approximately half of the pollen is liberated. The same anthers can thus be used to prepare a shed pollen culture from one half of the pollen and an isolated pollen culture from the other. The data of Table 7 clearly show the superiority of the shed pollen cultures. The argument that isolation and fractionation of the pollen is needed to achieve a suspension simulating a culture of a microorganism (31) is as irrelevant for barley as it is for tobacco.

EXPRESSION OF COMPETENCE *IN VITRO*

Dehiscence of anthers in float culture is accompanied by release of a specific conditioning factor (CF) into the medium. The factor is essential for development of the competent pollens *in vitro*. In barley, the competent pollens will not develop if anthers are cultured at the usual densities; for high yielding cultures producing many thousands of calluses, densities of 60-120 anthers per ml are needed (33). Labor of dissecting anthers can be minimized by resort to drops of medium preconditioned by anthers (34) (see Table 7). The data of Xu et al. (34) show that the level of CF increases with increasing anther age. Hence, better preconditioning is given by more advanced anthers than are needed for the induction of competence.

The CF is interpreted as the key to initiation and maintenance of cell division in competent pollens. In barley, exogenous hormones and other accessory growth compounds such as m-inositol have no influence unless sufficient CF is present. At high inoculation

Table 7. Callus Production in vitro from Shed and from Isolated Pollen of Hordeum vulgare cv Sabarlis (data from 29).

Medium for Float Culture	Removal of Anthers Days	No. of Calluses from Shed Pollen	Time of Pollen Isolation Days	No. of Calluses from Isolated Pollen	
				P-9H(IG)	P-9(IG)
			0	0	0
P-9H	0-1	145	1	0	0
	0-3	2635	3	2	62
	0-7	1412	7	82	88
P-9H(C)	0-1	515	1	0	0
	0-3	2225	3	6	42
	0-7	2303	7	132	212

P-9H: potato medium (38) 9% sucrose, kinetin 0.5 mg/l, 2,4-D 1.5 mg/l.
P-9H(C): same medium preconditioned for 7d at 25°C with stage 5 - 6 anthers (10 - 12 per ml) (34).
P-9H(IG): P-9H plus m-inositol 1000 mg/l, glutamine 160 mg/l.
P-9(IG): same as above minus hormones.

Anthers floated in 0.2 ml drops of medium (24 per drop). After removal of anthers, further medium added at intervals to give a final volume of 1 ml. Isolated pollen suspended in 1.0 ml aliquots of medium.

densities, hormones and inositol interact synergistically with the CF and enhance the growth rate of the calluses (35). Culture yields are thus seen as being regulated by the availability of the CF, whereas hormones and inositol act in a subsidiary capacity augmenting the influence of the CF and regulating expression of competence in the direction required. These findings have given much more credence to the model proposed in 1974 (1) in respect of an anther-wall contribution.

The hypothesis of a specific endogenous factor controlling division of competent pollens makes sense of the poor results obtained with many species of pollen isolated at day 0 (Table 7). The endogenous CF content is too low. Higher endogenous levels may be proposed to account for those species, such as the peonies (loc cit) and Datura innoxia (18) whose pollen does respond to isolation at day 0. Isolated pollen culture has been much more successful after a short period of anther preculture (15). After 3 - 7d in float culture, barley anthers contain abundant multicellular pollens (32). Such units exercise a greater degree of autonomy and thus are more likely to survive after isolation (see Table 7).

CONCLUSION

In my opinion the aims and aspirations of plant cell culture in crop improvement will in the long term be better served by resort to populations of cells that have already been induced into a state of competence. Much greater attention needs to be given to preculture conditions. The endogenous environment has also been too long neglected. It is all too easy to dismiss the argument as nothing more than a plea for empirical manipulation of endogenous hormone levels instead of exogenous levels. Stress pretreatments will certainly alter endogenous hormone levels. There is some evidence, however, that the barley conditioning factor may not be hormonal. The activity of preconditioned media is destroyed by heat as is the synergism with exogenous hormones and inositol (35). The possibility cannot be ignored that we may be dealing with an entirely different class of regulating molecules specific for each species. Species like oats, which exhibit dimorphism _in vivo_ (11), but nevertheless are unresponsive in culture, may lack the essential factor or contain overriding inhibitors.

I think the view that explants are merely groups of uniform cells to be arbitrarily manipulated in the hope that one or more cells may be shocked into compliance is far too narrow. In the case of pollen, a high frequency of both induction and expression is needed if a plant's heterozygosity is to be exploited to the full. To do this, there is no stereotyped formula that must be rigidly adhered to. One can only offer useful guidelines. Each species and cultivar presents its own particular challenges and the measure of success is gauged by the ingenuity of the individual worker in sorting out systematically by properly controlled experiments the best conditions to meet the specific demands.

Whatever our training and reading may have led us to believe, the lesson of pollen is that individual units are not synchronous with respect to the cell cycle. The degree of asynchrony varies from one species to another. Cereal pollens also vary in polarity during the spore cycle and differences may, therefore, be expected in the numbers and distribution of organelles at the time of excision. During the critical G1 interphase in an asynchronous system, there will be no uniformity in enzyme states or activities. Excision interrupts the normal ordered spore/tapetal relationship and under stress conditions the signalling of specific gene events may fail (5). The same heterogeneity of states may well apply to somatic tissues. Is it not that somatic tissues appear to be uniform because we cannot recognize competent cells in them until they start to divide?

Finally, I think we should resist having inflexible procedures imposed upon us. For the mass production of doubled haploids for breeding programs, float culture of anthers meets most demands. There are exceptions, as for instance, the peonies, which by virtue of slow growth in anther culture warrant agar culture. The procedure adopted to obtain high density suspensions of viable pollens for genetic studies should be determined by the size of anthers in question, the ease with which they can be dissected out and processed, the dehiscence properties and the magnitude of the E grain fraction induced. Peony anthers, like those of D. innoxia (21), do not dehisce readily so that shed pollen culture is impractical. For the smaller anthers of the cereals, on the other hand, shed pollen is the obvious choice. The Wernicke (17) fractionation has its place for the separation of competent from incompetent pollens when these differ greatly in size and buoyant density (as in mature anthers in vivo). The aim should be, however, to induce the whole population of cells into competence by inductive pretreatments. There is then little point in killing much of the induced pollen by mechanical isolation and then using density gradients simply to remove viable from inviable pollens.

REFERENCES

1. N. Sunderland, Anther culture as a means of haploid induction, in: "Haploids in Higher Plants: Advances and Potential," K.J. Kasha, ed., University of Guelph Press (1974).
2. M.S. Ramanna, The origin and in vivo development of embryoids in the anthers of Solanum hybrids, Euphytica 23:623-632 (1974).
3. N. Sunderland, Observations on anther culture of ornamental plants, in: "La culture des tissus et des cellules des végétaux. Travaux dédiés à la mémoire de Georges Morel," R.J. Gautheret, ed., Masson, Paris (1977).
4. Y.S. Murty and V. Kumar, In vivo production of plantlets from the anthers of Dianthus caryophyllus L., Acta Bot. Ind. 42:172-173 (1976).
5. N. Sunderland, Induction of growth in the culture of pollen, in: "Differentiation in vitro," M.M. Yeoman, D.E.S Truman, eds., Cambridge University Press (1982).
6. N. Sunderland and F.M. Wicks, Embryoid formation in pollen grains of Nicotiana tabacum, J. Exp. Bot. 22:213-226 (1971).
7. N. Sunderland, G.B. Collins and J.M. Dunwell, The role of nuclear fusion in pollen embryogenesis of Datura innoxia Mill, Planta 117:227-241 (1974).
8. N. Sunderland and J.M. Dunwell, Pathways in pollen embryogenesis, in: "Tissue Culture and Plant Science 1974," H.E.

Street, ed., Academic Press, London (1974).
9. P.J. Dale, Pollen dimorphism and anther culture in barley, Planta 127:213-220 (1975).
10. M. Horner and H.E. Street, Pollen dimorphism -- origin and significance in pollen plant formation by anther culture, Ann. Bot. 42:763-771 (1978).
11. N. Sunderland, Strategies in the improvement of yields in anther culture, in: "Proceedings of Symposium on Plant Tissue Culture," Science Press, Peking (1978).
12. N. Sunderland, Anther and pollen culture 1974-1979, in: "The Plant Genome," D.R. Davies and D.A. Hopwood, eds., John Innes Charity, Norwich (1980).
13. N. Sunderland and M. Roberts, Anther and pollen culture, in: "John Innes Annual Report #67," Norwich (1976).
14. M. Roberts and N. Sunderland, Pollen culture in Paeonia, in: "John Innes Annual Report #68," Norwich (1977).
15. C. Nitsch, Pollen culture -- a new technique for mass production of haploid and homozygous plants, in: "Haploids in Higher Plants: Advances and Potential," K.J. Kasha, ed., University of Guelph Press (1974).
16. E. Heberle-Bors and J. Reinert, Isolated pollen cultures and pollen dimorphism, Naturwiss 67:311-312 (1980).
17. W. Wernicke, C.T. Harms, H. Lorz and E. Thomas, Selective enrichment of embryogenic microspore populations, Naturwiss 65:540-541 (1978).
18. C. Nitsch and B. Norreel, Effet d'un choc thermique sur le pouvoir embyrogène de pollen de Datura innoxia cultivé dans l'anthère ou isolé de l'anthère, C R hebd Séanc Acad. Sci. Paris 276:303-306 (1973).
19. N. Sunderland and M. Roberts, Cold-pretreatment of excised flower buds in float culture of tobacco anthers, Ann. Bot. 43:405-414 (1979).
20. N. Sunderland and D.C. Wildon, A note on the pretreatment of excised flower buds in float culture of Hyoscyamus anthers, Plant Sci. Lett. 15:169-175 (1979).
21. N. Sunderland, Comparative studies of anther and pollen culture, in: "Plant Cell and Tissue Culture: Principles and Applications," W.R. Sharp, P.O. Larsen, E.F. Paddock and V. Raghavan, Eds., State University Press, Columbus, Ohio (1979).
22. S.S. Bhojwani, J.M. Dunwell and N. Sunderland, Nucleic-acid and protein contents of embryogenic tobacco pollen, J. Exp. Bot. 24:863-871 (1973).
23. J.M. Dunwell and N. Sunderland, Pollen ultrastructure in anther culture of Nicotiana tabacum. II. Changes associated with embryogenesis, J. Exp. Bot. 25:363-373 (1974).
24. B. Huang and N. Sunderland, Temperature-stress pretreatment in barley anther culture, Ann. Bot. 49:77-88 (1982).
25. W.A. Keller and K.C. Armstrong, Stimulation of embryogenesis and haploid production in Brassica campestris anther culture

by elevated temperature treatment, Theor. Appl. Genet. 55: 65-67 (1979).
26. N. Sunderland and L.J. Evans, Multicellular pollen formation in cultured barley anthers. II. The A, B, and C pathways, J. Exp. Bot. 31:501-514 (1980).
27. N. Sunderland and M. Roberts, New approach to pollen culture, Nature 270:236-238 (1977).
28. Y. Chen, R.F. Wang, W.H. Tian, Q. Zuo, S. Zheng, D.Y. Lu and G. Zhang, Studies on pollen culture in vitro and induction of plantlets in Oryza sativa subsp. Keng., Acta Gen. Sinica 7:46-47 (1980).
29. Z.H. Xu and N. Sunderland, Pollen culture in Hordeum vulgare, J. Exp. Bot. (in press).
30. M. Horner and H.E. Street, Problems encountered in the culture of isolated pollen of a Burley cultivar of Nicotiana tabacum, J. Exp. Bot. 29:217-226 (1978).
31. A. Rashid and J. Reinert, High-frequency embryogenesis in ab initio pollen culture of Nicotiana tabacum, Naturwiss 68: 378-379 (1981).
32. N. Sunderland, M. Roberts, L.J. Evans and D.C. Wildon, Multicellular pollen formation in cultured barley anthers. I. Independent division of the generative and vegetative cells, J. Exp. Bot. 30:1133-1144 (1979).
33. Z.H. Xu and N. Sunderland, Inoculation density in the culture of barley anthers, Scientia Sinic. (in press).
34. Z.H. Xu, B. Huang and N. Sunderland, Culture of barley anthers in conditioned media, J. Exp. Bot. 32:767-778 (1981).
35. Z.H. Xu and N. Sunderland, Glutamine, inositol and conditioning factor in the production of pollen callus in vitro, Plant Sci. Lett. 23:161-168 (1981).
36. T. Murashige and F. Skoog, A revised medium for rapid growth and bioassays with tobacco tissue cultures, Physiol. Plant. 15:473-497 (1962).
37. C.C. Chu, The N_6 medium and its application to anther culture of cereal crops, in: "Proceedings of Symposium on Plant Tissue Culture," Science Press, Peking (1978).
38. C.C. Chuang, T.W. Ouyang, H. Chia, S.M. Chou and C.K. Ching, A set of potato media for wheat anther culture, in: "Proceedings of Symposium on Plant Tissue Culture," Science Press, Peking (1978).

INDUCTION OF EMBRYOS IN *AB INITIO* POLLEN CULTURES OF *NICOTIANA*

A. Rashid

Department of Botany
University of Delhi
Delhi 110007
INDIA

INTRODUCTION

Angiosperm pollen grains, being haploid, may be considered analogous to microbes. Their totipotency (1) makes them potentially ideal for genetic studies. However, culture of pollen grains direct from the plant (*ab initio* culture) to give embryos in high frequency has so far proved difficult. This can be ascribed to lack of knowledge about pollen capable of embryo formation. An important contribution towards the identification of such pollen is the work of Horner and Street (2). These authors indicated that in a Burley cultivar of *Nicotiana tabacum* pollen grains capable of embryogenesis are smaller than those destined to form gametes. In the cultivar Badischer Burley their number can be increased by subjecting plants raised in long days and high temperature to short days and a lower temperature just prior to flowering (3). Nonetheless, it remained to be confirmed that the small pollen grains are capable of embryogenesis because in culture only a few embryos were formed.

More recently it has been possible to raise *ab initio* pollen cultures with a high frequency of embryo formation in *Nicotiana tabacum* cv. Badischer Burley (4 - 8). A comprehensive account of this work forms the subject matter of the present communication.

RESULTS AND DISCUSSION

To produce high yielding *ab initio* pollen cultures, young buds of an equal sepal-petal length (20 mm) from plants flowering at 18°C were subjected to a cold treatment at 10°C for 10 days. Embryogenic pollen grains were then selectively isolated from non-embryogenic pollen grains by density gradient centrifugation in percoll and sucrose (4). In culture, embryogenic pollen grains formed embryos at a frequency of 2% of the cultured pollen. The culture medium comprised a filter-sterilized mineral-sucrose medium (9), supplemented with glutamine and asparagine (pH 6.5). The choice of high pH was deliberate because it is known to favor divisions at low cell density (10). These results, the first successful attempts at *ab initio* pollen culture in *Nicotiana* were unconventional in a number of ways, namely (a) pollen was taken from plants flowering at 18°C, (b) buds were subjected to a cold treatment, 10°C for 10 days, (c) potentially embryogenic pollen was separated from non-embryogenic pollen, and (d) the embryogenic pollen was cultured in a medium of pH 6.5.

Further experiments indicated that high pH is indeed beneficial for pollen embryogenesis, and at pH 6.8 it is possible not only to have an increased frequency of embryogenesis (5) but also to dispense with the amides glutamine and asparagine (5) which are described as being essential for embryo formation from pollen isolated from precultured anthers (11). However, how pH affects the process of pollen embryogenesis remains to be ascertained. Possibly by its effect on permeability, pH exercises a control on the influx and efflux of substance(s) from the pollen and its new environment, the nutrient medium. This is pertinent because the division frequency of cells in a nutrient medium is described as being a function of an equilibrium between the substance(s) of the cells and nutrient medium (10). These experiments clearly indicated that nutrient requirements of pollen for embryo formation are very simple.

The experiments outlined above laid down the foundations of successful *ab initio* pollen culture but were far short of the goal of the high frequency embryogenesis required for application of the technique. However, further improvement resulted in high frequency embryogenesis (7). For this the pollen was taken from plants induced to flower at a still lower temperature (15°C). Also, the buds selected for cold treatment were slightly older (petal length *ca* 25 mm) than those employed in earlier experiments, and the separation of embryogenic pollen was done in a refrigerated centrifuge at 15°C. Such pollen formed embryos up to a frequency of 30% of the cultured pollen (6). High frequency embryogenesis was in part due to proper separation of the two types of pollen. At the slightly older stage of development, the gamete-forming pollen is at a more advanced stage and a better separation is possible; about 15% of the pollen grains form the embryogenic fraction, and of these about 30% regularly form embryos (i.e., 5% of the original population).

An inquiry into the causal factors contributing towards the success of these experiments indicated that cold treatment brings about the formation of embryogenic pollen as demonstrated by Sunderland (12). This became clear when pollen cultures were raised from older buds (petal length *ca* 35 mm) without any prior cold treatment. At this stage of development it is possible to separate two types of pollen. In cultures raised form plants induced to lfower at 18°C the embryos fialed to appear, but embryos appeared if the cultures were given a cold treatment at 10°C for 10 or more days (7). In contrast the cultures raised form plants flowering at 15°C formed embryos at a low frequency which, however, could be increased substantially by giving the cultures a cold treatment. These experiments indicated that production of embryogenic pollen is a sequential process which can be initiated on the plant and completed in culture. Cold treatment of buds was, however, more effective than cold treatment of pollen as it resulted in a higher frequency of embryogenesis.

How cold treatment brings about the production of embryogenic pollen remains to be resolved. In this context a subcellular study of embryogenic and non-embryogenic pollen from cold-treated buds of 15°C plants is revealing. Cold treatment possibly causes repression of gametophytic differentiation. Gametophytic pollen grains have a thick exine, a dense cytoplasm enriched with ribosomes and well-developed mitochondria. Such pollen grains can determine can germinate in the presence of metabolic inhibitors such as actinomycin-D and cycloheximide, indicating that they have pre-formed messengers. This is considered to be a remarkable example of structure-function relationship (13). In contrast, embryogenic pollen grains have a thin exine, and an attenuated cytoplasm that has condensed mitochondria and is poor in ribosomes (8). Such pollen neither germinates nor forms embryos in the presence of metabolic inhibitors (7).

ACKNOWLEDGEMENTS

This work was done at Free University of Berlin while the author held an Alexander von Humboldt Fellowship. The award of this fellowship and the grant of study leave from the University of Delhi are gratefully acknowledged.

REFERENCES

1. S. Guha and S.C. Maheshwari, *In vitro* production of embryos from anthers of *Datura*, *Nature* 204:497 (1964).
2. M. Horner and H.E. Street, Pollen dimorphism - origin and significance in pollen plant formation by anther culture, *Ann. Bot.* 42:763-771 (1978).
3. E. Heberle-Bors and J. Reinert, Isolated pollen cultures and pollen dimorphism, *Naturwiss.* 67:311 (1980).
4. A. Rashid and J. Reinert, Selection of embryogenic pollen from cold-treated buds of *Nicotiana tabacum* var. Badischer Burley and their development into embryos in culture, *Protoplasma*

105:161-167 (1980).
5. A. Rashid and J. Reinert, Differentiation of embryogenic pollen in cold-treated buds of Nicotiana tabacum var. Badischer Burley and the nutritional requirements of the isolated pollen to form embryos, Protoplasma 106:137-144 (1981a).
6. A. Rashid and J. Reinert, High frequency embryogenesis in ab initio pollen cultures of Nicotiana tabacum, Naturwiss. 68:378-379 (1981b).
7. A. Rashid and J. Reinert, In vitro differentiation of embryogenic pollen, control by cold treatment and embryo formation in ab initio pollen cultures of Nicotiana tabacum var. Badischer Burley, Protoplasma 109:285-294 (1981c).
8. A. Rashid, A.W. Siddiqui and J. Reinert, Ultrastructure of embryogenic pollen of Nicotiana tabacum var. Badischer Burley, Protoplasma 107:375-386 (1981).
9. J.P. Nitsch, Experimental androgenesis in Nicotiana, Phytomorphology 19:389-404 (1969).
10. H.E. Street, Single-cell clones - derivation and selection, in: "Plant Tissue and Cell Cultures," H.E. Street, ed., Blackwell Scientific Publications, Oxford (1977).
11. C. Nitsch, Culture of isolated microspores, in: "Applied and Fundamental Aspects of Plant Cell Tissue and Organ Culture," J. Reinert and Y.P.S. Bajaj, eds., Springer-Verlag, Berlin-Heidelberg-New York (1977).
12. N. Sunderland, Comparative studies of anther and pollen culture, in: "Plant Cell and Tissue Cultures - Principles and Applications," W.R. Sharp, P.O. Larson, E.F. Paddock and V. Raghavan, eds., Ohio University Press, Columbus (1979).
13. W.A. Jensen, Reproduciton in flowering plants, in: "Dynamic Aspects of Plant Ultrastructure," A.W. Robards, ed., McGraw-Hill Book Co., Ltd., United Kingdom (1974).

GENETIC STABILITY AND VARIABILITY OF POLLEN-DERIVED PLANTS

Hu Han

Institute of Genetics
Academia Sinica
Beijing
China

INTRODUCTION

A great deal of attention has been directed towards evaluating the genetic stability and variability of pollen plants obtained through anther culture. These factors are directly related to the use of such plants in genetic studies and breeding programs. This paper summarizes research in China on the characterization of both genetically stable and variable pollen-derived crop plants.

GENETIC STABILITY

Stable, homozygous diploid strains can be developed through chromosome doubling of haploids. In earlier investigations on the application of anther culture to breeding programs in China, a great deal of research concerning the genetic stability of pollen plants derived from anthers of F_1 hybrids and their parents was carried out in tobacco (1), rice (2, 3) and wheat (4). These studies revealed that the coefficiencies of variance of important genetic traits in pollen-derived strains was similar to or less than those of the parents. The genetic characteristics of these pollen-derived strains could, therefore, be considered to be relatively stable.

In recent years genetic analysis and cytological investigations were further carried out in natural populations of pollen-derived strains. Results obtained from such populations of pollen plants in wheat (5, 6), rice (7), maize (8) and tobacco (9) over a period of several years indicated that about 90% of the diploid lines were

Table 1. Genetic Stability in an H_2 Pollen-derived Wheat Population

No. of strains	Uniform strains No. (%)	Segregating strains No. (%)	Segregating Characteristics					
			Plant height	Spike shape	Awns	Grain color	Ferti-lity	Spring or Winter type
444	396 (89.2)	48 (10.8)	19	10	2	5	6	6

genetically uniform. An analysis of 444 pollen-derived lines of wheat obtained between 1972 and 1978 revealed that 90% were completely homozygous while 10% segregated for certain characteristics (Table 1). In addition, cytological investigations of root tip chromosomes from 54 H_1 plants were carried out (10). The pollen-derived plants could be classified into five types. Among them 88.9% of the regenerated plants were haploids and homozygous diploids (Table 2). The cytological examination of PMCs from 72 pollen-derived H_1 plants was also carried out (10). Table 3 shows that 87.5% of the regenerates were haploids (3x) and diploids (6x) while 12.5% were variants. These results indicated that diploids and heteroploids as well as haploids could be made available through anther culture. In 44 3x plants many of the PMCs had 21 univalents (Figure 1). In some cells two kinds of chromosome configurations, 19 I + 1 II and 17 I + 2 II were more frequently observed. Of 19 pollen-derived 6x plants all had bivalents (Figure 2), indicating they had the identical, stable chromosome configuration initially observed in their parents.

Using the maize varieties "Ba Tang Bai" and Qun Dan 105", Cao et al. (11) and Wu and Zhong (12) have recently obtained pollen-derived clones from cultured anthers. Within the last three years Gu et al. (8) subcultured the clones derived from "Ba Tang Bai" every four weeks. According to the degree of totipotency observed during this period, the clones could be classified into four categories. The calli of clone # 1 possessed a great capacity for morphogenesis and readily regenerated into green plantlets. Gu and

Table 2. Root Tip Chromosome Numbers in Pollen-derived Wheat Plants

No. of plants observed	3X	6X	5X	Mixoploids	Aneuploids
54	38	10	2	3	1
%	70.4	18.5	3.7	5.6	1.8

Table 3. Meiotic Chromosome Numbers in Pollen-derived Wheat Plants.

No. of plants observed	3X	6X	8X	Nullisomics (6X-2)	Mixoploids
72	44	19	1	2	6
%	61.1	26.4	1.4	2.8	8.3

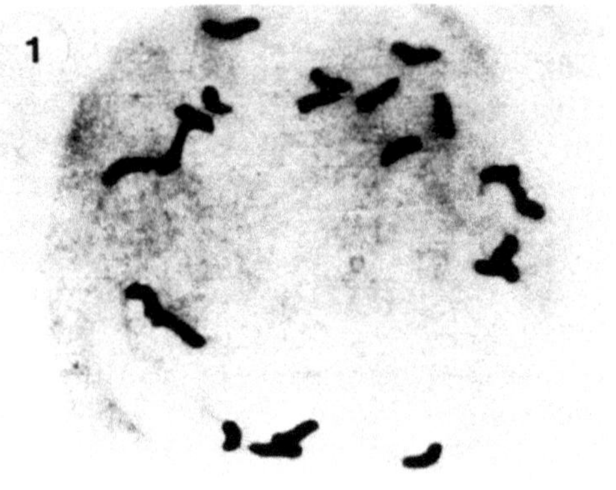

Figure 1. PMC of 3X wheat pollen plant with 21 I.

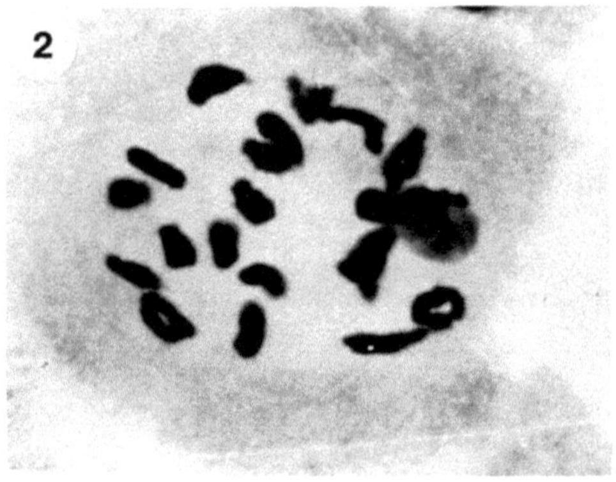

Figure 2. PMC (diakinesis) of 6X wheat pollen plant with 21 II.

Zhang (13) carried out a cytological study of calli and root tip cells of 35 pollen plants regenerated from this clone. The results showed that approximately 90% of the regenerates were haploids (Table 4). The karyotype analysis of calli and regenerated plants of clone # 1 was carried out through Giemsa banding. The chromosome C-banding patterns of callus cells and regenerated plants were similar. These results provided further evidence that anther culture methodology could be directly applied to plant breeding programs.

Table 4. Variation in Chromosome Numbers in Calli and Root-tip Cells of 35 Maize Plants Regenerated from Pollen-derived Clone # 1.

Materials	No. of Cells	Variation in chromosome number			
		Hypohaploid (10)	Haploid (=10)	Hypodiploid (20)	Diploid (=20)
Callus	427	34	383	2	8
	%	7.9	89.7	0.2	1.9
Root-tip	645	25	564	10	46
	%	3.9	87.4	1.6	7.1

VARIABILITY OF POLLEN-DERIVED PLANTS

Cultured cells of both animal and plant tissues are characterized by instability of chromosome number and structure. The same phenomenon has also been observed in pollen-derived plants. These variations, in addition to gene (point) mutations, consisted mainly of chromosomal changes and were generally expressed as changes in ploidy level and fertility in about 28 species belonging to 19 genera and 7 families of dicots and monocots. Table 5 comprises a list of monocots in which heteroploids have been obtained through anther culture. These heteroploids are useful because they could serve as the source of material for chromosome engineering studies.

In our previous studies on wheat anther culture, the majority (90%) of plants obtained were haploids and homozygous diploids but some pentaploids, octoploids and mixoploids were also detected (5, 10, 14). It should, therefore, be possible to routinely produce heteroploids and aneuploids through anther culture techniques.

Production of Aneuploids from Intervarietal Hybrids of Common Wheat

In recent years (1978-1980) aneuploids have been obtained from three types of common wheat (i.e., semi-winter, winter and spring types of intervarietal hybrids) via anther culture. A nullisomic (# 498) and non-typical nullisomic (# 453) were identified in 1978 in H_1 plants derived from pollen of a semi-winter intervarietal hybrid (Orofen x Xiaoyan 759). Furthermore, a monosomic 6B plant with a large satellite and lower arm ratio was identified in the H_3 population of nullisomic #453 (Figures 3a and 3b). In 1979 a monosomic line (A_{23}) was detected in pollen-derived plants of a winter intervarietal hybrid (Kotung 58 x Norin 10 F_1). In 1980 a nullisomic plant # J-87 (Figure 4) was isolated through anther culture of a F_1 spring wheat hybrid (Pitic 62 x Hua Pei #1). This plant was abnormally self-fertile and produced a few seeds.

Production of Aneuploids from Common Wheat Cultivars

For further verification of aneuploid production through anther culture, the common wheat cultivar Orofen served as the source of anthers and N6, Potato-II and Maihe 5 (W_5) basic media were used. In May 1980, 3,780 anthers were inoculated, and 96 green plants were obtained. Root tips and pollen mother cells of these plants (H_1) were cytologically evaluated. In addition to a great number of haploid and dihaploid plants, two tetrasomics ($2\underline{n} + 2 = 44$) and a trisomic ($2\underline{n} + 1 = 43$) were observed. One of the tetrasomics (plant J-92) had a pair of dicentric 6B chromosomes (Figure 5). One variant plant with a spelta-like spike was also observed (Figure 6). Furthermore, four nullisomic plants were obtained from two pollen-derived lines of the common wheat cultivar Jinghong # 5. These plants were characterized by long, narrow leaf blades and crimped

Table 5. Heteroploid Plants Regenerated from Anther Cultures of Various Monocot Species.

Species	Chromosome number of regenerated plants	Reference(s)
Agropyron repens	3*n*, 4*n*, or higher	20
Hordeum vulgare	4*n*, chimera	21, 22, 23
Oryza sativa	3*n*, 4*n*, 5*n*, chimera	24, 25, 26
Secale cereale	polyploids	27
Triticum aestivum	5*n*, 8*n*, 12*n* and aneuploids	5, 10, 14, 28, 29
Triticum vulgare x *Agropyron glaucum*	monosomic haploid	30
Triticale	mixoploids	31
Triticale (6*X*) x *Triticum aestivum* (6*X*)	various aneuploids and mixoploids	32

flag leaves. Fertility was poor but a few seeds were obtained.

Production of Aneuploids from Intergeneric Hybrids of Hexaploid Triticale and Hexaploid Wheat

In 1979 hexaploid triticale (cv. Rosner) was used as the maternal parent in a cross with hexaploid wheat (cv. Kotung 58). In 1980, anthers from the F_1 hybrid were cultured on B_5 medium. A total number of 2,881 anthers were inoculated and 62 green plantlets were produced from 2.2% of the cultured anthers. Root tip cells of 27 vigorous pollen-derived plants were analyzed cytologically. Different types of aneuploids, such as aneuhaploids, aneudihaploids and mixoploid plants were identified (Table 6). The aneuploids and mixoploids set seeds in varying degrees.

CONCLUSIONS

The above mentioned results indicated that fertile aneuploids, such as nullisomics, monosomics, trisomics and tetrasomics could be obtained from intervarietal hybrids and from wheat cultivars through

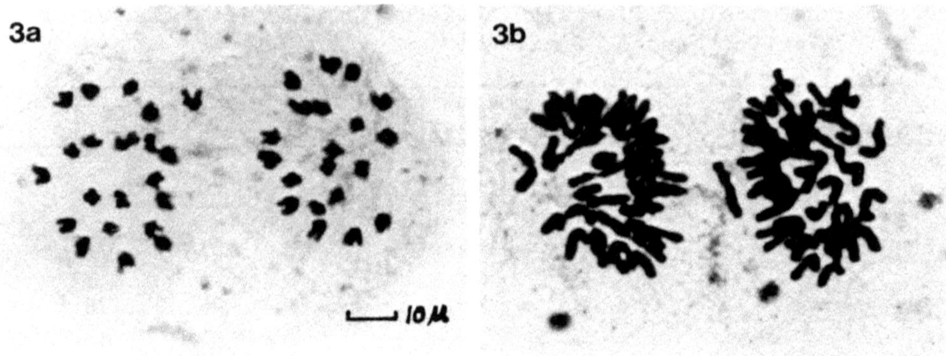

Figure 3. Pollen plant (# 453) derived from a semi-winter intervarietal F_1 hybrid (Orofen x Xiaoyan 759).
a) Monosomic anaphase I, showing lagging chromosome;
b) Monosomic telophase I, showing free 6B chromosome.

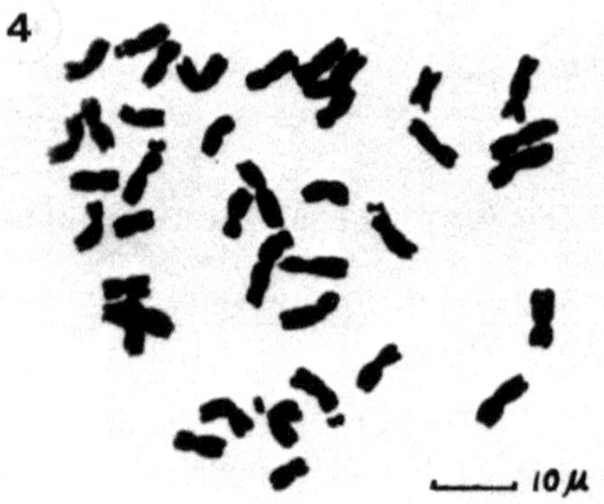

Figure 4. Pollen plant derived from a spring intervarietal F_1 hybrid (Pitic 62 x Hua pei #1). Root tip somatic chromosomes of a nullisomic (# J-87) $2\underline{n} = 6\underline{x}-2 = 40$.

anther culture. For example, plant J-92 derived from the cultivar Orofen was identified as a tetrasomic with dicentric 6B chromosomes. These observations indicate that anther culture will induce changes in chromosome structure as well as number. Translocation lines might be produced in this manner. Moreover, it is possible that different aneuploid plants (nullisomics, monosomics, etc.) without consanguinity to Chinese Spring could be obtained. Khush (15)

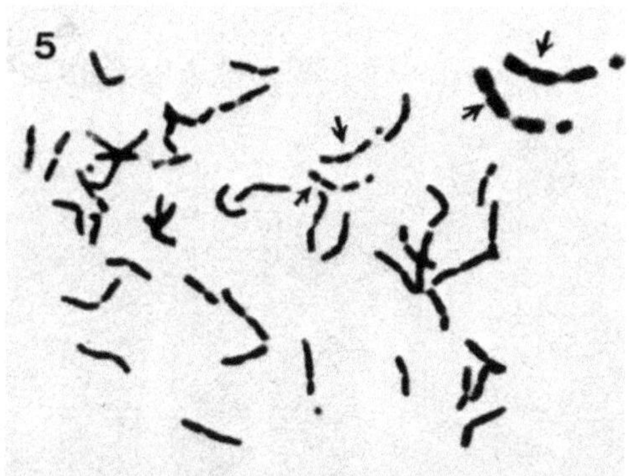

Figure 5. Root tip chromosomes of a tetrasomic wheat pollen plant (# J-92) derived from the cultivar Orofen. Note the pair of dicentric 6B chromosomes.

Figure 6. Spike of variant wheat pollen plant derived from the cultivar Orofen. Left: Orofen control plant; Right: variant plant.

Figure 7. Spikes of pollen aneuhaploids derived from Rosner x Kotung 58 F_1 plants. Left to right: Rosner (♀) Kotung (♂). Pollen aneuhaploid plants with 21-27 chromosomes.

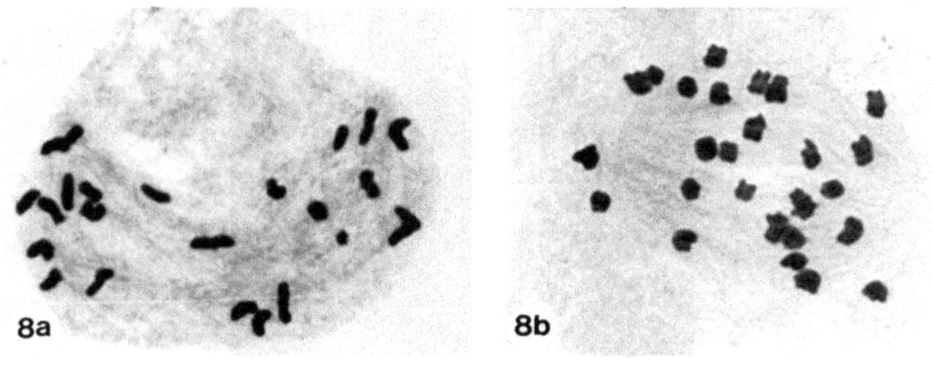

Figure 8. PMCs of aneuhaploid plants of intergeneric triticale (Rosner) x wheat (Kotung 58) F_1 hybrids.
a) PMC of an aneuhaploid with 24 univalents;
b) PMC of an aneuhaploid with 26 univalents.

Table 6. Types of Pollen-derived Plants Obtained from F_1 Hybrids of Triticale and Common Wheat

Chromosome number of the regenerated plants	21	22	23	24	25	26	27	46	Mixoploid
Number of plants	1	1	3	3	4	1	1	4	9

stated that nullisomic, monosomic, trisomic and tetrasomic plants originated from haploids or from progenies of hybrids between different ploidy levels or by treatment with physical and chemical factors. We have been able to demonstrate the direct production of aneuploids from hexaploid wheat through anther culture.

Research on different kinds of alien addition lines and alien substitution in wheat derived from hexaploid secondary triticale was recently reviewed by Muntzing (16). Triticale has been suggested as classical material for investigations on chromsome engineering (17, 18). In our work the maternal triticale parent (cv. Rosner) was identified as 14 IIW + 7 IIR by Shigenaga and Larter (19). For obtaining different alien addition and substitution lines of wheat, we used a small population of 27 pollen plants derived from cultured anthers of a triticale (cv. Rosner) x wheat (cv. Kotung 58) F_1 hybrid. A number of aneuploids and mixoploids were present in this population (Figure 7). The PMCs of the aneuhaploids possessed only univalents (Figures 8a and 8b) indicating that their genomes might be ABDR. Such plants may be valuable since, through chromosome doubling, alien addition lines, substitution lines and mixed amphiploids might be produced. Although the investigations were carried out in different years (from 1978 to 1980), using different varieties and hybrids as materials, a number of fertile or partially fertile aneuploids were repeatedly obtained through anther culture. Anther culture should, therefore, be considered as a potentially useful tool for chromosome engineering in higher plants.

REFERENCES

1. Institute of Tobacco, Academy of Agricultural Sciences; Institute of Botany, Academia Sinica, The evaluation of the progenies of the pollen plants of tobacco, Acta Genet. Sinica 1:26-39 (1974).
2. C. Hsu, K.C. Yin, C.Y. Chu, F.Y. Bi, S.D. Wang, D.Y. Lio, F.L. Wang, N.F. Chien and C. Chu, Inducing pollen plants of rice and observations of their progenies, Acta Genet. Sinica 2: 294-301 (1975).
3. 302 Research Group, Institute of Genetics, Academia Sinica, Genetic studies on pollen plants in rice (*Oryza sativa* L.),

Acta Genet. Sinica 3:277-285 (1976).

4. T.W. Ouyang, H. Hu, C.C. Chuang and C.C. Tseng, Induction of pollen plants from anthers of Triticum aestivum L. cultured in vitro, Scientia Sinica 16:79-95 (1973).

5. H. Hu, Z. Xi, J. Zhuang, J. Ouyang, J. Zeng, S. Jia, X. Jia, J. Jing and S. Zhou, Genetic investigation on pollen-derived plants in wheat (Triticum aestivum), Acta Genet. Sinica 6: 322-330 (1979).

6. D. Yang and J. Gao, Observation of genetic stability and viability of the progenies pollen-derived plants in wheat, Ann. Rep. Inst. Agric. Sci. Shandong Chang Wei:23-32 (1979).

7. Y. Chen and L. Li, Investigation and utilization of pollen-derived haploid plants in rice and wheat, in: "Proc. of Symp. on Plant Tissue Culture," Science Press, Peking (1978).

8. M. Gu, X. Zhang, Z. Cao and C. Guo, The totipotency and genetic stability of pollen callus of maize in subcultures, Ann. Rep. Inst. Genetics Acad. Sinica 1980:96-97 (1981).

9. H. Xu, S. Ai, Z. Chen and X. Jia, Report on heredity and viability of progenies of pollen-derived tobacco, Zhongguo Yancao 1:8-10 (1980).

10. H. Hu, T. Hsi and S. Chia, chromosome variation of somatic cells of pollen calli and plants in wheat (Triticum aestivum L.), Acta Genet. Sinica 5:23-30 (1978).

11. Z. Cao, C. Guo and J. Hao, A study of embryogenesis in pollen callus of maize (Zea mays L.), Acta Genet. Sinica 8:269-274 (1981).

12. J. Wu and Q. Zhong, Embryogenesis in corn culture, Acta Photo-physiol. Sinica 6:221-224 (1980).

13. M. Gu and X. Zhang, Studies on the karyotype of pollen callus clone of maize in long-term subcultures, Ann. Rep. Inst. Genetics Acad. Sinica 1980:97-98 (1981).

14. H. Hu, Z. Xi, J. Ouyang, S. Hao, M. He, Z. Xu and M. Zou, Chromosome variation of pollen mother cell of pollen-derived plants in wheat (Triticum aestivum L.), Scientia Sinica 23:905-912 (1980).

15. G.S. Khush, "Cytogenetics of Aneuploids," Academic Press, New York (1973).

16. A. Muntzing, Triticale results and problems, in: "Advances in Plant Breeding," Supplement to Journal of Plant Preeding, Paul Parey, Berlin (1979).

17. M. Feldman and E.R. Sears, The wild gene resources of wheat, Sci. Amer. 244:102-113 (1981).

18. J.G. O'Mara, Cytogenetic studies on Triticinae I.A. method for determining the effects of individual Secale chromosomes on Triticum, Genetics 25:401-408 (1940).

19. S.E. Shigenaga and E.N. Larter, Karyotype analysis of hexaploid triticale, Can. J. Cenet. Cytol. 13:585-591 (1971).

20. M. Zenkteler, E. Misiura and A. Ponitka, Induction of androgenetic embryoids in the in vitro cultured anther of several species, Experientia 31:289-291 (1975).

21. P.J. Dale, Pollen dimorphism and anther culture in barley, Planta 127:213-220 (1975).
22. B. Foroughi-Wehr, G. Mix, H. Gaul and H.M. Wilson, Plant production from cultured anthers of Hordeum vulgare, Z. Pflanzenzuchtg. 77:198-204 (1976).
23. H.M. Wilson, Culture of whole barley spikes stimulates high frequencies of pollen calluses in individual anthers, Plant Sci. Lett. 9:233-238 (1977).
24. T. Nishi and S. Mitsuoka, Occurrence of various ploidy plants from anther and ovary culture of rice plant, Jap. J. Genet. 44:341-346 (1969).
25. H. Niizeki and K. Oono, Rice plants obtained by anther culture, Colloq. Intl. C.N.R.S. #193 Les cultures de tissue de plantes:251-257 (1971).
26. Second Division 3rd Laboratory, Institute of Genetics, Academia Sinica, Investigations on the induction and genetic expression of rice pollen plants, Scientia Sinica 17:209-226 (1974).
27. G. Wenzel and E. Thomas, Observations on the growth in culture of anthers of Secale cereale, Z. Pflanzenzuchtg. 72:89-94 (1974).
28. Y. Henry et J. de Buyser, Androgènese sur des blés tendres en cours de sélection. I. L'obtention des plantes in vitro, Z. Pflanzenzuchtg. 83:49-56 (1979).
29. J. de Buyser and Y. Henry, Androgènese sur des blés tendres en cours de sélection. 2. L'obtention de grains, Z. Pflanzenzuchtg. 84:9-17 (1980).
30. C.C. Wang, Z.C. Chu, C.S. Sun, C. Hsu, K.C. Yin and F.Y. Bi, Induction of pollen plants from the anther culture of Triticum vulgare-Agropyron glaucum hybrid, Acta Genet. Sinica 2:72-80 (1975).
31. V.B. Vnuchkova, Growing haploids from triticale anthers and their cytological characteristics, Dokl. Vaskhnil. 10:8-10 (1979).
32. X.Z. Wang and H. Hu, Production of aneuploids with intergeneric hybrids between hexaploid triticale and hexaploid wheat, Ann. Rep. Inst. Genet. Acad. Sinica 1980:76 (1981).

RECENT PROGRESS IN RICE ANTHER CULTURE STUDIES

S.K. Raina

Genetics Division
Indian Agricultural Research institute
New Delhi 110012
INDIA

Raina and Iyer (1) identified Crossa-2, a purple pigmented cultivar, as the highest responding from among 25 _indica_ rice cultivars screened for anther response. Subsequently, using Crossa-2 as the test material, Raina (2) carried out studies on the cultural requirements for eliciting increased frequency of anther response and regeneration of shoots and roots from the pollen calli. Anthers at mid- to late-uninucleate pollen stage inoculated on Linsmaier and Skoog's (3) medium supplemented with 2,4-D (2 mg/l), sucrose 3%, pH 6.0 and incubated at 25°C $\pm$ 1°C in dark or under low light intensity (about 200 lux), gave the maximum response. Addition of zeatin (0.01 mg/l) or yeast extract (3 g/l) enhanced the response slightly. Other than that, auxins (IAA, IBA, 2,4-D and NAA) and cytokinins (kinetin and zeatin) individually or in several combinations, or some of the complex growth substances (coconut water, yeast extract and casein hydrolysate) individually or in several combinations with an auxin and/or a cytokinin, in most of the cases induced a much lower response than 2,4-D (2 mg/l) alone. Studies carried out similarly on organ regeneration showed that Linsmaier and Skoog's medium (LS) + coconut water (CW) (15%) as the most effective medium for inducing shoot-root regeneration from maximum number of pollen calli.

The present report deals briefly with some of the more recent studies. Anther culture technique is being tried with potentially desirable hybrids with the view to the isolation of promising recombinants directly in a homozygous state, after screening from among colchicine-doubled pollen-derived haploids or spontaneously-doubled haploids. The latter form the majority (about 80%) of the pollen-derived plant population, having undergone chromosome doubling during the course of _in vitro_ culture. Side by side, ancillary basic studies continue for further understanding/improvement of the technique.

Two sets of F_1 hybrids were tried. One involved Rassi x CO-40 and Vaigai x CO-40. Rassi and Vaigai are rather low yielders (4-5 tons/hectare), have a spreading plant type but are of short duration (125 days); whereas CO-40 is high yielding (6-7 tons/hectare), has a compact high tillering plant type, but of long duration (165 days). The objective being to isolate a high yielding short/medium duration and compact plant type. The other set of hybrids: P-150 x IR-36 and P269 x IR-36, have P-150 and P-269 as fine grain donors and IR-36 for boosting their yield potential and introduction of other desirable agronomic characteristics. Medium used for anther culture was that of N_6 (4), since it is found to be significantly better than LS and certain others (Table 1), supplemented with 2,4-D (2 mg/l). However, for inducing organogenesis in pollen calli, LS basal medium with CW (15%) was used. Sucrose was maintained at 3%.

Both the hybrids in the first set responded rather well (Table 2). The hybrid involving Vaigai responding better than the one having Rassi. The hybrid involving Vaigai also showed a higher percentage of shoot producing anther calli (Table 3). Anther response in the second set of hybrids was much lower (Table 2), and the organogenesis pattern was similar to the first set. In all about 950 plants regenerated from the four hybrids, but only 347 could be transplanted in soil, the rest being albinos. Among the 231 green plants derived from Vaigai x CO-40 and Rassi x CO-40 hybrids, 27 diploid plants (all of Vaigai x CO-40 origin) were selected for further study. Of these, seven plants (R_1 - R_7) derived from a single callus of an anther were particularly interesting owing to their promising characteristics: compact plant type (like CO-40), intermediate plant height (about 90 cms), matured in 135 days (compared to 125 days of Vaigai and 165 of CO-40) and good panicle length and spikelet fertility. Apparently all the seven plants looked alike, and the plant population derived from each of them not only looked uniform in itself but also resembled the others. Analysis of data for plant height, flowering date, number of productive tillers, panicle length, spikelet fertility and yield per plant showed that there were no significant differences for any of these characters within the population derived from each of the seven plants. However, there were significant differences for some characters between the plants (Table 4) indicating that the single callus was in fact genetically mosaic. Coalescence of adjacent pollen embryoids during the early stages of callusing seems to be the most probably reason for this mosaicism. R_4 proved to be the most promising of all, and in a subsequent study out-yielded R_1, R_2, Vaigai and IR-8 in performance trials carried out recently in Delhi (unpublished data). R_{10}, another promising one from the selected plants, matures very early (125 days) like Vaigai, has a higher number of productive tillers and more per plant yield than Vaigai. However, the plant is a

Table 1. Effect of Certain Basal Media on Anther Response.

cv. Crossa-2

Basal medium*	No. of Anthers: Inoculated	No. of Anthers: Responded	% Response	Average Response %
N_6 - Experiment 1	224	50	22.32	19.06
- Experiment 2	296	47	15.80	
LS - Experiment 1	245	31	12.65	11.48
- Experiment 2	221	24	10.31	
**SK-8 - Experiment 1	142	11	7.75	7.20
- Experiment 2	240	16	6.66	
**Chaleff's R-2 - Experiment 1	238	1	0.42	0.39
- Experiment 2	270	1	0.37	

* In all cases supplemented with 2,4-D (2 mg/1).
** From Rush and Shao (9).

Table 2. Anther Response of the Rice Hybrids.

Medium	Material	Cold Pre-Treatment °C/Duration	No. of Anthers Inoculated	No. of Anthers Responded	% Response
N_6 + 2,4-D (2 mg/1) Suc. (30 g/1)	P-269 x IR-36	-	5645	109	1.93
	P-150 x IR-36	-	5515	73	1.37
	Rassi x CO-40	-	676	15	2.21
	Vaigai x CO-40	-	857	149	17.38
N_6 + 2,4-D (2 mg/1) Suc. (30 g/1)	Rassi x CO-40	5°C for 24 h	751	73	9.72
	Vaigai x CO-40	5°C for 24 h	653	112	17.15
N_6 + 2,4-D (2 mg/1) Suc. (30 g/1) CW 15%	Rassi x CO-40	5°C for 24 h	514	42	8.15
	Vaigai x CO-40	5°C for 24 h	464	73	15.73

Table 3. Organ Regeneration from Anther Calli of the Rice Hybrids.

Material	Callus induction medium and pre-treatment if any	Anther calli subcul-tured #	Showing shoot & root gen-eration %	Green plants only %	Green and albino %	Albinos only %	Roots only %
Rassi x CO-40	N_6 + 2,4-D (2 mg/1)	12	25.0	-	-	25.0	37.5
Vaigai x CO-40		142	24.6	2.8	-	21.8	28.8
Rassi x CO-40	N_6 + 2,4-D (2 mg/1) Cold pre-treatment at 5°C for 24 h	67	31.3	1.5	1.5	28.3	23.8
Vaigai x CO-40		112	42.0	2.6	1.8	37.5	24.0
Rassi x CO-40	N_6 + 2,4-D (2 mg/1) + CW 15% and cold pre-treatment at 5°C for 24 h	37	35.1	2.7	2.7	29.7	21.6
Vaigai x CO-40		73	52.0	5.5	-	46.5	19.1

Table 4. Comparison of Mean Values of R_1 - R_7 with Reference to Certain Economic Traits.

Culture No.	Plant height (cms)	Productive tillers/ plant (#)	Panicle length (cms)	Days for 50% flowering (#)	Spikelets/ panicle (#)	Spikelet fertility (%)	Yield (q/ha)	5% level	1% level
R_1	89.0	12.0	27.7	109	147	75	53.43		
R_2	89.0	12.0	27.8	112	159	78	61.88		
R_3	88.0	11.0	27.0	111	174	84	46.41		
R_4	89.0	12.0	21.3	115	119	79	79.78		
R_5	89.0	9.0	21.1	112	101	78	75.00		
R_6	90.0	10.0	22.2	112	140	84	66.25		
R_7	90.0	9.0	21.3	111	103	79	58.12		
F Value	1.62	5.61**	1.06	0.00	6.53**	2.89*	8.42**	2.66	4.01
CD at 5%		1.55			15.32	5.90	1.07		

* Significant at 5% probability level.
** Significant at 1% probability level.

rather spreading type like Vaigai.

Out of the 27 plants which were studied in some detail, two of them, R_{18} and R_{40}, were found unstable. The type of segregation exhibited in both the cases suggested that the instability was due to a single gene mutation. It seems likely that this would have occurred after the pollen calli had diploidized. Whereas in R_{18} there was a small number (4 out of 91) of late maturing tall plants, in the case of R_{40} there were almost as many self-sterile plants (5 out of 87) in an otherwise uniform population in each case. Out of the 106 normal green plants recovered from the second set of hybrids, R_{409} of P-150 x IR-36 origin was found to be the most promising and has bred true. It has inherited the fine grain characteristics from P-150 and high tillering ability from IR-36.

Sunderland (5), working with Nicotiana tabacum, for the first time observed that only the lightly staining smaller (S) grains are potentially embryogenic and not the normal (N) grains which comprise most of the pollen population. This observation was further substantiated by fractionated pollen culture studies in N. tabacum by Wernicke et al. (6). Assuming that it might be possible to increase the frequency of S-grains through stress conditions imposed on the donor plants, water stress was tried. Using the cultivar Crossa-2, water stress was imposed by light irrigation once or twice a week, just enough to keep the plants growing; whereas the control plants remained almost all through under 2-5 cms of water. Contrary to the expectations, no pollen dimorphism was observed. The average pollen size in water stressed as well as control plants remained the same. However, anther response from water stressed donor plants was better than the control plants (Table 5).

Albinism continues to be a major problem so far as the application of the technique is concerned. It seems unlikely that these chlorophyll-deficient plants could be mutants since in certain cultivars the frequency of such plants produced is sometimes 100%. The possibility of reduced gene dosage being the cause for their high frequency occurrence can be ruled out, as on cytological screening most of the albinos analyzed were found to be diploids and not haploids. The problem could be due to deficient pollen cytoplasm, as it appears so from the recent reports of the production of comparatively higher frequency of normal green plants from egg cells in unpollinated ovary cultures (7, 8). Working with the high responding cultivar Crossa-2, temperature during anther incubation was found to affect the frequency of albinos produced (Table 6). When anthers were cultured at 23°C-24°C as against 25°C-26°C, the frequency of albinos produced was subsequently lower by about 20%. Perhaps temperature affects the orientation of the first division(s) in pollen which could be significant in that the embryoid initial so produced may not be adequate in cytoplasmic contents. Nevertheless, albinism in anther cultures needs to be studied in more detail and

Table 5. Effect on Anther Response Using Anthers from Water Stressed Donor Plants.

Treatment	No. of Anthers		Response (%)	Average Response (%)
	Inoculated	Responded		
Water stress				
Experiment 1	194	44	22.68	21.97
Experiment 2	221	47	21.26	
Control				
Experiment 1	240	46	19.16	17.98
Experiment 2	232	39	16.81	

Table 6. Effect of Temperature During Anther Incubation on the Frequency of Albino Plant Production.

Temperature during anther incubation	No. of anther calli: Subcultured	No. of anther calli: Showing shoot and root regeneration (%)	Green plants only (%)	Albinos (%)	Average frequency of albinos (%)
25°C–26°C					
Experiment 1	98	35 (35.71)	14 (40.00)	21 (60.00)	65.71
Experiment 2	66	21 (31.90)	6 (28.57)	15 (71.42)	
23°C–24°C					
Experiment 1	78	27 (34.61)	14 (59.25	13 (48.14)	46.98
Experiment 2	81	24 (29.62)	13 (54.16)	11 (45.83)	

attempts made accordingly to significantly lower their frequency, if not completely eliminate them. Unless this is achieved, "anther culture breeding" particularly in rice and wheat, where the albinism frequency is generally high, cannot be utilized as a dependable and more efficient alternative breeding technology.

ACKNOWLEDGEMENT

The author is grateful to S. Ramanujam, Genetics Division, IARI, New Delhi, for providing facilities and to Drs. S. Sree Rangaswamy and R.B. Ranganathan of the Tamil Nadu Agricultural University, Coimbatore, for providing some of their rice hybrid material and laboratory facilities at Coimbatore. Thanks are due to Suman Bala for technical assistance.

REFERENCES

1. S.K. Raina and R.D. Iyer, A study of response to anther culture, Indian J. Genet. 34A:283-285 (1974).
2. S.K. Raina, "Studies on induction of haploids through anther culture," Ph.D. Thesis, Agra University, Agra, India (1977).
3. E.M. Linsmaier and F. Skoog, Organic growth factor requirements of tobacco tissue culture, Physiol. Plant. 18:100-127 (1965).
4. C.C. Chu, The N_6 medium and its applications to anther culture of cereal crops, in: "Proc. Symposium on Plant Tissue Culture," Science Press, Peking (1978).
5. N. Sunderland, Anther culture as a means of haploid induction, in: "Proc. Intl. Symposium on Haploids in Higher Plants," University of Guelph Press, Guelph (1974).
6. W. Wernicke, C.T. Harms, H. Lörz and E. Thomas, Selective enrichment of embryogenic microspore populations, Naturwiss. 65:540-541 (1978).
7. Z.C. Zhu, H.S. Wu, Induction of haploid plantlets from the unpollinated ovaries of Nicotiana tabacum, Acta Genetica Sinica 6:181-185 (1979).
8. Z. Chang and H.Y. Yuan, Induction of haploid rice plantlets by ovary culture, Plant Sci. Lett. 20:231-237 (1981).
9. M.C. Rush and Q.Q. Shao, Rice improvement through cell and tissue culture, in: "Proc. Symposium on Rice Research Strategies for the Future," IRRI Press, Manila (1980).

THE PRODUCTION AND UTILIZATION OF MICROSPORE-DERIVED HAPLOIDS IN BRASSICA CROPS

W.A. Keller, K.C. Armstrong and A.I. de la Roche

Ottawa Research Station
Agriculture Canada
Ottawa, Ontario K1A 0C6
Canada

INTRODUCTION

The large scale availability of haploids could offer plant breeders specific advantages for developing new crop cultivars (1). Although haploids may be obtained through a number of techniques, a great deal of attention is being given to anther culture as a universal method of haploid production. Under appropriate conditions, the microspores within the anthers of some species can be induced to undergo embryogenesis resulting in the development of differentiated embryos which resemble zygotic seed embryos and from which plants may be regenerated. Haploids have been obtained from anther cultures in a number of crop species (2), but in many instances their yield and reliability of production have not been satisfactory to permit application to breeding programs.

The genus *Brassica* contains a number of economically important crop species including oilseed rape which might be more efficiently improved through the use of haploids. Studies have been undertaken in our laboratory as well as in other laboratories to develop procedures for haploid *Brassica* production and to identify factors which influence microspore embryogenesis (3 - 7). We have evaluated potential factors influencing *Brassica* microspore embryogenesis on the basis of embryo yield per 1,000 cultured anthers (4, 8). These yield values are simply calculated from observations of anther response (percent of anthers producing embryos) and anther productivity (number of embryos produced by responding anthers). In this paper we will report on recent studies initiated to identify optimal conditions for microspore embryogenesis in *Brassica* species. The procedures and media utilized to establish the anther cultures have previously been described (3, 9). We shall also describe preliminary experiments designed to evaluate the potential uses of haploids in breeding.

FACTORS INFLUENCING MICROSPORE EMBRYOGENESIS

Pollen Developmental Stage

A number of studies have shown that embryogenesis can only be obtained from pollen at a specific developmental stage. In most species the uninucleate microspore stage is optimal but in some cases mitotic or binucleate microspores respond more effectively in culture (10). Our studies with *Brassica* have indicated that the uninucleate stage appears to be the most responsive. We have been able to develop an external staging system which permits rapid identification of suitable anthers; in *B. campestris* the optimal stage occurred when the length of the petal was 1/4 - 1/2 of the length of the anther (Table 1). This staging procedure has been successfully employed with other *Brassica* species.

Composition of the Culture Medium

Medium composition is a key factor influencing microspore embryogenesis. In *Brassica* anther culture we have observed that sucrose concentration is critical with levels of at least 4% (w/v) required for induction of embryogenesis. The optimal concentration for *B. campestris* has been considered to be 10% (3), although Lichter (7) has recently identified 8% as optimal for *B. napus*.

Table 1. Effect of Developmental Stage on Embryogenesis in Anther Cultures of *Brassica campestris* (Torch x R500 F_1)

Length of petal in comparison to length of anther	No. of anthers cultured*	Anther response %	Anther productivity	Embryo yield / 1000 anthers
Between 1/4 and 1/2	888	12.50	2.52	315
Approx. 1/2	954	6.81	1.78	121
More than 1/2	732	2.05	1.73	36

* Anthers were cultured at 35°C for 3 days prior to transfer to 25°C; the culture procedures and medium employed were as described by Keller et al. (3) and Keller and Armstrong (9).

Supplementation with auxins and reduced nitrogen in the form of glutamine has been shown to be beneficial (3). As a result of the observations on the positive effect of serine in isolated tobacco micro spore cultures (11), we have also added this amino acid to our Brassica anther culture medium (9).

The employment of liquid medium for anther culture has been advantageous for a number of species including tobacco (12) and barley (13). Liquid medium has recently been used for B. napus anther culture (7). Preliminary tests with liquid medium in our laboratory have revealed that in some cases a 100% increase in yield of B. napus embryos was obtained (unpublished results).

Physiological Condition of the Anther Donor

Several investigators have demonstrated that the conditions under which donor plants are grown subsequently influences the behavior of microspores in culture (14 - 16). We have made similar observations in the case of Brassica. Greater success has been obtained with plants grown in controlled climate chambers than with field-grown material (Table 2). Investigations on the effect of growth temperature revealed that higher embryo yields were obtained if B. napus donor plants were grown in a 15°C/10°C day/night cycle rather than at higher temperatures (17). However, more recently a similar study with B. campestris F_1 hybrids revealed that greater embryo yields were obtained if the donor plants were grown at a 20°C/15°C cycle (Table 3). In B. campestris a study on the influence of photoperiod indicated that superior anther donors could be obtained under continuous light (Table 4). Another study on the role of light intensity demonstrated that the best B. campestris donors were grown under 3,000 foot candles, which was the maximal intensity available (Table 5). On the basis of these results, further tests involving the employment of higher light intensities are warranted. The significance of light quality was also evaluated and included the use of Cool White fluorescent (Sylvania), Gro-Lux Wide Spectrum (Sylvania) and Vitalite (Duro-Test) tubes used independently or in combination with incandescent bulbs. Embryos were obtained from anthers of plants grown under all conditions but slightly higher yields were obtained with a combination of Gro-Lux Wide Spectrum with 45 watt incandescent bulbs supplying 16% of the intensity (total intensity at source was 1,800 fc).

The studies carried out thus far with Brassica have revealed that manipulation of the donor growth conditions will influence the subsequent response of anthers in culture. Conditions which either completely inhibited microspore embryogenesis or caused dramatic embryo yield increases were not identified. We did not evaluate the role of donor plant nutrition; however, investigations on this topic are warranted since nitrogen nutrition has been shown to affect tobacco microspore embryogenesis (15).

Table 2. Embryogenesis in Cultured Anthers Obtained from Field-grown and Chamber-grown *B. campestris*.

Donor growth condition	Genotype	No. of anthers cultured*	Anther response %	Anther productivity	Embryo yield / 1000 anthers
Chamber (16 h photo-period; 20°C/15°C day/night cycle)	Torch x R500	1224	8.74	2.02	176
	R500 x Torch	1842	9.66	2.30	223
Field	Torch x R500	924	3.35	2.94	98
	R500 x Torch	1218	3.28	1.35	44

* Anthers were cultured one day at 35°C followed by one day at 30°C prior to transfer to 25°C; the culture procedures and medium employed were described by Keller et al. (3) and Keller and Armstrong (9).

Table 3. Effect of Donor Growth Temperature on Embryogenesis in Anther Cultures of *B. campestris*.

Donor growth temp. (°C) Light (16h)	Donor growth temp. (°C) Dark (8h)	Genotype	No. of anthers cultured*	Anther response %	Anther productivity	Embryo yield/1000 anthers
15	10	Torch x R500 F_1	1404	7.48	2.10	157
		R500 x Torch F_1	1014	4.34	2.00	87
20	15	Torch x R500 F_1	1746	13.69	1.67	229
		R500 x Torch F_1	1068	16.95	2.96	501
25	20	Torch x R500 F_1	1716	5.48	2.23	122
		R500 x Torch F_1	900	8.44	1.49	126

* Anthers were cultured 1 day at 35°C followed by 1 day at 30°C prior to transfer to 25°; the culture procedures and medium employed were as described by Keller et al. (3) and Keller and Armstrong (9).

Table 4. Effect of Photoperiod Received by Donor Plants on Embryogenesis in Anther Cultures of *Brassica campestris*.

Photoperiod (h) Light (20°C)	Photoperiod (h) Dark (15°C)	Genotype	No. of anthers cultured*	Anther response %	Anther productivity	Embryo Yield / 1000 anthers
16	8	Torch x R500 F_1	1182	6.18	1.34	83
		R500 x Torch F_1	1158	7.69	1.85	142
24	0	Torch x R500 F_1	876	14.73	1.67	247
		R500 x Torch F_1	1704	12.56	1.88	236

* Anthers were cultured 1 day at 35°C followed by 1 day at 30°C prior to transfer to 25°C; the culture procedures and medium employed were as described by Keller et al. (3) and Keller and Armstrong (9).

Table 5. Effect of Light Intensity Received by Donor Plants on Embryogenesis in Anther Cultures of *Brassica campèstris*.

Light intensity (f.c.)	Genotype	No. of anthers cultured*	Anther response %	Anther productivity	Embryo Yield / 1000 anthers
750	Torch x R500 F_1	384	3.64	1.57	57
	R500 x Torch F_1	1866	4.34	3.57	155
1500	Torch x R500 F_1	522	2.87	1.40	40
	R500 x Torch F_1	1176	5.70	1.87	106
3000	Torch x R500 F_1	1224	8.74	2.02	176
	R500 x Torch F_1	1842	9.66	2.30	223

* Anthers were cultured 1 day at 35°C followed by 1 day at 30°C prior to transfer to 25°C; the culture procedures and medium employed were as described by Keller et al. (3) and Keller and Armstrong (9).

Table 6. Effect of Sequential Increase and Decrease of Culture Temperature on Embryogenesis in Anthers of *Brassica campestris* (Torch x R500 F_1).

Treatment	No. of anthers cultured*	Anther response %	Anther productivity	Embryo Yield / 1000 anthers
35(1d)--30(1d)--25	852	15.49	2.13	330
25(1d)--30(1d)--35(1d)--30(1d)--25	1002	1.50	1.67	25

* The culture procedures and medium employed were as described by Keller et al. (3) and Keller and Armstrong (9).

Culture Temperature

In Brassica spp., anther culture temperature has been shown to be a highly significant factor influencing microspore embryogenesis (4, 8, 18). An initial period of culture at elevated temperature (e.g. 30°C, 35°C) followed by maintenance at 25°C greatly increased the yield of microspore-derived embryos. There appeared to be species differences in optimal temperature requirements with spring B. napus giving maximal embryo yields after culture at 30°C for 14 days (4), while B. campestris and B. oleracea gave better yields after a 1 - 3 day initial culture period at 35°C. Preliminary tests had indicated that the sequential reduction of culture temperature in 5°C steps from 40 or 35°C to 25°C was superior to a one-step reduction. However, further investigations with B. campestris revealed that embryo yields were not significantly higher. An experiment was also undertaken to evaluate a sequential increase in culture temperature from 25°C to 35°C followed by a subsequent reduction to 25°C. However, this treatment decreased the yield of embryos in comparison to immediate culture at 35°C (Table 6). The foregoing experiment (Table 6) indicated that it was important to expose the anthers to elevated culture temperatures as quickly as possible in order to maximize embryo yields. A study in which B. campestris buds were surface sterilized and floated in water at room temperature for periods of 0.5 to 6 h prior to anther excision and plating supported this supposition; embryo yields dropped considerably after four hours of bud maintenance under these conditions (Table 7).

Although the importance of employing elevated culture temperature has been recognized, information regarding the molecular or cellular events occurring during this treatment is still lacking. Both biochemical and histological studies could provide valuable information on the site(s) and mechanism(s) of action of the treatment.

Anther Pretreatments

The beneficial effect of pretreating buds, or inflorescences at low temperatures prior to anther culture has been demonstrated for many species (10, 19). In B. campestris we evaluated a single low temperature pretreatment of buds maintained in sealed plastic bags in darkness at 5°C for one week but observed that such a treatment in fact reduced embryo yield (Table 8). It is, however, possible that a systematic study of different temperatures and lengths of treatment would reveal conditions more favorable for Brassica microspore embryogenesis.

We have tested the effect of high temperature pretreatments in B. oleracea var. italia (broccoli) and have found them to be effective in stimulating microspore embryogenesis (Keller and Armstrong, manuscript in preparation). Maximal yields (more than 100 embryos/

Table 7. Effect of Delay Between Bud Excision and Anther Planting on Microspore Embryogenesis in *Brassica campestris* R500 x Torch F_1.

Time between excision and planting (h)	No. of anthers cultured*	Anther response %	Anther productivity	Embryo Yield / 1000 anthers
0.5	276	4.35	1.75	76
1.0	840	7.86	1.42	112
2.0	798	10.53	1.42	149
4.0	540	5.93	1.41	83
6.0	600	3.17	1.21	38

* Anthers were cultured at 35°C for 3 days prior to transfer to 25°C; the culture conditions and medium employed were as described by Keller et al. (3) and Keller and Armstrong (9).

Table 8. Effect of Low Temperature Bud Pretreatment on Embryogenesis in Cultured Anthers of *Brassica campestris* Torch x R500 F_1.

Pretreatment	No. of anthers cultured*	Anther response %	Anther productivity	Embryo Yield / 1000 anthers
None	2442	7.2	2.2	155
5°C for 7 days	1092	1.1	2.0	22

* Anthers were cultured at 35°C for 3 days prior to transfer to 25°C; the culture procedures and medium employed were as described by Keller et al. (3) and Keller and Armstrong (9).

1,000 anthers) were obtained when excised inflorescences maintained in water were pretreated at 45°C for one hour followed by 40°C for three hours after which the buds were surface sterilized and the anthers cultured at 35°C for one day, and 30°C for one day prior to transfer to 25°C. In *B. campestris*, however, high temperature pretreatments were not stimulatory (unpublished results).

Genotype

The role of donor genotypes in microspore embryogenesis has been shown to be significant in a number of crop species (19, 20). In Brassica we have obtained a response with all species and cultivars tested although there were differences in degree of response amongst lines. For example, in B. campestris anthers from F_1 hybrids yielded more embryos than those from either parental cultivar (17). Recent tests with B. campestris have shown that the cultivars R500 and Torch yielded more embryos than Candle (unpublished results). Genotypic differences in optimal culture temperature requirements for microspore embryogenesis have also been observed. Anthers of winter B. napus cultivars gave maximal yields of microspore-derived embryos after an initial culture period of 35°C for two days (unpublished results) whereas the spring B. napus cultivar Tower responded best to a 14 day initial culture temperature of 30°C (4).

REGENERATION AND CHARACTERIZATION OF MICROSPORE-DERIVED PLANTS

In order to regenerate plants from microspore-derived Brassica embryos, it is essential to transfer them to an embryo culture medium in which auxins have been eliminated and the sucrose concentration reduced (3, 4). Immature, globular embryos, which in some cases comprise more than 50% of the population generally do not survive on the embryo culture medium which contains 2% sucrose. Such embryos may require a more complex medium or a gradual step-wise reduction of sucrose level to facilitate their survival. Of those embryos which survive, less than 20% develop directly into plantlets; the majority undergo elongation and/or thickening of the hypocotyl but fail to develop a normal shoot system. However, plants can be obtained from such embryos through the induction of shoot organogenesis in hypocotyl explants (9).

Analyses of regenerates of B. napus, B. campestris and B. oleracea have revealed the presence of diploids as well as haploids (4, 8, 18). By employing the procedures developed for B. campestris (8), anther-derived regenerates have also been obtained from B. hirta, B. juncea and B. nigra. Cytological evaluation has revealed haploids amongst the B. hirta and B. juncea regenerates but only diploids amongst the B. nigra regenerates (unpublished results). Brassica haploids can be readily identified by their small, sterile flowers. On this basis we have screened large populations of anther-derived regenerates and have observed that the frequency of haploids varies amongst species and may range from 20% to more than 50% (4, 8, 18).

Genetic analyses of diploids obtained from anthers of heterozygous donors have shown them to be homozygous (3). These plants must have arisen from haploid microspores presumably as a result of endomitosis or nuclear fusion during anther culture (3).

For application of haploids in breeding it is essential to double the chromosome number to restore fertility. Wenzel et al. (5) have described the use of colchicine for doubling B. napus haploids. The induction of organogenesis in petiole explants of haploid tobacco can result in chromosome doubling (21). Similar procedures could possibly be developed for Brassica since the induction of shoot regeneration on hypocotyl explants of microspore-derived B. oleracea embryos increased the diploid frequency in the regenerates in comparison to regenerates obtained directly from embryos (18). Spontaneously-doubled haploids which have been routinely obtained through anther culture could, of course, be directly used in breeding programs.

UTILIZATION OF HAPLOIDS IN CROP IMPROVEMENT

A major application of haploid production relates to greatly increased effectiveness in selection of genetic recombinants (22, 23). This should result in a significant reduction in the time required to develop new crop cultivars. In B. napus, field-grown doubled haploids have been evaluated (17) and lines showing potential improvements in yield and quality characteristics have been selected for further testing (24). Homozygous lines obtained from spontaneously-occurring B. napus haploids have demonstrated superior yields to the parental varieties (25), providing further evidence for the value of haploids in the development of new cultivars.

Microspore-derived Brassica embryos often possess differentiated cotyledons which contain storage lipids identical to those present in zygotic seed embryos (26). In oilseed rape breeding it has been possible to perform chemical analysis on single cotyledons and directly select for traits such as erucic acid content which are controlled by the sporophyte (27). It should, therefore, be possible to select for certain seed traits in a non-lethal manner in microspore-derived embryo populations. We crossed the "zero" erucic B. campestris cultivar Torch with the high erucic cultivar R500 and produced F_1 seed with intermediate levels of erucic acid. Single cotyledons from embryos derived from cultured F_1 anthers were sampled according to the procedures described by de la Roche and Keller (26). A range of erucic acid levels was detected in the cotyledons of these embryos (Table 9). In cases where "zero" levels of erucic acid were present, the seed of plants grown from these embryos also contained low erucic acid levels indicating that effective selection for this trait could be made in microspore-derived embryo populations. Such selection procedures are especially advantageous in B. campestris where self-incompatibility may create an additional problem in obtaining seed for chemical analysis. It should be noted

Table 9. Erucic Acid Content of Single Cotyledons of Microspore-derived Embryos and of Seed of Plants Regenerated from Embryos which were Obtained from Anther Cultures of *B. campestris* F_1 Hybrids (Torch x R500).*

	Percent erucic acid	
Pedigree	Cotyledons in vitro	Derived seed
824-46	21.4	46.8
900-17	20.7	28.1
920-10-A-1	49.3	38.4
920-10-A-6	trace	trace
824-20	trace	trace
900-10	41.3	38.0
920-7-A-1	trace	trace

* Culture procedures and medium employed were those described by Keller et al. (3) and Keller and Armstrong (9).

that selection for high erucic acid levels cannot be made as effectively at the embryo level since the embryos are not dormant and have likely utilized some storage lipid.

Another important application of *Brassica* haploid production relates to its use in the rapid synthesis of homozygous lines for the development of hybrid cultivars. This approach should be feasible in vegetable *Brassicas* where self-incompatibility systems have made the production of inbred lines time-consuming and expensive. The recent production of large numbers of broccoli haploids (Keller and Armnstrong, manuscript in preparation) would suggest it is already possible to evaluate the haploid approach in the development of hybrid seed in *Brassica* crops.

REFERENCES

1. J.G.T. Hermsen and M.S. Ramanna, Haploidy and plant breeding, *Phil. Trans. R. Soc. London B* 292:499-507 (1981).
2. I.K. Vasil, Androgenetic haploids, *Int. Rev. Cytol.* Suppl. 11A: 195-223 (1980).

3. W.A. Keller, T. Rajhathy and J. Lacapra, In vitro production of plants from pollen in Brassica campestris, Can. J. Genet. Cytol. 17:655-666 (1975).
4. W.A. Keller and K.C. Armstrong, High frequency production of microspore-derived plants from Brassica napus anther cultures, Z. Pflanzenzuchtg. 80:100-108 (1978).
5. G. Wenzel, F. Hoffmann and E. Thomas, Anther culture as a breeding tool in rape. I. Ploidy level and phenotype of androgenetic plants, Z. Pflanzenzuchtg. 78:149-155 (1977).
6. M. Renard et F. Dosba, Etude de l'haploidie chez le Colza (Brassica napus L. var. oleifera Metzger), Ann. Amelior. Plantes 30:191-209 (1980).
7. R. Lichter, Anther culture of Brassica napus in a liquid culture medium, Z. Pflanzenphysiol. 103:299-337 (1981).
8. W.A. Keller and K.C. Armstrong, Stimulation of embryogenesis and haploid production in Brassica campestris anther cultures by elevated temperature treatments, Theor. Appl. Genet. 55:65-67 (1979).
9. W.A. Keller and K.C. Armstrong, Embryogenesis and plant regeneration in Brassica napus anther cultures, Can. J. Bot. 55: 1383-1388 (1977).
10. N. Sunderland, Anther and pollen culture 1974-1979, in: "The Plant Genome," D.R. Davies and D.A. Hopwood, eds., The John Innes Charity, Norwich (1980).
11. C. Nitsch, La culture de pollen isolé sur milieu synthétique, C R Acad. Sci. Paris 278:1031-1034 (1974).
12. N. Sunderland and M. Roberts, New approach to pollen culture, Nature 270:236-238 (1977).
13. Z.H. Xu, B. Huang and N. Sunderland, Culture of barley anthers in conditioned media, J. Exp. Bot. 32:767-778 (1981).
14. J.M. Dunwell, Comparative study of environmental and developmental factors which influence embryo induction and growth in cultured anthers of Nicotiana tabacum, Env. Exp. Bot. 16:109-118 (1976).
15. N. Sunderland, Strategies in the improvement of yields in anther culture, in: "Proc. of Symposium on Plant Tissue Culture," Science Press, Peking (1978).
16. B. Foroughi-Wehr and G. Mix, In vitro response of Hordeum vulgare L. anthers cultured from plants grown under different environments, Env. Exp. Bot. 19:303-309 (1979).
17. W.A. Keller and G.R. Stringam, Production and utillization of microspore-derived haploid plants, in: "Frontiers of Plant Tissue Culture 1978," T.A. Thorpe, eds., The International Association for Plant Tissue Culture, Calgary (1978).
18. W.A. Keller and K.C. Armstrong, Production of anther-derived haploid plants in autotetraploid marrowstem kale (Brassica oleracea var. acephala), Can. J. Genet. Cytol. 23:250-269 (1981).
19. S.C. Maheshwari, A.K. Tyagi, K. Malhotra, Induction of haploidy from pollen grains in angiosperms - the current status,

Theor. Appl. Genet. 58:193-206 (1980).

20. G. Wenzel, Recent progress in microspore culture of crop plants, in: "The Plant Genome," D.R. Davies and D.A. Hopwood eds., The John Innes Charity, Norwich (1980).
21. M.J. Kasperbauer and G.B. Collins, Reconstitution of diploids from leaf tissue of anther-derived haploids in tobacco, Crop Sci. 12:98-101 (1972).
22. G. Melchers, Haploid higher plants for plant breeding, Z. Pflanzenzuchtg. 67:19-32 (1972).
23. T. Rajhathy, Haploid flax revisited, Z. Pflanzenzuchtg. 76:1-10 (1976).
24. G.R. Stringam, An assesment of the anther culture technique for plant improvement in Brassica napus L., in: "The Plant Genome," D.R. Davies and D.A. Hopwood, eds., The John Innes Charity, Norwich (1980).
25. K.F. Thompson, Superior performance of two homozygous diploid lines from naturally occurring polyhaploids in oilseed rape (Brassica napus), Euphytica 28:127-135 (1979).
26. A.I. de la Roche and W.A. Keller, The morphogenetic control of erucic acid synthesis in Brassica campestris, Z. Pflanzenzuchtg. 78:319-326 (1977).
27. R.K. Downey and B.L. Harvey, Methods of breeding for oil quality in rape, Can. J. Plant Sci. 43:271-275 (1963).

PLANT PROTOPLASTS: AN EXPERIMENTAL SYSTEM FOR CELL BIOLOGISTS

L.C. Fowke[1] and F. Constabel[2]

[1]Department of Biology
University of Saskatchewan
Saskatoon, Saskatchewan
Canada S7N 0W0

[2]National Research Council of Canada
Prairie Regional Laboratory
Saskatoon, Saskatchewan
Canada S7N 0W9

INTRODUCTION

Protoplasts are spherical naked plant cells produced by the removal of the cell wall with digestive enzymes (Fig. 1). They are usually derived from either leaf tissue or from cell suspension cultures and have been isolated from a wide variety of plant species. Plant protoplasts can be cultured in defined media and will form a new cell wall, divide, and in many cases regenerate complete plants (e.g. 1, 2, 3, 4). Protoplasts provide a unique experimental system for studies of the structure and function of plant cells. The removal of the cell wall from plant cells, for example, has created new and exciting possibilities for exploring the structure, chemistry and function of the plant plasma membrane. Some of the more important applications in cell biology are listed in Table 1. This table includes only recent selected references and, therefore, does not contain a comprehensive list of relevant literature.

Three of the applications listed in Table 1 will be examined in order to illustrate the usefulness of plant protoplasts as an experimental system for cell biologists: 1) plant protoplasts for studying cytoplasmic organelles associated with the plasma membrane, 2) plant protoplasts for studying idioblasts and secondary metabolism of plants, 3) protoplast fusion as a possible technique for investigating the interaction of plasma membranes.

Table 1. Applications of Protoplasts for Studies in Plant Cell Biology

Method	Application	Selected References
1. Protoplast isolation	(a) structure and chemistry of the plasma membrane	(5), (6)
	(b) <u>cytoplasmic organelles associated with the plasma membrane</u>	(7), (8)
	(c) uptake and membrane transport by protoplasts	(9), (10), (11)*
	(d) primary metabolism (photosynthetic parenchyma)	(12), (13)*
	(e) <u>secondary metabolism (idioblasts)</u>	(14)
	(f) organelle isolation	
	(i) nuclei	(15), (16)
	(ii) chloroplasts	(17)*
	(iii) vacuoles	(18), (19)
2. Protoplast culture	(a) cell wall formation	(11)*, (20)*
	(b) plant regeneration	(2)*, (3)*
3. Protoplast fusion	(a) <u>plasma membrane interaction</u>	(21), (22), (23),
	(b) behavior of organelles in heterokaryocytes and hybrid cells	(24), (11)*, (25)*
	(c) cell cycle parameters	(26)

* Review articles.

The underlined applications are discussed in the text.

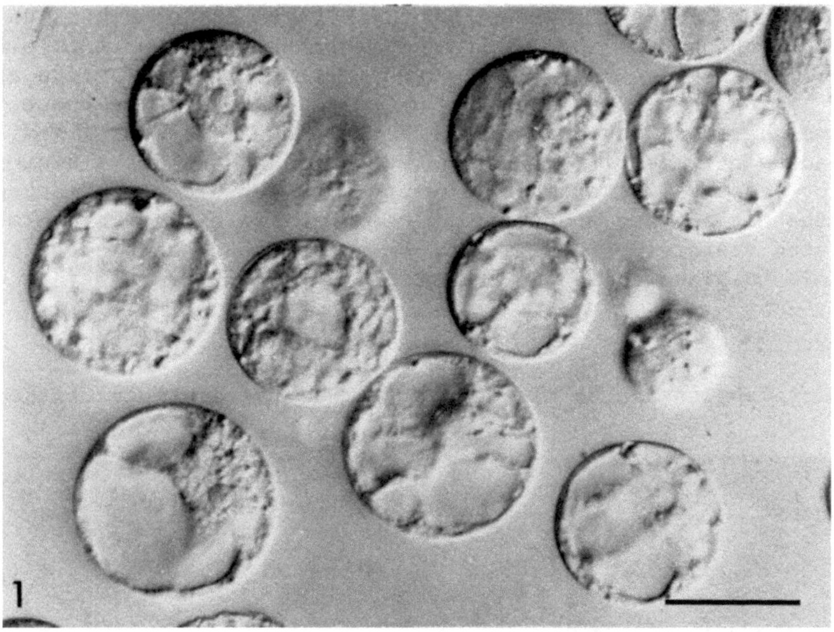

Figure 1. Light micrograph of protoplasts isolated from tobacco cell suspension culture. Nomarski optics. Bar = 30 μm.

CYTOPLASMIC ORGANELLES ASSOCIATED WITH THE PLASMA MEMBRANE

Plant protoplasts are particularly useful for investigating characteristics of the plant plasma membrane. Studies using specific stains have confirmed the presence of carbohydrates in the plant plasma membrane while electron microscope studies using lectins have demonstrated the mobility of sugar groups within the membrane (see review 11). The results of the lectin studies demonstrate the fluid nature of the plant plasma membrane.

The plant plasma membrane appears to be the site of synthesis of the cellulose microfibrils which constitute the major structural components of the cell wall. Enzyme complexes capable of polymerizing the microfibrils are believed to move in the plane of the plasma membrane (27). Cytoplasmic organelles are also thought to be involved in the process of cell wall formation. Microtubules, for example, seem to orient cellulose microfibrils while a variety of Golgi derived vesicles contribute carbohydrates and possibly protein to the developing wall (28, 29). Recently, techniques have been developed which permit the study of these organelles and their association with the plasma membrane of protoplasts. Plant protoplasts are ideally suited to studies of cell wall formation because they rapidly regenerate a new wall when cultured. Relevant techniques developed by Marchant (30) involve attaching plant protoplasts to either glass coverslips or electron microscope grids with polylysine and then bursting them osmotically. The resulting plasma membrane fragments which remain attached to the substrate are washed, fixed and examined using immunofluorescence techniques at the light microscope level or negatively stained and observed in the electron microscope. These techniques have successfully been used to examine microtubules and coated vesicles, organelles which are believed to play an important role in plant cell wall formation (7, 8, 31, 32).

Microtubules

Microtubules can be visualized on the inner surface of the

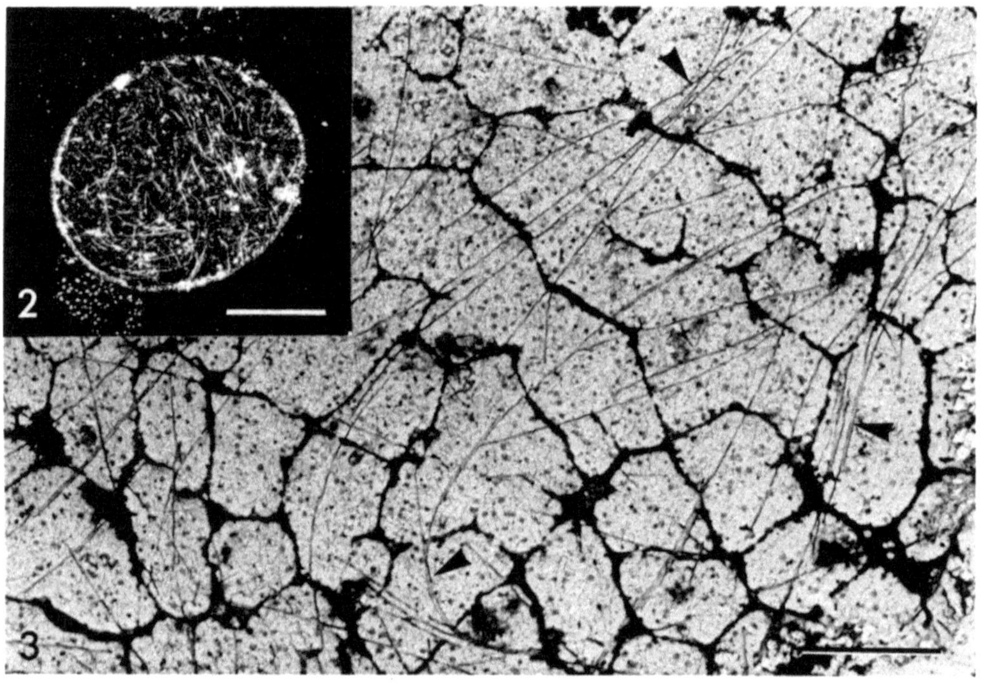

Figure 2. Light micrograph showing fluorescent microtubules on the inner surface of a tobacco protoplast plasma membrane fragment. The membrane was treated with antibodies to porcine brain tubulin followed by a second antibody labelled with FlTC. Bar = 20 μm.

Figure 3. Electron micrograph of a membrane fragment similar to the one in Figure 2 stained with uranyl acetate. Many microtubules (arrows) are apparent on the membrane surface. (Kindly provided by Dr. P. Van der Valk). Bar = 2 μm.

plasma membrane fragments by staining them with fluorescent antitubulin and examining them with a fluorescence microscope (Fig. 2). This technique reveals the general frequency and distribution of microtubules. More detailed information is available when membrane fragments are examined by transmission electron microscopy (Figs. 3, 4). Ultrastructural studies provide data concerning number and size of microtubule bundles, microtubule lengths and microtubule frequencies (7). To obtain similar information via conventional thin sectioning techniques is a laborious task. Membrane fragments can also be prepared from cultured protoplasts during early stages of wall regeneration and future research will hopefully clarify the role of

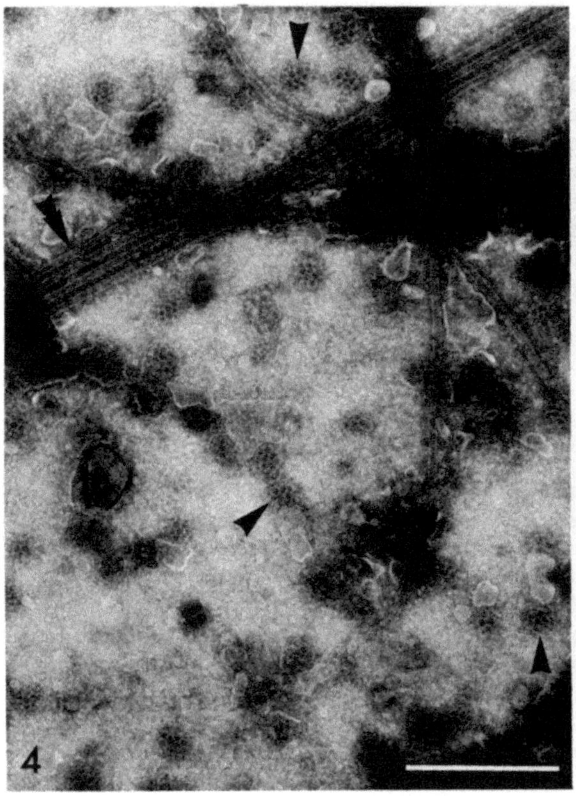

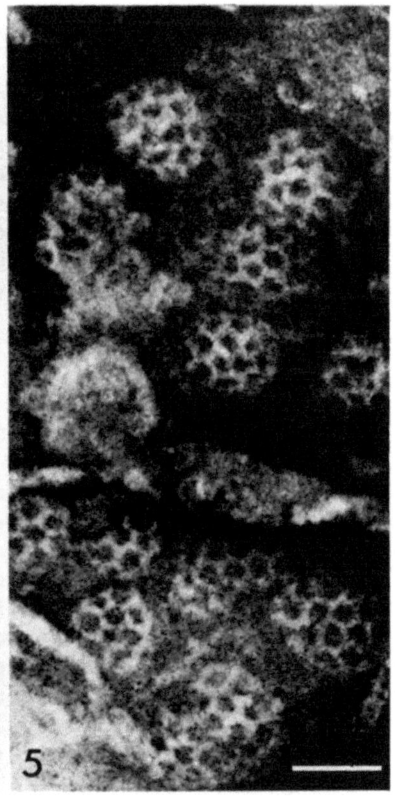

Figure 4. Electron micrograph of a negatively stained membrane fragment similar to those in Figs. 2 and 3. Numerous coated vesicles (single arrows) are present on the membrane. Note the bundle of microtubules (double arrow). Bar = 400 nm.

Figure 5. Electron micrograph of negatively stained coated vesicles on the inner surface of a plasma membrane fragment from tobacco. (Kindly provided by Dr. P. Van der Valk). Bar = 100 nm.

microtubules in wall formation and the establishment of cell shape.

Coated Vesicles

Coated vesicles are small vesicles (100 nm diameter) consisting of an internal membranous sac enclosed within a distinctly patterned coat composed mainly of the protein clathrin. They have been observed in a wide variety of plant and animal tissues (e.g. 33, 34). While in animal cells coated vesicles are involved in selective endocytosis and membrane recycling, their function in plants is not clearly understood. It has been suggested that they are produced by Golgi bodies, travel to and fuse with the plasma membrane, thus contributing materials to the growing cell wall (33).

Membrane fragments prepared as described above (1a) provide a unique system for examining coated vesicles at the cell surface with the electron microscope (Figs. 4, 5). They permit the rapid visualization of large areas of the inner surface of the protoplast plasma membrane. Coated vesicles have been examined in membranes prepared from tobacco protoplasts capable of rapid cell wall synthesis. This study has provided information concerning the distribution, structure and fate of coated vesicles (8). The results are consistent with the suggested role for coated vesicles in the process of cell wall formation.

SECONDARY METABOLISM AND IDIOBLASTS

The study of secondary metabolism in plants derives its fascination from the almost infinite variety of products, its taxonomic and phylogenetic implications, and the physiological activity of some of its products. While the chemistry of secondary products is well advanced, their biology is little understood (cf. 35). The site of synthesis and of accumulation of terpenoids, flavonoids, steroids and alkaloids, the specialization of cells harboring secondary products (idioblasts), their non-random distribution in plants, the transport of and reaction of cells to their own products, and, most strikingly, the molecular genetics of product formation, while receiving occasional attention have rarely acquired the status of a major research program. Modern biotechnology programs aimed at *in vitro* production of industrially important metabolites would significantly profit from research in the biology of product formation. The methodology of plant tissue culture has permitted some progress in defining conditions for synthesis, accumulation and degradation of secondary products (36, 37). The methodology of protoplast isolation and culture appears to offer novel approaches to studies of the biology of idioblasts and secondary metabolism.

Investigation of idioblasts and secondary metabolism by means of protoplast methodology have, so far, been modest but noteworthy.

Idioblasts

Burgeff (1883-1976) is credited with the suggestion that isolation of young latex vessels of papaya fruits, for instance, might allow for a better understanding of factors which determine the growth of non-articulated cells and latex formation. While the isolation of latex cells as protoplasts has been achieved with periwinkle leaves (Fig. 6), conditions for prolonged culture and growth have not been determined. Nutrient requirements of these idioblasts differ from those of leaf parenchyma protoplasts; the latter regenerated cell walls and formed callus.

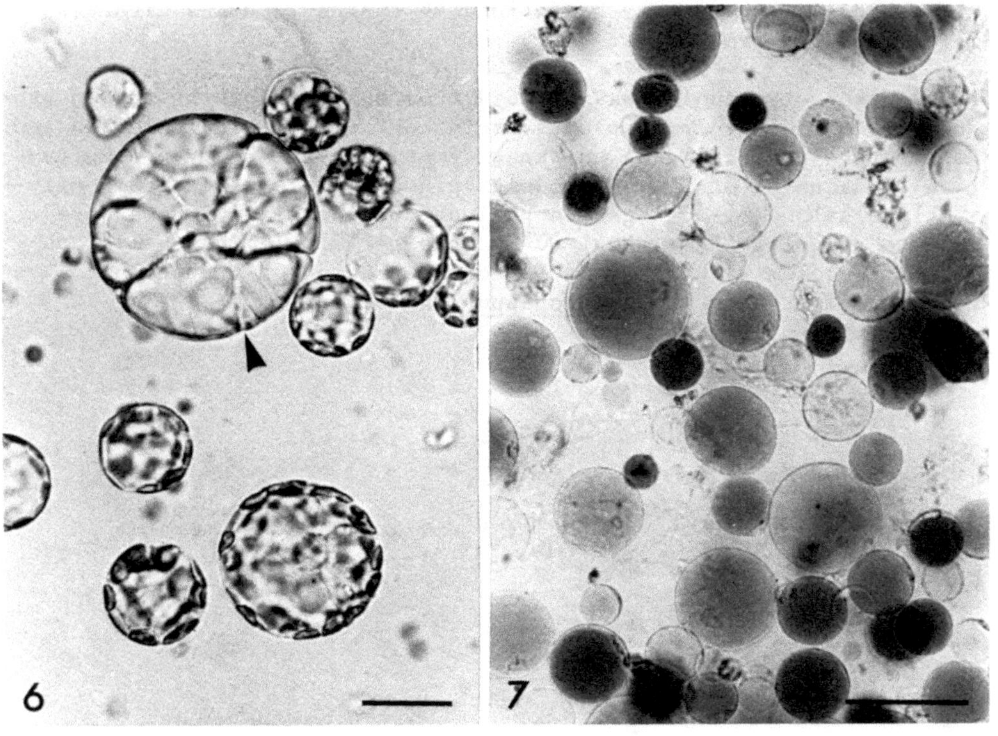

Figure 6. Light micrograph showing leaf protoplasts from periwinkle. The large protoplast (arrow) is derived from a latex cell (idioblast). Bar ≈ 20 μm.

Figure 7. Light micrograph of multicolored vacuoles isolated from protoplasts of red beet parenchyma. Bar ≈ 10 μm.

Product Accumulation

Studies of the accumulation of secondary metabolites have substantially been refined by methodology to isolate viable vacuoles from protoplasts (Fig. 7). The first report on vacuole isolation showed that remarkable efficiency had been accomplished; the percentage of intact vacuoles was extremely high and the vacuoles remained intact for 20 hr. (38). The anthocyanin concentration in tulip petal vacuoles was found to be 7.5 mM (39). Vacuoles of *Sorghum bicolor* leaf cells were recognized as the site of accumulation of the cyanogenic glucoside dhurrin, 11 μmoles/10^7 vacuoles (14). Similarly, vacuoles of *Nicotiana rustica* leaf cells were confirmed to store nicotine, 41 μg/10^6 vacuoles (40). These observations would suggest that experimentation with isolated vacuoles to elucidate the biology of secondary products is feasible. For example, vacuole isolation combined with spectrophotometry may permit study of the formation and fluctuations in products during idioblast development, and metabolic studies may be refined by fusing vacuoles with vesicles carrying precursors.

Fusion of vacuoles was deduced from observations of fusion products obtained with protoplasts isolated from periwinkle leaves and *Haplopappus gracilis* cell cultures (Figs. 8, 9). Heavy precipitation in vacuoles of fusion products presumably resulted from periwinkle alkaloids reacting with *Haplopappus* phenolics. Data were used to calculate that about 90% of the periwinkle leaf cells contained vacuoles with alkaloids at concentrations high enough to cause precipitation with *Haplopappus* phenolics (41).

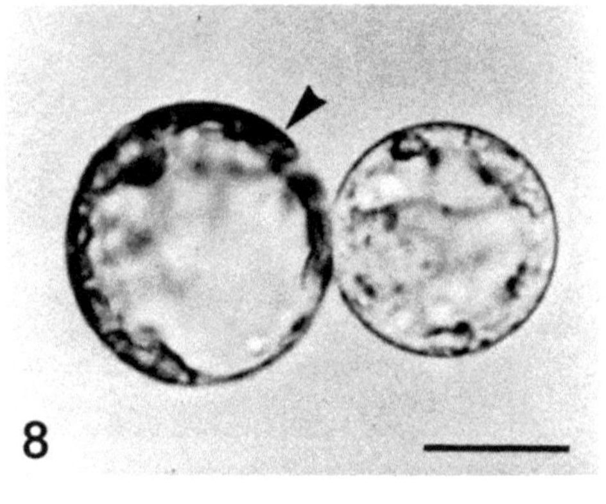

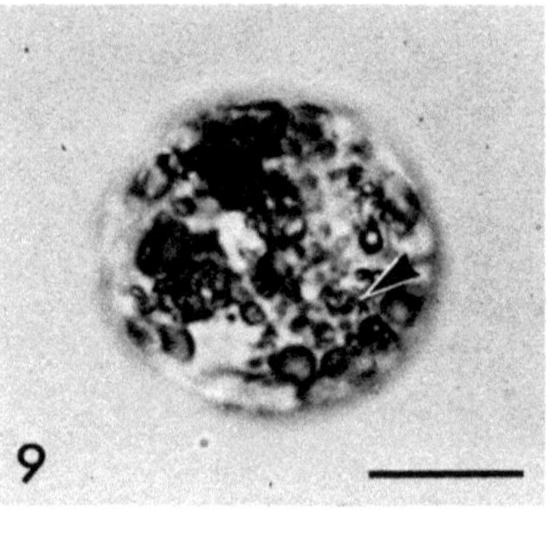

Figure 8. Light micrograph of protoplasts from *Haplopappus gracilus* cell culture and periwinkle leaf (arrow). Bar ≃ 20 μm.

Figure 9. Light micrograph showing fusion product from protoplasts of *Haplopappus gracilus* cell culture and periwinkle leaf. Note the heavy vacuolar precipitate (arrow). Bar ≃ 20μm.

Cell cloning of a periwinkle leaf by way of protoplast isolation and culture led to the establishment of many cell lines all of which showed alkaloid accumulation (42). Variation of the alkaloid profiles among the clones was comparatively low. The objective of cell cloning, i.e., the creation of variation for selection of high yielding strains, was approached when the isolation and culture of protoplasts from one particular periwinkle cell line furnished a subline with an increased capacity for catharanthine production (10 mg/100 g dry weight).

The greatest impact of protoplast methodology is seen in enhancing somatic cell genetics. Apart from cell cloning, enucleation, hybridization, and transformation are techniques which have already widened the scope of cell biology substantially (e.g. 2). Secondary products, i.e., pigments, have occasionally served as markers. Somatic cell genetics of cells producing secondary metabolites, in particular idioblasts, it is hoped, will receive proper attention soon. The combination of ergot alkaloid profiles of two Claviceps strains by protoplast fusion would demonstrate the feasibility of such experimentation (43).

PROTOPLAST FUSION AND PLASMA MEMBRANE INTERACTIONS

Plant protoplasts can be induced to fuse by a variety of techniques. The most successful and reproducible method involves the agglutination of protoplasts with high concentrations of high molecular weight polyethylene glycol (PEG) and the gradual elution of the PEG to induce fusion (44). PEG-induced fusions have been achieved between protoplasts of a wide variety of plant species as well as between plant protoplasts and animal cells (see reviews 2, 25, 45).

Considerable descriptive information is available concerning the agglutination and subsequent fusion of plant protoplasts in a number of different protoplast combinations. The processes have been described at both the light and electron microscope level (11, 46 and Figs. 10, 11). Despite the availability of this descriptive information the nature of the fusion process itself and the fate of the individual plasma membranes remains to be elucidated.

Protoplast fusion should provide a powerful tool for studying the interaction and fate of plasma membranes and may, therefore, furnish fundamental information concerning the nature of the plant plasma membrane. It should be possible, for example, to attach different labels to two populations of protoplasts, fuse them and follow the fate of labels during and after fusion. Fluorescent dyes such as fluorescein isothyocyanate (FITC) are available for light microscope studies, while a variety of labels (e.g. ferritin, hemocyanin, colloidal gold) could be used to investigate membrane interactions with the electron microscope. Preliminary work by Burgess

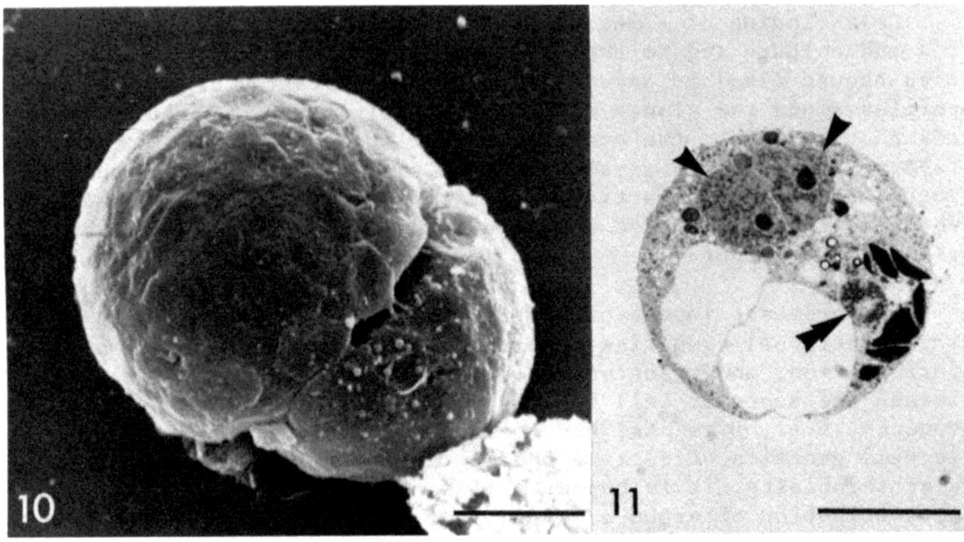

Figure 10. Scanning electron micrograph of agglutinated pea leaf protoplasts. Bar = 10 µm.

Figure 11. Light micrograph showing fusion product from protoplasts of pea leaf and Vicia cell cultures. The pea nucleus (double arrow) is surrounded by chloroplasts. Leucoplasts are detectable in the cytoplasm near the two Vicia nuclei (single arrows). Bar = 20 µm. (From Fowke et al. - 22).

et al. (21) indicates that protoplasts labelled with colloidal gold are capable of agglutinating. This approach should be particularly useful in determining the extent of membrane mixing which directly reflects the fluidity of the plasma membrane. Experiments of this nature have been completed with animal cells (e.g. 47).

An alternative technique which might be used to investigate the fusion of membranes is freeze-etching. This technique permits the visualization of the internal structure of the plasma membrane and has been successfully used to demonstrate movement of intramembranous particles away from the zone of fusion in animal cells (e.g. 48). A similar mobility of intramembranous particles seems to occur in the plasma membranes of fusing plant protoplasts (49), but further research is necessary to confirm and augment these observations.

A rather interesting model system which may prove particularly useful for investigating membrane interactions employs higher plant protoplasts and algal protoplasts. Fusion of these protoplasts permits the insertion of the tiny algal plasma membrane into the surface of the higher plant protoplast. The most promising combination studied to date involves *Chlamydomonas* and carrot protoplasts (24 and Figs. 12, 13). *Chlamydomonas* is particularly well suited to such studies because cell wall-less mutants (e.g. CW15) are easily cultured as a ready source of protoplasts and the basal bodies associated with the *Chlamydomonas* plasma membrane serve as a useful marker for membrane fusion at the electron microscope level. Future research with this system might involve a study of membrane fusion utilizing surface labels, as described above, or an investigation of the nature of cell walls synthesized by cultured fusion products. Since basal bodies are not a normal component of higher plant cells, it would also be interesting to try to follow the fate of the *Chlamydomonas* basal bodies located on the surface of the carrot proto-

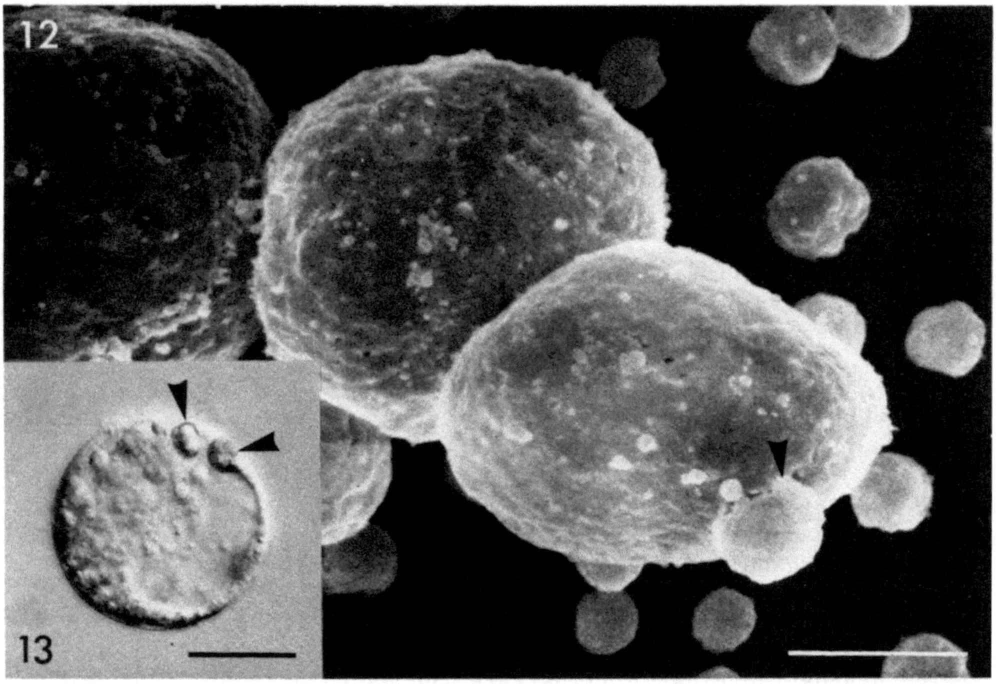

Figure 12. Scanning electron micrograph showing agglutinated carrot protoplasts. A *Chlamydomonas* protoplast (arrow) is attached to one carrot protoplast. Bar = 10 µm.

Figure 13. Light micrograph showing contents of two *Chlamydomonas* protoplasts (arrows) within a carrot protoplast. Nomarski optics. Bar = 20 µm.

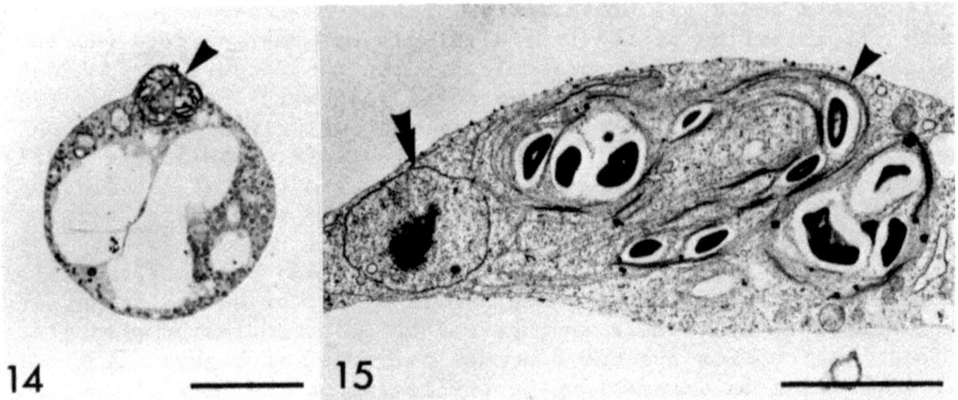

Figure 14. Light micrograph showing Stigeoclonium protoplast (arrow) fusing with a carrot protoplast. Bar = 20 μm.

Figure 15. Electron micrograph showing Stigeoclonium chloroplast (single arrow) and nucleus (double arrow) within the carrot cytoplasm following fusion. Bar = 3 μm.

plasts following fusion. With the variety of Chlamydomonas mutants available it might be possible to introduce algal membranes with known transport characteristics into the surface of carrot protoplasts.

Recently, protoplasts of the filamentous green alga Stigeoclonium have been successfully fused with carrot protoplasts (50 and Figs. 14, 15). The fusion process appears similar to that observed with the Chlamydomonas-carrot combination. The algal plasma membrane seems to be incorporated into the carrot plasma membrane while the algal organelles are released into the carrot cytoplasm.

Protoplasts have a broad range of applications for studies of plant cells (Table 1). The three applications outlined in the foregoing discussion clearly illustrate the usefulness of plant protoplasts as an experimental system for cell biologists. Studies of the inner surface of the protoplast plasma membrane have provided information concerning the structure and distribution of associated cytoplasmic organelles and future research should clarify their role in cell wall formation. Protoplasts are currently being used to study secondary product formation in plants. Results of this research will directly benefit biotechnology programs as well as provide basic knowledge concerning the structure and chemistry of latex cells and plant vacuoles. Protoplast fusion should prove to be a particularly useful tool for investigating properties of the plant plasma membrane.

ACKNOWLEDGEMENTS

We wish to express our appreciation to Pieter Van der Valk, Harvey Marchant and Peter Gresshoff for their research contributions. We also acknowledge the excellent technical assistance of Pat Rennie, Sandy Rambold, Jim Kirkpatrick and Jerry Shyluk. The research was supported by the Natural Sciences and Engineering Research Council of Canada. L.F. gratefully acknowledges travel funds provided by the Commonwealth Foundation, London, England.

REFERENCES

1. E.C. Cocking, Protoplasts: past and present, in: "Advances in Protoplast Research," Proceedings, 5th Intl. Protoplast Symposium, Hungarian Academy Sciences, Szeged (1979).
2. O.L. Gamborg and P.J. Bottino, Protoplasts in genetid modifications of plants, in: "Advances in Biochemical Engineering," A. Fiechter, ed., Springer-Verlag, Berlin (1981).
3. I.K. Vasil and V. Vasil, Isolation and culture of protoplasts, in: "Perspectives in Plant Cell and Tissue Culture," Intl. Rev. Cytol., Suppl. 11B, I.K. Vasil, ed., Academic Press, New York (1980).
4. L. Wetter and F. Constabel (eds.), "Plant Tissue Culture Methods," 2nd rev. ed., National Research Council of Canada, Ottawa, NRCC #19876 (1982).
5. J. Burgess and P.J. Linstead, Membrane mobility and the concanavalin A binding system of the plasmalemma of higher plant protoplasts, Planta 136:253-259 (1977).
6. A.R.D. Taylor and J.L. Hall, An ultrastructural comparison of lanthanum and silicotungstic acid/chromic acid as plasma membrane stains of isolated protoplasts, Plant Sci. Lett. 14: 139-144 (1979).
7. P. Van der Valk, P.J. Rennie, J.A. Connolly and L.C. Fowke, Distribution of cortical microtubules in tobacco protoplasts. An immunofluorescence microscopic and ultrastructural study, Protoplasma 105:27-43 (1980).
8. P. Van der Valk and L.C. Fowke, Ultrastructural aspects of coated vesicles in tobacco protoplasts. Can. J. Bot. 59: 1307-1313 (1981).
9. M. Suzuki, I. Takebe, S. Kajita, Y. Honda and C. Matsui, Endocytosis of polystyrene spheres by tobacco leaf protoplasts, Exptl. Cell Res. 105:127-135 (1977).
10. D.P. Briskin and R.T. Leonard, Ion transport in isolated protoplasts from tobacco suspension cells, Plant Physiol. 64:

959-962 (1979).

11. L.C. Fowke and O.L. Gamborg, Applications of protoplasts to the study of plant cells, Intl. Rev. Cytol. 68:9-51 (1980).
12. M. Nishimura and T. Akazawa, Biosynthesis of ribulose-1, 5-biphosphate carboxylase in spinach leaf protoplasts, Plant Physiol. 62:97-100 (1978).
13. E. Galun, Plant protoplasts as physiological tools, Ann. Rev. Plant Physiol. 32:237-266 (1981).
14. J.A. Saunders and E.E. Conn, Presence of the cyonogenic glucoside dhurrin in isolated vacuoles from sorghum, Plant Physiol. 61:154-157 (1978).
15. K. Ohyama, L.E. Pelcher and D. Horn, A rapid, simple method for nuclei isolation from protoplasts. Plant Physiol. 60:179-181 (1977).
16. H. Lorz and I. Potrykus, Investigations on the transfer of isolated nuclei into plant protoplasts, Theor. Appl. Genet. 53:251-256 (1978).
17. P.H. Quail, Plant cell fractionation, Ann. Rev. Plant Physiol. 30:425-484 (1979).
18. M. Nishimura and H. Beevers, Hydrolysis of protein in vacuoles isolated from higher plant tissue, Nature 277:412-413 (1979).
19. T. Boller and H. Kende, Hydrolytic enzymes in the central vacuole of plant cells, Plant Physiol. 63:1123-1132 (1979).
20. J.H.M. Willison and A.S. Klein, Cell wall regeneration by protoplasts isolated from higher plants, in: "Cellulose and Other Natural Polymer Systems: Biogenesis, Structure and Degradation," R.M. Brown, ed., Plenum Press, New York (1982).
21. J. Burgess, P.J. Linstead and V.E.L. Fisher, Studies on higher plant protoplasts by scanning electron microscopy, Micron 8:113-122 (1977).
22. L.C. Fowke, P.J. Rennie, J.W. Kirkpatrick and F. Constabel, Can. J. Bot. 53:272-278 (1975).
23. L.C. Fowke, F. Constabel and O.L. Gamborg, Fine structure of fusion products from soybean cell culture and pea leaf protoplasts, Planta 135:257-266 (1977).
24. L.C. Fowke, P.M. Gresshoff and H.J. Marchant, Transfer of organelles of the alga Chlamydomonas reinhardii into carrot cells by protoplast fusion, Planta 144:341-347 (1979).
25. F. Constabel, Development of protoplast fusion products, heterokaryocytes and hybrid cells, in: "Frontiers of Plant Tissue Culture 1978," T.A. Thorpe, ed., Intl. Assoc. Plant Tissue Cult., Calgary (1978).
26. L. Szabados and D. Dudits, Fusion between interphase and mitotic plant protoplasts, Exptl. Cell Res. 127:441-446 (1980).
27. S.C. Mueller and R.M. Brown, Evidence for an intramembrane component associated with a cellulose microfibril-synthesizing complex in higher plants, J. Cell Biol. 84:315-326 (1980).
28. M.J. Chrispeels, Biosynthesis, intracellular transport, and

secretion of extracellular macromolecules, Ann. Rev. Plant Physiol. 27:19-38 (1976).

29. D.G. Robinson, Plant cell wall synthesis, in: "Advances in Botanical Research," H.W. Woolhouse, ed., Academic Press, New York (1977).
30. H.J. Marchant, Microtubules associated with the plasma membrane isolated from protoplasts of the green alga Mougeotia, Exptl. Cell Res. 115:25-30 (1978).
31. H.J. Marchant and E.R. Hines, The role of microtubules and cell wall deposition in elongation of regenerating protoplasts of Mougeotia, Planta 146:41-48 (1979).
32. M.E. Doohan and B.A. Palevitz, Microtubules and coated vesicles in guard-cell protoplasts of Allium cepa L., Planta 149:389-401 (1980).
33. E.H. Newcomb, Coated vesicles: their occurrence in different plant cell types, in: "Coated Vesicles," C.J. Ockleford and A. Whyte, eds., Cambridge University Press, New York (1980).
34. M.F. Pearse and M.S. Bretscher, Membrane recycling by coated vesicles, Ann. Rev. Biochem. 50:85-101 (1981).
35. K. Mothes, Historical introduction, in: "Secondary Plant Products," E.A. Bell and B.V. Charlwood, eds., Encyclopedia of Plant Physiology, Springer-Verlag, Berlin (1980).
36. H. Böhm, Die bildung sekundärer naturstoffe durch pfanzliche zellkulturen, Biol. Rdsch 19:138-154 (1981).
37. W. Barz and B.E. Ellis, Plant cell cultures and their biotechnological potential, Ber. Deutsch. Bot. Ges. 94:1-26 (1981).
38. G.J. Wagner and H.W. Siegelman, Large-scale isolation of intact vacuoles and isolation of chloroplasts from protoplasts of mature plant tissues, Science 190:1298-1299 (1975).
39. G.J. Wagner, Content and vacuole/extravacuole distribution of neutral sugars, free amino acids and anthocyanin in protoplasts, Plant Physiol. 64:88-93 (1979).
40. J.A. Saunders, Investigations of vacuoles isolated from tobacco I. Quantitation of nicotine, Plant Physiol. 64:74-78 (1979).
41. F. Constabel, H. Koblitz, J.W. Kirkpatrick and S. Rambold, Fusion of cell sap vacuoles subsequent to protoplast fusion, Can. J. Bot. 58:1032-1034 (1980).
42. F. Constabel, S. Rambold, K.B. Chatson, W.G.M. Kurz and J.P. Kutney, Alkaloid production in Catharanthus roseus (L) G. Don. VI. Variation in alkaloid spectra of cell lines derived from one single leaf, Plant Cell Reports 1:3-5 (1981).
43. C. Spalla, Production of ergot alkaloids by fermentation, in: "Indole and Biogenetically Related Alkaloids," J.D. Phillipson and M.H. Zenk, eds., Academic Press, New York (1980).
44. K.N. Kao and M.R. Michayluk, A method for high-frequency intergeneric fusion of plant protoplasts, Planta 115:355-367 (1974).
45. O. Schieder and I.K. Vasil, Protoplast fusion and somatic hybridization, in: "Perspectives in Plant Cell and Tissue Cul-

ture," Intl. Rev. Cytol., Suppl. 11B, I.K. Vasil, ed., Academic Press, New York (1980).

46. L.C. Fowke, Ultrastructure of isolated and cultured protoplasts, in: "Frontiers of Plant Tissue Culture 1978," T.A. Thorpe, ed., Int. Assoc. Plant Tissue Cult., Calgary (1978).
47. L.D. Frye and M. Edidin, The rapid intermixing of cell surface antigens after formation of mouse-human heterokaryons, J. Cell Sci. 7:319-335 (1970).
48. L. Orci, A. Perrelet and D.S. Friend, Freeze-fracture of membrane fusions during exocytosis in pancreatic B-cells, J. Cell Biol. 75:23-30 (1977).
49. H. Robenek and E. Peveling, Beobachtungen am plasmalemma wahrend der fusion isolierter protoplasten von Skimmia japonica Thunb., Ber. Deutsch. Bot. Ges. 91:351-359 (1978).
50. L.C. Fowke, H.J. Marchant and P.M. Gresshoff, Fusion of protoplasts from carrot cell cultures and the green alga Stigeoclonium, Can. J. Bot. 59:1021-1025 (1981).

PROTOPLAST, CYTOPLAST AND SUB-PROTOPLAST FROM RIPENING TOMATO FRUITS: THEIR NATURE AND FUSION PROPERTIES

Deepesh N. De and D. Swain

Applied Botany Section
Indian Institute of Technology
Kharagpur 721302
INDIA

INTRODUCTION

Since the achievement of the fusion of heterotypic protoplasts to bring about somatic hybridization (1), the feasibility of using liposomes (2), vacuoplasts (3) and various types of vesicles (4) for chromosome or organelle transfer is being gradually appreciated. Such protoplast units may be variously produced by differential centrifugation (3) or by synthetic methods (5). The ripening fruits of tomato normally produce various types of protoplasts which may be equally useful (6). The present investigation is addressed to the structure and diversity of these protoplasmic units and their spontaneous and induced fusion properties. The staining reactions with vital basic and acidic dyes also reveal their potential use as marker or fusogen.

MATERIALS AND METHODS

At Kharagpur it takes about 8 weeks for the complete ripening of the tomato fruits from the day of opening of the flower in autumn. After six weeks of growth, the fruit begins to ripen; the increase in fruit weight reaches its peak and the color of the fruit is orange-red. The ripening tomato fruit was cut open and the locular content was gently squeezed out, filtered through a fine nylon mesh and centrifuged at 100 rpm for 5 min. The pellet was resuspended in 7% mannitol. The process was repeated three more times to obtain a clean suspension of protoplasmic units in 7% mannitol which is used as the initial material for all experiments.

For the isolation of mesophyll protoplasts the third leaf from the stem apex of vigorously growing plants of Solanum nigrum was used. Protoplasts were isolated following the "one step" method of Power and Cocking (7). PEG-induced fusion of protoplasts and protoplasmic units was carried out following the method of Kao (8). Isolated protoplasts were treated with various concentrations of methylene blue (Gurr's England Lot # 18633), neutral red (Riedel De Haenag Hannover, Germany Lot # 907746), and Evan's blue (Gurr's England Lot # 9446), prepared from 0.2% stock solution in 7% mannitol and incubated for various durations at room temperature (28°C-30°C). A Carl Zeiss Standard WL POL interference microscope was used for the study.

RESULTS AND DISCUSSION

Protoplasmic Units of Tomato Fruit

The locular sap of 6-7 week old ripening tomato fruits contains (a) protoplasts complete with nucleus, cytoplasm and one or more vacuoles (Figure 1); (b) cytoplasts without a nucleus but with or without vacuoles (Figures 2, 3, 4); (c) sub-protoplasts (Figure 3); (d) vacuoplasts consisting of a vacuole and a thin layer of cytoplasm but lacking a nucleus; and (e) free vacuoles, as have been described by Cocking and Gregory (6). None of these units contains any cell wall material as indicated by the absence of fluorescence with 0.05% Calcofluorwhite (9).

Fusion of the Protoplasmic Units

In a freshly isolated drop of locular sap, a certain degree of spontaneous fuion of various protoplasmic units takes place. The fusion of the various types of protoplasmic units appears to be non-random (Figures 4, 5, 6). The large plastid- and chromoplast-containing nucleate protoplasts fuse with each other (35%). The large cytoplasts also fuse with each other (25%). Similar homotypic fusions of vacuoplasts and vacuoles occur at a much lower frequency (18%). In addition to these homotypic fusions, heterotypic fusions are also seen between cytoplasts and vacuoplasts. As the protoplasmic units are progressively purified, their apparent fusion frequency gradually decreases. When the pellet is mounted on to a slide and observed after one hour of incubation, fusion of the order of 1.5% was noticed. With PEG 1540 the frequency of fusion increases to 22%. Unlike the spontaneous fusion, the induced fusion process appears to be random between various types of units. However, bigger cytoplasts tended to undergo fusion more frequently, possibly because of their larger surface.

The fusion frequency of fruit protoplasts and mesophyll protoplasts occuring spontaneously and as induced by PEG is given in

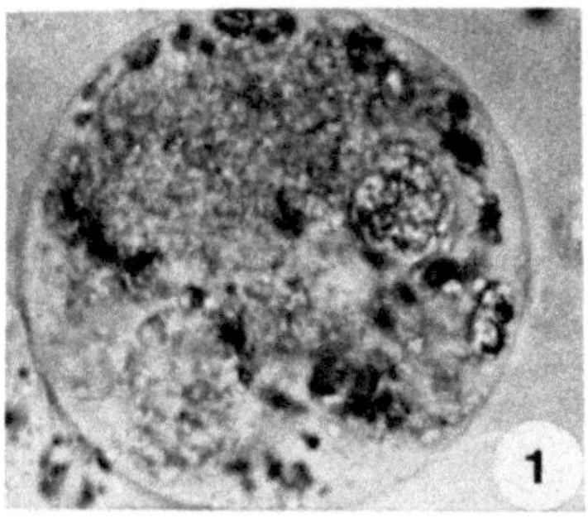

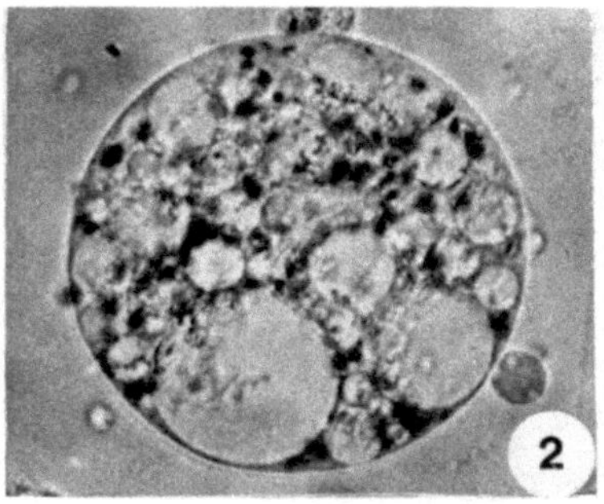

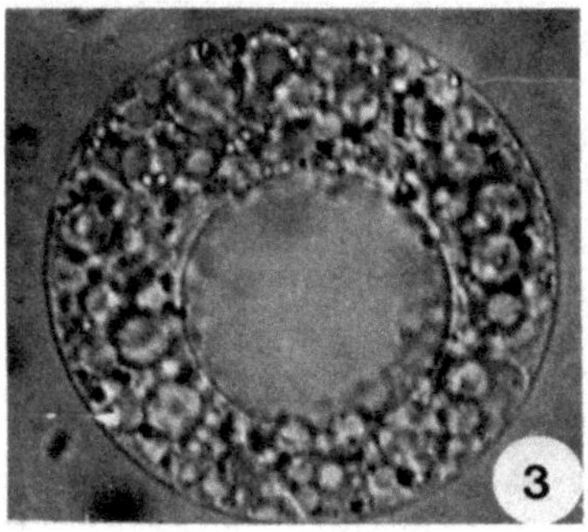

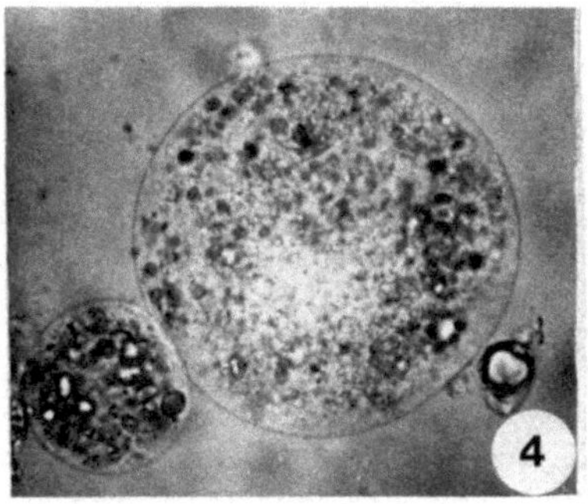

Figure 1. A typical nucleated protoplast.

Figure 2. A typical anucleate protoplast with numerous small vacuoles.

Figure 3. A protoplast with a large central vacuole and numerous small vesicles

Figure 4. Aggregation of a large nonvacuolate cytoplast with a small cytoplast containing starch grains.

Table I. Although their spontaneous fusion frequencies were different, there was little difference in the PEG-induced frequencies.

Staining and Fusion by Vital Dyes

The protoplasmic units were stained with various concentrations of dyes without changing the osmolar properties of the medium. The staining pattern and the induction of aggrregation and fusion were observed after 1 hr of incubation of the protoplasmic units in the dye solution in a humid chamber at room temperature. The pH was adjusted to 6.0 in view of the basic nature of methylene blue and neutral red. The solution of Evan's blue was adjusted to pH 7.0.

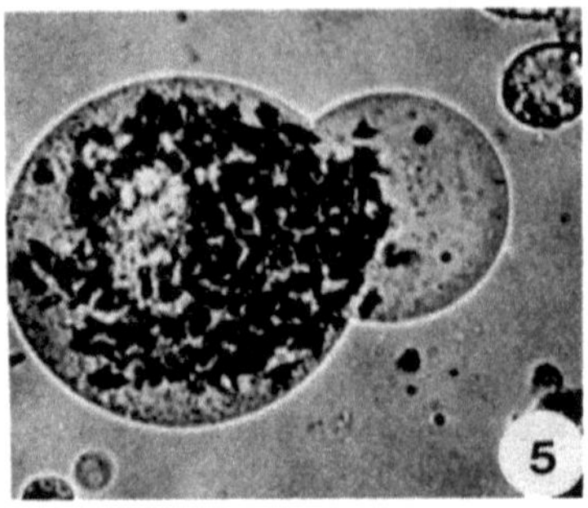

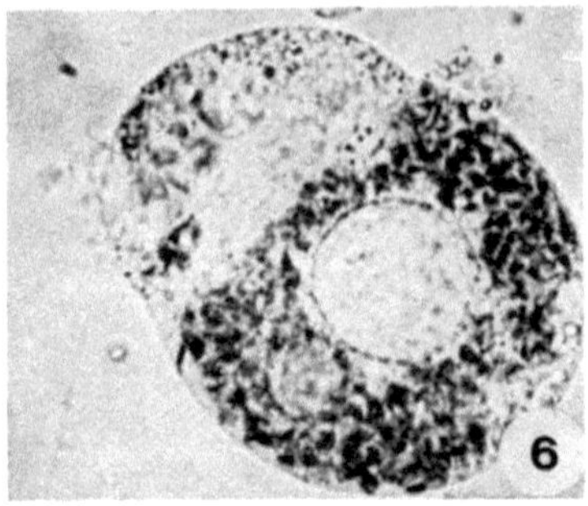

Figure 5. Spontaneous fusion of a chromoplast-containing cytoplast with a small non-vacuolate cytoplast.

Figure 6. An advanced state of spontaneous fusion of a protoplast with a vacuole with a non-vacuolate protoplast.

With a very low concentration (0.001%) of methylene blue, the staining intensity of the preparation was very poor and the dye did not induce any appreciable aggregation or fusion. However, with 0.01% of the dye, most of the protoplasmic units were unstained, but a small fraction of the units did take up the dye and become stained. These protoplasmic units are considered to be dead in view of their loss of membrane property, permitting the dye to penetrate. However, the free vacuoles took up the dye. In addition, the dye also induced aggregation and fusion (Figures 6, 7, 8 and Table IIA).

When 0.001% of neutral red was used, all the vacuoles whether inside the protoplasmic unit or free, were stained. The protoplasmic units without a vacuole were not stained. No appreciable amount of aggregation or fusion was noticed. However, when 0.01% dye was used, a similar pattern of staining was observed but with higher

Table I. Comparison of Spontaneous and PEG-induced Fusion of Fruit and Mesophyll Protoplasts. The protoplasts were incubated with PEG for 45 min. followed by washing for 1 hr. For spontaneous fusion no PEG was added.

Protoplasts	Spontaneous fusion %	PEG-induced %
Tomato protoplasmic units	1.26	22.01
Mesophyll protoplasts	2.93	25.61

Table II. Response of Protoplasts to the Dyes as Expressed by Percentage of Aggregation and Fusion. The protoplasts wer incubated in 0.01% dye solution at room temperature (28°C-30°C) for 1 hour. About 300-400 protoplasts were counted for each treatment.

A: Observations as made directly in the dye medium after 1 hour of incubation.

	Tomato protoplasmic units (%)				*S. nigrum* mesophyll protoplasts (%)			
Treatment	Aggregation of two	Aggregation of three	Multiple of aggregates	Fusion	Aggregates of two	Aggregates of three	Multiple of aggregates	Fusion
Control (spontaneous)	11.39	1.26	1.26	1.26	4.78	1.43	0.95	2.93
Methylene blue	18.42	0	2.63	5.25	8.39	0.76	2.96	2.96
Neutral red	16.66	0	2.80	3.08	9.40	1.48	0.99	3.96
Evan's blue	7.64	0.80	0.80	1.62	3.48	0.99	0.49	1.90

B: Observations in mannitol after 3 washings preceeded by 1/2 hour incubation in the dyes.

Methylene blue	11.67	1.05	1.23	0.51	3.07	0.90	0.27	2.28
Neutral red	10.93	1.28	3.03	0.77	2.72	0.03	0.18	1.97
Evan's blue	7.84	0	1.27	0	3.73	0.74	0.75	2.21

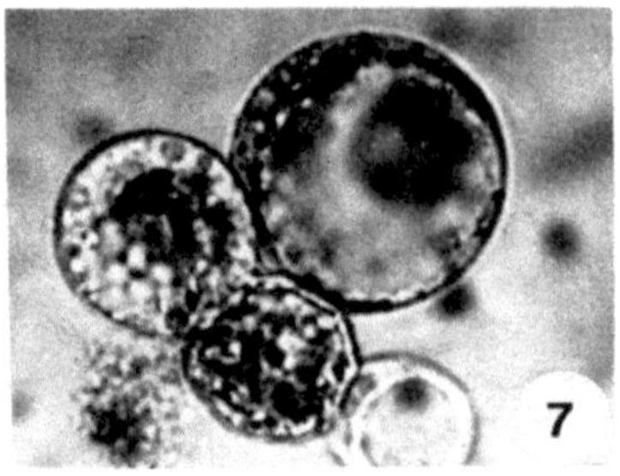

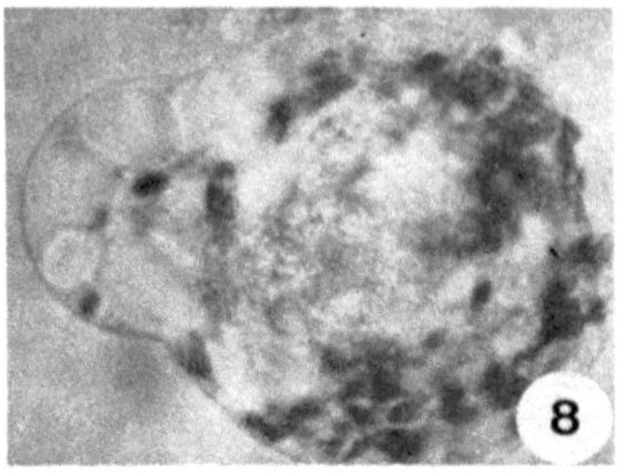

Figure 7. Aggregation of diverse types of protoplasts induced by 0.01% methylene blue for 1 hour.

Figure 8. Complete fusion of a large chromoplast-containing protoplast with a highly vacuolate protoplast.

intensity. This concentration also induced a considerable rate of aggregation and fusion of the protoplasmic units (Table IIA).

In the case of Evan's blue, with 0.001% concentration, no staining of the protoplasmic units was observed. With 0.01% almost all the protoplasmic units were unstained and a few of them were light blue in color. The vacuolar units were stained readily. The aggregation and fusion frequency induced by Evan's blue is less than those produced spontaneously. Hence, it should be considered as an inhibitor of fusion.

From the experiments recorded above, it is apparent that the various types of protoplasmic units are different from those artificially obtained from various tissues. In order to make a direct comparison, mesophyll protoplasts of Solanum nigrum were chosen. Both fruit and mesophyll protoplasts were obtained in as identical condition as possible and were studied for their spontaneous fusion, as well as for their frequency of aggregation and fusion induced by the three vital dyes at the concentration of 0.01% in 7% mannitol medium. The results are sumamrized in Table II. It is seen that the tomato protoplast units show a higher rate of aggregation under all conditions, but the resultant fusions occur at approximately the same frequency as with mesophyll. On the whole methylene blue seems to produce a little higher frequency of fusion in tomato (Table IIA). If however, after 1/2 hour of incubation with the dyes, the protoplasts were washed with mannitol, the difference between the two types of protoplasts became more prominent. The results are summarized in Table IIB. The frequency of aggregates of tomato protoplasts remain high, but that of mesophyll decreases sharply. In contrast, the fusion frequency of fruit protoplasts with all the dyes is very low, but that of the mesophyll remain virtually unchanged. Although the mechanism of the origin of the protoplasmic units is not known, it is likely that most of the units which are fragments of the original protoplast would be surrounded by the plasma membrane. The vacuoles are surrounded by the tonoplast. Although true lysosomes may be absent in plant cells, there may be vacuoles which are products of autophagy in the degenerating cells (D.N. De and H.S. Gupta, unpublished results) or of lipid utilization (10). In addition, the possibility exists that some of the protoplasts or vacuoles may be outlined by the unit membrane of endoplasmic reticulum as suggested by Cocking and Gregory (6). Thus, it is obvious that membrane properties of the various units are not identical. The staining reactions of the three dyes support this contention. The units which take up the dyes readily, by definition, have lost their semipermeable nature to the ionized state of the dye and can be considered as damaged, if not dead. All free vacuoles readily take up the dye. In many cases the vacuoles within the protoplast also readily take up the dye, whereas cytoplasts without vacuoles remain unstained.

The induction of aggregation and fusion of both fruit protoplast and mesophyll protoplasts by the vital dyes opens up a new dimension to protopalst research. The possibility of using the vital dyes both as markers and as fusogens for non-random heterotypic fusion is presently being explored and will be reported elsewhere.

REFERENCES

1. P.S. Carlson, H.H. Smith and R.D. Dearing, Parasexual interspecific plant hybridization, Proc. Natl. Acad. Sci. USA 69:2292-2294 (1972).
2. A.B. Mukherjee, S. Orloff, J. Butter, B. De, T. Triche, P. Lalley and J.D. Schulman, Entrapment of metaphase chromosomes into phospholipid vesicles (lipochromosomes): carrier potential in gene transfer, Proc. Natl. Acad. Sci. USA 75:1361-1365 (1978).
3. H.H. Lorz, C.T. Harms and I. Potrykus, Isolation of "vacuoplasts" from protoplasts of higher plants, Biochem. Physiol. Pflanzen. 169S:617-620 (1976).
4. K.L. Giles, Uptake by plant protoplast, Abstract, 13th Intl. Bot. Congress, Sydney (1981).
5. E.D. Korn, ed., "Methods in Membrane Biology I," Plenum Press, New York (1974).
6. E.C. Cocking and D.W. Gregory, Organized protoplasmic units of plant cell: their occurrence, origin and structure, J. Exp. Bot. 14:504-511 (1963).

7. J.B. Power and E.C. Cocking, Isolation of leaf protoplasts: macromolecule uptake and growth substance response, J. Exp. Bot. 21:64-70 (1970).
8. K.N. Kao, A method for fusion of plant protoplasts with polyethylene glycol, in: "Plant Tissue Culture Methods," O.L. Gamborg and L.R. Wetter, eds., National Research Council of Canada, Saskatoon (1975).
9. T. Nagata and I. Takebe, Cell wall regeneration and cell division in isolated tobacco mesophyll protoplasts, Planta 92:301-308 (1970).
10. S.C. Maitra and D.N. De, Ultrastructure of root cap cells: formation and utilization of lipid, Cytobios 5:111-118 (1972).

EMBRYOIDS FROM MESOPHYLL PROTOPLASTS OF VINGA MUNGO L. HEPPER, A SEED LEGUME CROP PLANT

R.R. Sinha, K. Das and S.K. Sen

Programme in Genetical Research
Botany Department
Bose Institute
Calcutta 700009
INDIA

INTRODUCTION

Isolation and culture of seed legume protoplasts of the most important group of crop plants which supply the bulk of food protein to the vegetarian population of the world, have proved difficult. There is only one report where limited success in anther derived cultured tissue protoplasts of Dolichos biflorus has been achieved (1). The present communication deals with our success of inducing embryogenesis in calli from leaf mesophyll protoplasts of Vinga mungo, a widely cultivated seed legume crop of the Indian subcontinent.

MATERIALS AND METHODS

Seeds of V. mungo CV T9, T27 and T65 obtained from the Agricultural Research Institute Dholi, Bihar were surface sterilized for 5-7 min. in 0.2% mercuric chloride, 5 min. water, 5 min. 70% ethanol and 5 changes of water. Seeds were subsequently grown in 500 ml Erlenmeyer flasks containing 8-fold diluted SS-B-8 medium (2) with 2 g/l sucrose at pH 6.5 and solidified with 0.8% agarose. The flasks were maintained in dark at 30°C $\pm$ 2°C till germination is completed. They were afterwards exposed to white fluorescent light (3000 lux) for 16 h. Leaves of these aseptically grown 20-25 day old plant were incubated for 6 h in 0.3 M mannitol in dark. After removal of lower epidermis with the help of a fine forceps, leaves were cut into small pieces and incubated in an enzyme mixture containing 0.5% hemicellulase, 3% cellulase, 1% macerozyme in 0.7 M mannitol (710 mOsmol/kg of H_2O) at pH 6.0. Digestion was allowed to proceed on cell production roller apparatus at 30°C ± 2°C. After 3 h of incubation, the mixture was passed through a 67 M stainless steel sieve and purified protoplasts were obtained by repeated centrifugation on 0.6 M sucrose cushions (100 g for 5 min.). Viability of protoplasts were determined by FDA staining technique of Larkin (3). Eleven inorganic media formulations established in the past for cell, tissue and protoplast culture (1) and a new medium SS-K-8 formulated for cell suspension culture of legumes (2) in their full and half concentration with six phytohormone combinations were screened through modified MDA screening technique (4) under the same organic and osmotic conditions. Protoplasts were cultured in 50 1 hanging droplets and 50 1 normal fresh mannitol free medium with other carbohydrates, namely xylose, fructose and mannose, were added after 15 d and then cultured in sitting position. Gibberellic acid and five auxins, namely IAA, IBA, NAA, 2,4-D and Picloram, in combination with a cytokinin were tested.

The SS-K-8 medium (2) standardized earlier for growth of legume cells in suspension culture consisted of (in mg/l): NH_4NO_3, 500; KNO_3, 2000; KH_2PO_4, 200; $CaCl_2.2H_2O$, 300, $NaH_2PO_4.2H_2O$, 50; $MgSO_4.7H_2O$, 300; $FeSO_4.7H_2O$, 27.8; Na_2 EDTA, 37.3; H_3BO_3, 5; KI, 1; $MnSO_4.H_2O$, 10; $ZnSO_4.H_2O$, 2; $CuSO_4.5H_2O$, 0.075; $Na_2MoO_4.2H_2O$, 4; $CoCl_2.6H_2O$, 0.075. All the media formulations were supplemented with organic constituents as described elsewhere (1) and were modified with mannitol to give a final osmotic pressure of 710 mOsmol/kg of H_2O.

For histological studies, differentiating tissues at various stages of embryogenesis were fixed in FAA (formalin:acetic acid:30% ethanol, 1:1:10) for 24 h, dehydrated through ethanol series and embedded in parafin. Sections were cut by a rotary microtome at 12-15 m, stained with safranin (1%, dissolved in 50% ethanol) light green (1% dissolved in clove oil) and mounted in canada balsam.

RESULTS

Out of the three cultivars tested, leaves of CV T9 yielded about 4-5 x 10^6 protoplasts/g tissue and 70-80% of these protoplasts were intact and viable (Figure 1). The other two cultivars yielded less number of viable protoplasts. The principal source of technical difficulty associated with production of protoplasts was the removal of the lower epidermis.

Amongst the 24 media formulations tested with six phytohormone combinations, only SS-K-8 and V47 (5) media in their half concentration enabled 5-7% single division, plus some sustained division. Addition of 50 µl full concentration of SS-K-8 and V47 media to their respective droplets after 15 d of culture induced unaccount-

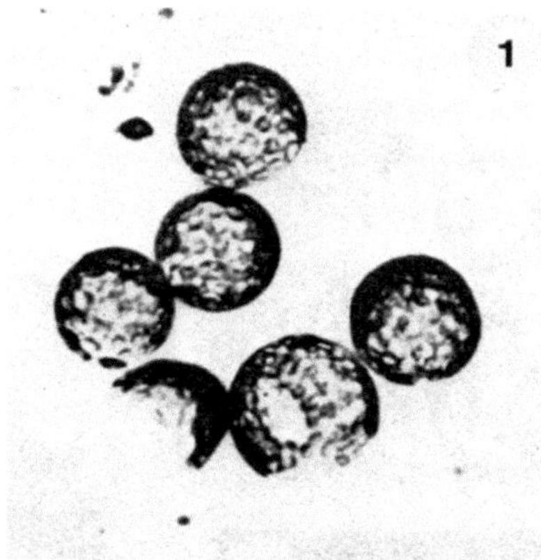

Figure 1. Freshly isolated protoplasts from leaf mesophyll cells of Vinga mungo CV T9.

able sustained division. Maximum division frequency was observed in SS-K-8 medium follwoed by V47 medium. Amongst six auxins Pic was found to be more potent followed by 2,4-D. The other four auxins failed to induce division.

Freshly isoalted protoplasts were cultured in SS-K-8 medium with 0.5 and 1.0 mg.1 of Pic and BAP respectively. Maximum division frequency was observed at population density of 1.6 - 2.0 x 10^4 protoplasts/ml, i.e., 800-1000 protoplasts/50 µl droplet. 60-70% protoplasts regenerated their cell wall within 72 h. The chloroplasts rearranged with increase in cell volume accompanied by wall formation. The first cell division was observed within 5 d and within 8 d some of the dividing cells had completed their second division (Figure 2). Most of the cells stopped division either after wall formation or after first division. However, addition of normal fresh media enabled these cells to resume further division. Large number of cell colonies were observed in the droplets after 27 d. The cell colonies were collected and transferred over agar solidified SS-K-8 medium containing different combinations of phytohormones. Some tiny organized globular structures were observed after 45 d of growth (Figure 3) on the medium containing GA_3, Pic, IAA and zeatin at the level of 5 x 10^6 M, 1 x 10^{-7} M, 5 x 10^{-7} M, and 10^{-6} M respectively. After 60 d, these calli were picked up and cultured on medium containing BAP, NAA and Pic at the level of 3 x 10^{-6} M, 5 x 10^{-7} M and 1 x 10^{-6} M, respectively. Adenine sulfate was used at the level of 10 mg/l. These globular structures differentiated further into heart and torpedo shaped embryoids within 40 d (Figure 4). Subculturing these embryoids on the same medium resulted in slightly enlarged embryoids with dark green tips within 15 d. We failed to regenerate plantlets from these embryoids either on hormone depleted

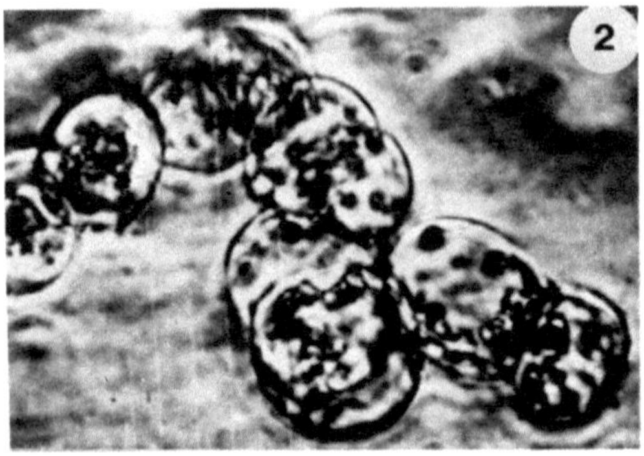

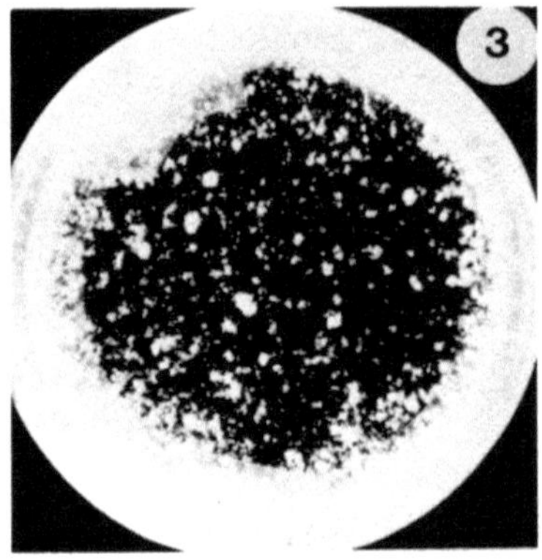

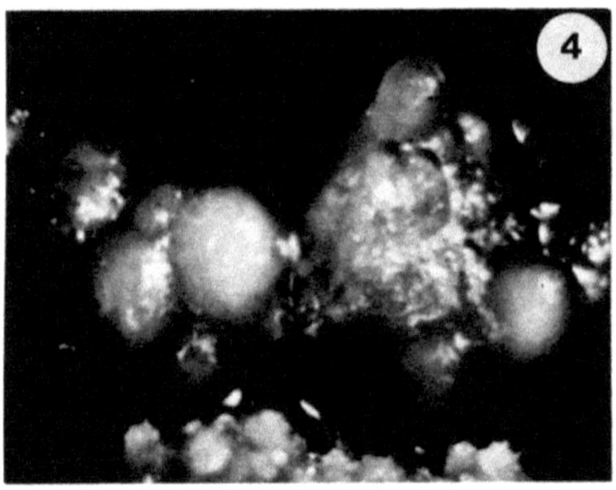

Figure 2. Several divisions in protoplasts after 7 d in culture.

Figure 3. Cell colonies showing tiny globular structures after 45 d of growth on agar media.

Figure 4. Heart and torpedo shaped embryoids.

medium or on other media containing different combinations of phytohormones.

To confirm that differentiation had actually occurred through true embryoid formation, materials were fixed at different stages of development for histological studies. Examinations indicated that embryoids generally formed form peripheral layer of callus (Figures 5, 6) and as the rate of differentiation increased, embryoids started to initiate from the inner regions as well.

DISCUSSION

Success was achieved in isolation and culture of protoplasts from V. mungo mesophyll cells which resulted in formation of heart

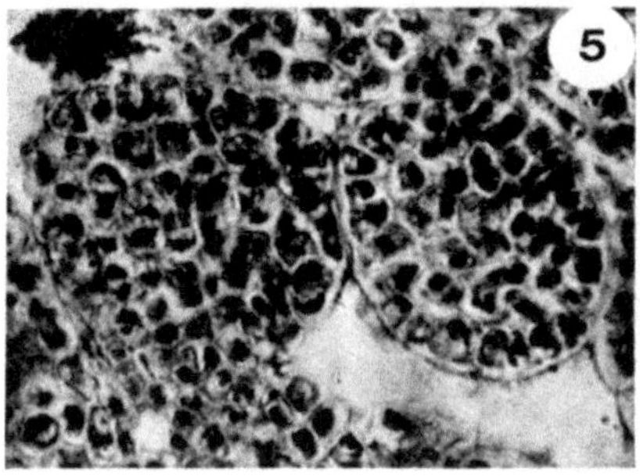

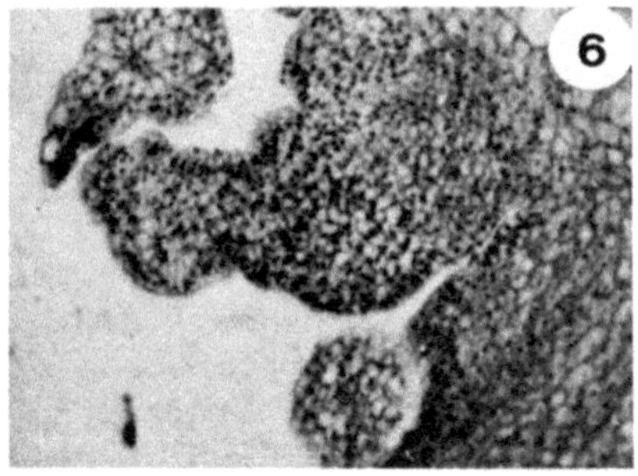

Figure 5. A microtome section through cell colonies showing emergence of globular embryoids.

Figure 6. A microtome section showing torpedo shaped embryoid.

and torpedo shaped embryoids. The genotype and the physiological condition of the leaf material for protoplast isolation were critical for achieving cell regeneration and division. Regeneration of cell wall and induction of first cell division required one set of conditions, whereas sustained division required another, as earlier experienced by Kameya (6). The optimum condition for wall regeneration is two-fold dilution of V47 and SS-K-8 media and for colony formation, normal strength media is required. Concentration and composition of mineral salts in the medium was of great significance in inducing division of Vinga mungo leaf mesophyll cell protoplasts. Carbohydrates also played an important role. Protoplasts showed first division and occasional second division in medium containing sucrose. However, sustained division was induced by supplementing various other carbohydrates simultaneously.

Embryogenesis was possible to be induced in a three step fashion. The last two steps called for two or more than two auxins and a cytokinin, where a natural cytokinin, zeatin, is essential during the second step. It appears that the regeneration of plants form protoplast of seed legumes is a tricky affair. Our inability to regenerate plants from embryoids of Dolichos biflorus (1) additionally points to that direction.

ACKNOWLEDGEMENT

The present investigation has been carried out under the financial support of a research project (# HCS/DST/310/76) from the Department of Science and Technology, New Delhi, India. The authors gratefully acknowledge the help of Mr. Harbilash Das.

REFERENCES

1. R.R. Sinha and S.K. Sen, Dolichos biflorus an ideal material for legume protoplast technology, in: "Proc. Intl. Tissue Culture Workshop," A.S. Islam, ed., Dacca (1981).
2. R.R. Sinha, "Application of Plant Cell Culture Techniques in Certain Pulse Crops," Ph.D. Thesis, Calcutta University (1980).
3. P.J. Larkin, Purification and viability determinations of plant protoplasts, Planta 128:213-216 (1976).
4. I. Potrykus, H. Lorz and C.T. Harms, On some selected problems and results concerning culture and genetic modification of higher plants, in: "Plant Tissue culture and Its Biotechnological Application," W. Barz, E. Reinhard and M.H. Zenk, eds., Springer-Verlag, Berlin (1977).
5. H. Binding, Regeneration von haploiden und diploiden Pflanzen aus protoplasten von Petunia hybrida L., Z. Pflanzenphysiol. 74:327-356 (1974b).
6. T. Kameya, Induction of hybrids through somatic cell fusion with dextran sulfate and gelatin, Jap. J. Genet. 50:235-246 (1975).

FUSION OF PLANT PROTOPLASTS: RECENT ADVANCES IN STUDIES ON REGULATION OF CELL CYCLE, GENE EXPRESSION AND IN PARASEXUAL GENE TRANSFER

Denes Dudits

Institute of Genetics
Biological Research Center
Hungarian Academy of Sciences
Szeged, HUNGARY

INTRODUCTION

Since the early 1970s the technique of fusion between plant protoplasts (1 - 4) has become a powerful tool in genetic manipulation of higher plants. Experiments performed by fusion of protoplasts have proved that hybridization of plant species can be parasexually accomplished. Several excellent reviews of the plant somatic hybridization field have discussed different aspects of production of somatic hybrids with special emphasis on selection systems (5 - 12). As shown by the continuously increasing number of examples, intra- and inter-specific hybridization through protoplast fusion can result in integration of the parental diploid genomes with basically similar gene expression patterns as in the sexual hybrids (13 - 18). At present, the somatic hybridization with very few exceptions is restricted only to species belonging to Solanaceae family (12). Because of this limitation, any suggestion or general conclusion about the potential use of somatic cell hybridization with application in plant breeding requires that the spectrum of species included in fusion experiments be widened.

As the results of extensive studies on the fate of cytoplasmic organelles after protoplast fusion, significant progress has been achieved in genetic analysis of cytoplasmic traits. Biparental organelle transmission mediated by protoplast fusion has offered a unique opportunity for studies on extrachromosomal inheritance such as chloroplast function and segregation (19 - 24), cytoplasmic male sterility (25 - 27) and mitochondrial recombination (28, 29).

In studies on expression of nuclear genes, isoenzyme analysis was frequently used for the characterization of hybrid cell lines or plants originating from fused protoplasts (15 - 18, 23, 30 - 34). In addition to several nuclear chlorophyll deficient mutants (14, 16, 18, 32, 33, 35 - 37), the morphogenic potential (17, 38) and the nitrate reductase activity (39) were found to be restored in somatic hybrids. Fusion between different amino acid analogue resistant lines resulted in appearance of double-resistant colonies. These observations can be interpreted as an indication of a dominant or semidominant nature of resistance against the tested analogues (40, 41). The tumor markers of crown gall cells have been found to be dominantly expressed in intraspecific hybrids (42). Hormone autotrophy was also previously detected in genetic tumors produced by protoplast fusion (13, 43).

Contrary to the situation with somatic hybrids between related species, the progress in production of intergeneric hybrid plants through cell fusion is still slow and limited to only a very few fusion combinations. In the light of experimental findings it is likely that intergeneric hybrid plants possessing all the chromosomes of both parental species (22, 44, 45) represent a minority as compared with the large number of fusion experiments in which protoplast fusion failed to result in intergeneric hybrid plants (8, 35, 46 - 48). Accumulation of observations about the restrictions caused by somatic incompatibility reactions at different stages of somatic hybridization between non-related species have focused the attention on the need of methods for transferring only a small part of plant genomes (32, 36).

CELL CYCLE CONTROL IN FUSED SOMATIC PLANT CELLS

The mechanism by which the cell or chromosome cycle is controlled in plant cells is poorly understood. Fusion between protoplasts with different cell cycle stages can offer a new approach to analyze the factors regulating the chromosome condensation. Now an appropriate method is available for the isolation of mitotic protoplasts from partially synchronized cell suspension cultures (49). Spontaneous or polyethylene glycol (PEG) induced fusion can form heterophasic fusion products. Cytological examination of mitotic-interphase homo- or heterokaryons revealed that the mitotic partner could induce premature chromosome condensation (PCC) in an interphase nucleus within a few hours after fusion (49). Similar to that of the mammalian cells, the morphology of prematurely condensed chromatin reflects the actual cell cycle stage of the interphase partner at the time of fusion. Because the interphase chromosomes of G_1 nuclei are much more extended than the normal metaphase ones (Figure 1A), they offer better possibilities for the identification of rearrangements in plant chromosomes. The appearance of pulverized or fragmented nuclei (Figure 1B) after interphase-mitotic fusion shows that PCC can generate small chromosome fragments which might be inte-

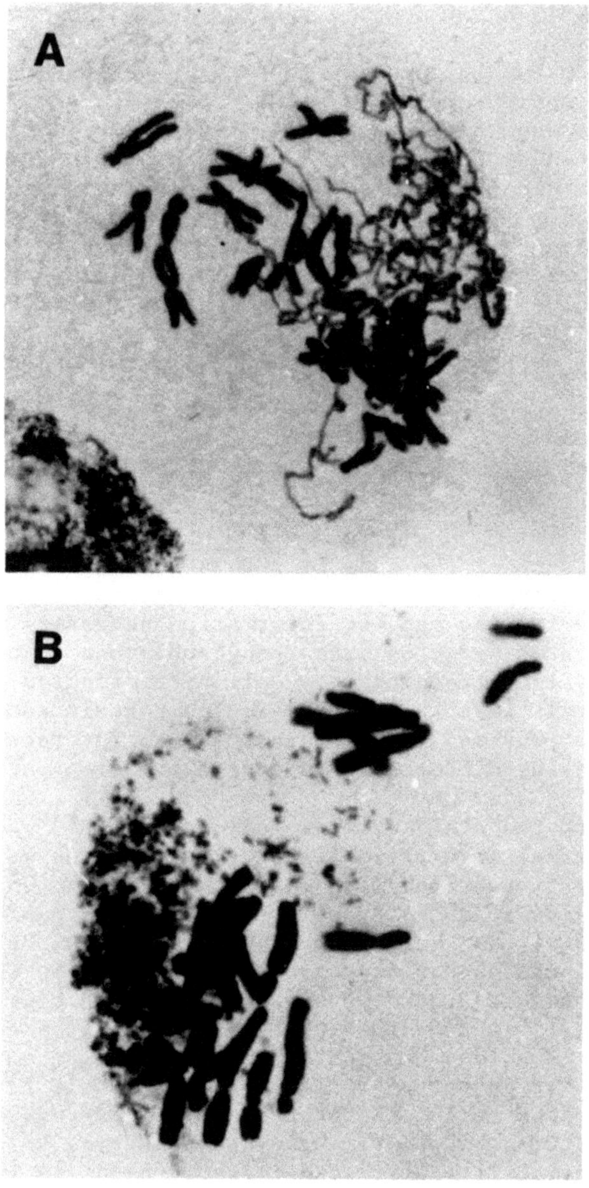

Figure 1. Induction of premature chromosome condensation in G_1 (A) and S (B) interphase nucleus of wheat (Triticum monococcum) in heterophasic homokaryons. (G. Bisztray, T. Praznovszky and G. Hadlaczky, unpublished.)

grated into the genome during subsequent division as it was proposed on the basis of the genetic constitution of mammalian cell hybrids (50 - 54) as well as of somatic hybrid plants (36, 55).

Studies on PCC in somatic plant cells suggest the existence of an inducer molecule which is synthesized in dividing cells and can initiate chromosome condensation or mitosis in other non-dividing partners after mixing the two cells by protoplast fusion (56). Mitotic activation was previously observed in fused plant cells 3 - 5 days after fusion (57, 58). The chemical nature of the suggested inducer(s) has not yet been identified in plants. It is interesting to note, however, that the factor involved seems not to be species specific as shown by intergeneric (49) or by interkingdom heterokaryons (56, 59). Further characterization of the mitotic inducer is likely to help overcome the difficulties in culture of plant protoplasts.

REGULATION IN EXPRESSION OF CYCLOHEXIMIDE RESISTANCE IN SOMATIC HYBRIDS

The questions of the genetic regulation in expression of differentiated functions can be approached through fusion between somatic plant cells having the genetic information necessary to carry out the differentiated function even though only one of the parental cells is actually expressing it. Cell hybridization experiments with carrot cell lines that differ in their resistance to cycloheximide (CH) have provided a way to analyze the interaction between genomes possessing differences in gene expression pattern.

In carrot, the cycloheximide resistance (CH^r) was shown to be a function expressed in differentiated plant tissues but not in unorganized callus tissues (60). A variant line, WCH 105, expressing the CH^r in callus cultures as well as in somatic embryos was isolated by Sung (61). The mechanism of CH^r in embryos of wild carrot and in WCH 105 calluses and embryos was found to be CH inactivation (60). Fusion between carrot protoplasts from CH sensitive albino callus cells (A1) lacking the inactivation function and from CH resistant cells (WCH 105) expressing it, resulted in somatic hybrids with CH sensitive phenotype in callus cells (Table 1) (62). The reappearance of CH^r colonies through somatic segregation in cell suspension cultures of somatic hybrids exhibiting increased chromosome number variability indicated that CH^r based on inactivation was a recessive trait. On the basis of these results, it was hypothesized that the CH sensitive cells produce a diffusible substance which blocks the expression of the inactivation function.

Table 1. Karyotypic and Cycloheximide Resistance Analysis of the Parental Lines, Hybrid # 25-12, and Three CH^r Somatic Segregants (62).

Cell Lines	*Growth in liquid medium containing 10 μg/ml CH (% of control)	Colony forming ability on agar plates containing 10 μg/ml CH (% of control)	**Cycloheximide inactivation ability	Chromosome numbers
A_1	0	0		18-46
WCH 105	95	92	+	18
Hybrid Line # 25-12 at the first test	0	0	-	36
Hybrid Line # 25-12 after subculture for one year	18	0.002	±	32-72
# $25\text{-}12^{s\text{-}11}$	89	18	+	20-70
# $25\text{-}12^{s\text{-}12}$	82	29	+	32-58
# $25\text{-}12^{s\text{-}30}$	99	9	+	30-100

* Growth of liquid cultures were measured by dry weight increase after 10 days cultivation with and without CH.
Colony forming ability was estimated on the basis of number of colonies per plate. Numbers represent averages of triplicate samples from one experiment.

** CH inactivation was performed by <u>Saccharomyces</u> agar-gel diffusion test after 20 hr incubation of cells with 10 μg/ml CH.

\+ means CH is inactivated. There is no zone of yeast inhibition; less than 0.1 μg CH is present.

\- means CH is not inhibited; 1.6-1.8 cm of zone of inhibition is present.

± means results were variable, but the inhibition zone never exceeded 1 cm.

INTERGENERIC GENE TRANSFER MEDIATED BY PROTOPLAST FUSION

In attempts at overcoming the somatic incompatible barriers expressed after fusion of protoplasts from *Umbellivefeae* species (48), fusion with mitotically inactivated or X-ray irradiated protoplasts was successfully used for introduction of a few nuclear genes into the genome of an albino carrot mutant (32, 36). Cell fusion mediated gene transfer was achieved by fusion between protoplasts from cell cultures of the albino carrot mutant (2n = 18) (Figure 2A) and mitotically inactive leaf protoplasts of *Aegopodium podagraria* (2n = 42) (Figure 2B). The green plants selected after fusion (designated CAPO) possessed *Aegopodium* specific characters but lacked cytologically detectable *Aegopodium* chromosomes (36, 55). Despite the apparent elimination of the *Aegopodium* genome, the selected plants showed serious morphological and developmental abnormalities (Figure 2C). Therefore, it became necessary to introduce an additional carrot genome by the parasexual means of somatic cell hybridization (63).

The leaf protoplasts of CAPO plants were fused back with protoplasts isolated from cell suspension cultures of albino carrot. The green plants originated from back fusion exhibited carrot phenotype (Figure 2D). In comparison with the original CAPO plants they showed normal growth and formed well developed root systems. While the CAPO plants could only be maintained in complete culture medium, the newly regenerated green plants were successfully transplanted into soil. In contrast with CAPO plants lacking any sex organs in their underdeveloped flowers, these plants (2n = 35), recovered from backfusion, formed normal flowers. However, the cytological analysis of meiotic cells frequently showed chromosomal abnormalities such as bridge formation (63).

In sexually distant hybridization, backcross with one of the parents is a frequently used method for elimination of abberations. The same approach can be applied parasexually through backfusion. In the present experiment a significant improvement in viability of CAPO plants was achieved by introduction of an additional albino carrot genome. This finding provides further support for the hybrid origin of CAPO plants in which the presence of foreign gene or genes is responsible for the observed developmental defects. Since in other fusion combinations the somatic hybrid plants resulting from intergeneric protoplast fusion exhibit abnormalities (32, 44, 45) a somatic back hybridization program might have wide application in the genetic manipulation of plants.

ISOLATED PLANT CHROMOSOMES AS POTENTIAL VECTORS FOR TRANSFERRING PLANT GENES

In searching for methods of transferring plant genes one of the

Figure 2. Parental and somatic hybrid plants used in somatic back hybridization experiments.
A: Nuclear albino mutant from carrot (Daucus carota, 2n = 18).
B: Aegopodium podagraria plant (2n = 42).
C: Somatic hybrid plant (CAPO 2n = 18) regenerated after fusion of protoplasts from albino carrot and from leaves of Aegopodium.
D: Flowering plant (2n = 35) regenerated after back-fusion with albino carrot protoplast.

possible ways could be the use of isolated plant chromosomes as vectors. The cell-free preparation of mammalian chromosomes is extensively used to transfer nuclear genes (64). In this system, whole chromosomes are taken up and integrated into the genome of recipient cells as additional chromosomes or as subchromosomal fragments. On the basis of these experiments one can expect that the similar structure of plant chromosomes will also make possible their isolation and transfer.

Chromosomes from plant cells can be released from mitotic cell protoplasts by disrupting the plasmalemma with detergent (65, 66). Subsequently, the chromosomes are purified by differential centrifugation in chromosome isolation medium (67) supplemented with sucrose (66). The efficiency of purification and structural integrity of the isolated chromosomes is strongly affected by the following: cell type, metabolic stage of the cells, degree of mitotic synchrony, and composition of isolation medium.

For chromosome-mediated gene transfer, the isolated plant chromosomes have to be introduced into the recipient protoplasts. Recently, cytological studies have proved that PEG treatment can result in uptake of plant chromosomes (66). Since the frequency of chromosome uptake is low, further studies are needed to improve the efficiency of transfer. Simultaneously, experiments are in progress to detect the expression of genes carried by the introduced chromosomes.

REFERENCES

1. J.B. Power, S.E. Cummins, E.C. Cocking, Fusion of isolated plant protoplasts, Nature 225:1016-1018 (1970).
2. W.A. Keller and G. Melchers, The effect of high pH and calcium on tobacco leaf protoplast fusion, Z. Naturforsch. 28c:737-741 (1973).
3. K.N. Kao and M.R. Michayluk, A method for high-frequency intergeneric fusion of plant protoplasts, Planta 115:355-367 (1974).
4. A. Wallin, K. Glimelius and T. Eriksson, The induction of aggregation and fusion of *Daucus carota* protoplasts by polyethylene glycol, Z. Pflanzenphysiol. 74:64-80 (1974).
5. O.L. Gamborg, F. Constabel, L.C. Fowke, K.N. Kao, K. Ohyama, K.K. Kartha and L.E. Pelcher, Protoplast and cell culture methods in somatic hybridization in higher plants, Can. J. Genet. and Cytol. 16:737-750 (1974).
6. G. Melchers, Plant hybrids by fusion of protoplasts, in: "Recombinant Molecules: Impact on Science and Society," R.F. Beers and E.G. Bassett, eds., Raven Press, New York (1977).
7. K.N. Kao, Plant protoplast fusion and somatic hybrids, in: "Proc. Symposium on Plant Tissue Culture," Science Press, Peking (1978).
8. E.C. Cocking, Selection and somatic hybridization, in: "Frontiers of Plant Tissue Culture," T.A. Thorpe, ed., International Association for Plant Tissue Culture, Calgary (1978).
9. R.G. Butenko, Cultivation of isolated protoplasts and hybridization of somatic plant cells, Intl. Rev. Cytol. 59:323-373 (1979).
10. E. Thomas, P.J. King and I Potrykus, Improvement of crop plants via single cells *in vitro*, an assessment, Z. Pflanzenzucht. 82:1-30 (1979).

11. D.J. Cove, The uses of isolated protoplasts in plant genetics, Heredity 43:295-314 (1979).
12. O. Schieder and I.K. Vasil, Protoplast fusion and somatic hybridization, Intl. Rev. Cytol. Suppl. 11B:21-42 (1980).
13. P.S. Carlson, H.H. Smith and R.D. Dearing, Parasexual interspecific plant hybridization, Proc. Natl. Acad. Sci. USA 69:2292-2294 (1972).
14. G. Melchers and G. Labib, Somatic hybridization of plants by fusion of protoplasts. I. Selection of light resistant hybrids of "haploid" light sensitive varieties of tobacco, Molec. Gen. Genet. 135:277-294 (1974).
15. J.B. Power, E.M. Frearson, C. Hayward, D. George, P.K. Evans, S.F. Berry and E.C. Cocking, Somatic hybridization of Petunia hybrida and P. parodii, Nature 263:500-502 (1976).
16. D. Dudits, G. Hadlaczky, E. Levi, O. Fejer, Z. Haydu and G. Lazar, Somatic hybridization of Daucus carota and D. capillifolius by protoplast fusion, Theor. Appl. Genet. 51:127-132 (1977).
17. P. Maliga, G. Lazar, F. Joo, A.H. Nagy and L. Menczel, Restoration of morphogenetic potential in Nicotiana by somatic hybridization, Molec. Gen. Genet. 157:291-296 (1977).
18. D.A. Evans, L.R. Wetter and O.L. Gamborg, Somatic hybrid plants of Nicotiana glauca and Nicotiana tabacum obtained by protoplast fusion, Physiol. Plant. 48:225-230 (1980).
19. K. Chen, S.G. Wildman and H.H. Smith, Chloroplast DNA distribution in parasexual hybrids as shown by polypeptide composition of fraction I protein, Proc. Natl. Acad. Sci. USA 74:5109-5112 (1977).
20. G. Belliard, G. Pelletier, F. Vedel and F. Quatier, Morphological characteristics and chloroplast DNA distribution in different cytoplasmic parasexual hybrids of Nicotiana tabacum, Molec. Gen. Genet. 165:231-237 (1978).
21. Y.Y. Gleba, Extranuclear inheritance investigated by somatic hybridization, in: "Frontiers of Plant Tissue Culture," T.A. Thorpe, ed., International Association for Plant Tissue Culture, Calgary (1978).
22. G. Melchers, M.D. Sacristan and A.A. Holder, Somatic hybrid plants of potato and tomato regenerated from fused protoplasts, Carlsberg Res. Commun. 43:277-294 (1978).
23. P. Medgyesy, L. Menczel and P. Maliga, The use of cytoplasmic streptomycin resistance: chloroplast transfer from Nicotiana tabacum into Nicotiana sylvestris, and isolation of their somatic hybrids, Molec. Gen. Genet. 179:693-698 (1980).
24. L. Menczel, F. Nagy, Z. Kiss and P. Maliga, Streptomycin resistant and sensitive hybrids of Nicotiana tabacum + Nicotiana knightiana: correlation of resistance with N. tabacum plastids, Theor. Appl. Genet. 59:191-195 (1981).
25. A. Zelcher, D. Aviv and E. Galun, Interspecific transfer of cytoplasmic male sterility by fusion between protoplasts of normal Nicotiana sylvestris and X-ray irradiated protoplasts

of male-sterile N. tabacum, Z. Pflanzenphysiol. 90:397-407 (1978).

26. S. Izhar, J.B. Power, Somatic hybridization in Petunia: a male sterile cytoplasmic hybrid, Plant Sci. Lett. 14:49-55 (1979).

27. D. Aviv and E. Galun, Restoration of fertility in cytoplasmic male sterile (CMS) Nicotiana sylvestris by fusion with X-irradiated N. tabacum protoplasts, Theor. Appl. Genet. 58: 121-127 (1980).

28. G. Belliard, F. Vedel and G. Pelletier, Mitochondrial recombination in cytoplasmic hybrids of Nicotiana tabacum by protoplast fusion, Nature 281:401-403 (1979).

29. F. Nagy, I. Torok and P. Maliga, Extensive rearrangements in the mitochondrial DNA in somatic hybrids of Nicotiana tabacum and Nicotiana knightiana, Molec. Gen. Genet. 183:437-439 (1981).

30. J.B. Power, E.M. Frearson, C. Hayward and E.C. Cocking, Some consequences of the fusion and selective culture of Petunia and Parthenocissus protoplasts, Plant Sci. Lett. 5:197-207 (1975).

31. L.R. Wetter, Isoenzyme pattern in soybean-Nicotiana somatic hybrid cell lines, Molec. Gen. Genet. 150:231-235 (1977).

32. D. Dudits, O. Fejer, G. Hadlaczky, C. Koncz, G.B. Lazar and G. Horvath, Intergeneric gene transfer mediated by plant protoplast fusion, Molec. Gen. Genet. 179:283-288 (1980a).

33. P. Maliga, Z.R. Kiss, A.H. Nagy and G. Lazar, Genetic instability in somatic hybrids of Nicotiana tabacum and N. knightiana, Molec. Gen. Genet. 163:145-151 (1978).

34. Y.Y Gleba and F. Hoffmann, Hybrid cell lines Arabidopsis thaliana + Brassica campestris: no evidence for specific chromosome elimination, Molec. Gen. Genet. 165:257-264 (1978).

35. O. Schieder, Hybridization experiments with protopalsts from chlorophyll-deficient mutants of some Solanaceaous species, Planta 137:253-257 (1977).

36. D. Dudits, G. Hadlaczky, G. Bajszar, D. Koncz, G. Lazar and G. Horvath, Plant regeneration from intergeneric cell hybrids, Plant Sci. Lett. 15:101-112 (1979).

37. E.C. Cocking, D. George, M.J. Price-Jones and J.B. Power, Selection procedures for production of inter-species somatic hybrids of Petunia parodii. 2. Albino complementation selection, Plant Sci. Lett. 10:7-12 (1977).

38. O. Schieder, Somatic hybrids of Datura innoxia Mill. and Datura discolor Beruh., and Datura innoxia Mill. and Datura stramonium L. var. tatula L., Molec. Gen. Genet. 167:113-119 (1978).

39. K. Glimelius, T. Eriksson, R. Grafe and A.J. Muller, Somatic hybridization of nitrate-reductase deficient mutants of Nicotiana tabacum by protoplast fusion, Physiol. Plant. 44: 273-277 (1978).

40. D.W.R. White and I.K. Vasil, Use of amino acid analogue-resistant cell lines for selection of Nicotiana sylvestris somatic cell hybrids, Theor. Appl. Genet. 51:161-167 (1979).
41. C.T. Harms, I. Potrykus and J.M. Widholm, Complementation and dominant expression of amino acid analogue resistance markers in somatic hybrid clones from Daucus carota after protoplast fusion, Z. Pflanzenphysiol. 101:377-390 (1981).
42. G.J. Wullems, L. Molendijk and R.A. Schilperoort, The expression of tumor markers in intraspecific somatic hybrids of normal and crown gall cells from Nicotiana tabacum, Theor. Appl. Genet. 56:3208 (1980).
43. H.H. Smith, K.N. Kao and N.C. Combatti, Interspecific hybridization by protoplast fusion in Nicotiana, confirmation and extension, J. Hered. 67:124-128 (1976).
44. Y.Y. Gleba and F. Hoffmann, "Arabidobrassica": A novel plant obtained by protoplast fusion, Planta 149:112-117 (1980).
45. G. Krumbiegel and O. Schieder, Selection of somatic hybrids after fusion of protoplasts from Datura innoxia Mill. and Atropa belladonna L., Planta 145:371-375 (1979).
46. N. Zenkteler and G. Melchers, In vitro hybridization by sexual methods and by fusion of somatic protoplasts, Theor. Appl. Genet. 52:81-90 (1978).
47. F. Hoffmann, H. Cshenck, H.W. Kohlenbach and Y.Y. Gleba, Regeneration and fusion of protoplasts from important crop plants of the Brassiceae, in: "Advances in Protoplast Research," L. Ferenczy and G.L. Farkas, eds., Akademiai Kiado, Budapest (1980).
48. D. Dudits, G. Hadlaczky, G. Lazar and Z. Haydu, Increase in genetic variability through somatic cell hybridization of distantly related plant species, in: F. Sala, B. Parisi, R. Cella and O. Ciferri, eds., Elsevier-North Holland (1980b).
49. L. Szabados and D. Dudits, Fusion between interphase and mitotic plant protoplasts, Exp. Cell Res. 127:441-446 (1980).
50. A.G. Schwartz, P.R. Coock and H. Harris, Correction of genetic defect in a mammalian cell, Nature New Biol. 230:5-8 (1971).
51. Y.L. Boyd and H. Harris, Correction of genetic defects in mammalian cells by the input of small amounts of foreign genetic material, J. Cell Sci. 13:841-861 (1973).
52. J.M. Graves, G.K. Chew, D.W. Cooper and P.G. Johnston, Marsupial-mouse cell hybrids containing fragments of the marsupial X chromosome, Somatic Cell Genet. 5:481-489 (1979).
53. A. Rodgers, Detection of small amount of human DNA in human-rodent hybrids, J. Cell Sci. 38:391-403 (1979).
54. K. Tsumamoto, R. Klein, M. Hatanaka, Insertion of muntjac gene segment into hamster cells by cell fusion, J. Cell. Physiol. 104:225-232 (1980).
55. D. Dudits, C. Koncz, G. Bajszar, G. Hadlaczky, G. Lazar and G. Horvath, Intergeneric transfer of nuclear markers through fusion between dividing and mitotically inactive plant proto-

plasts, in: "Advances in Protoplast Research," L. Ferenczy and G.L. Farkas, eds., Akademiai Kiado, Budapest (1980c).
56. D. Dudits, L. Szabados and G. Hadlaczky, Premature chromosome condensation in plant cells and its potential use in genetic manipulation, in: "Premature Chromosome Condensation: Application in Basic, Clinical and Mutation Research," P.N. Rao, R.T. Johnson and K. Sperling, eds., Academic Press, New York (in press) (1981).
57. K.N. Kao, F. Constabel, M.R. Michayluk and O.L. Gamborg, Plant protoplast fusion and growth of intergeneric hybrid cells, Planta 120:215-227 (1974).
58. D. Dudits, K.N. Kao, F. Constabel and O.L. Gamborg, Fusion of carrot and barley protoplasts and division of heterokaryocytes, Can. J. Genet. Cytol. 18:263-269 (1976).
59. B. Von der Haar, K. Sperling and D. Gregor, Maturing Xenopus oocytes induce chromosome condensation in somatic plant nuclei, Exp. Cell Res. 134:477-481 (1981).
60. Z.R. Sung, G.B. Lazar and D. Dudits, Cycloheximide resistance in carrot culture: a differentiated function, Plant Physiol. 68:261-264 (1981).
61. Z.R. Sung, Mutagenesis fo cultured plant cells, Genetics 84: 51-57 (1976).
62. G.B. Lazar, D. Dudits and Z.R. Sung, Expression of cycloheximide resistance in carrot somatic hybrids and their segregants, Genetics 98:347-356 (1981).
63. D. Dudits, Backfusion with somatic protoplasts as a method in genetic manipulation of plants, Acta Biologica Acad. Sci. Hungaricae (in press) (1982).
64. L.A. Klobutcher and F.H. Ruddle, Chromosome mediated gene transfer, Ann. Rev. Biochem. 50:533-554 (1981).
65. R. Malmberg and R.J. Griesbach, The isolation of mitotic and meiotic chromosomes from plant protoplasts, Plant Sci. Lett. 17:141-147 (1980).
66. L. Szabados, G. Hadlaczky and D. Dudits, Uptake of isolated plant chromosomes by plant protoplasts, Planta 151:141-145 (1981).
67. P.G.N. Jeppensen, A.T. Bankier and L. Sanders, Non-histone proteins and the structure of metaphase chromosomes, Exp. Cell Res. 115:293-302 (1978).

MECHANISMS OF UPTAKE INTO PLANT PROTOPLASTS

Kenneth L. Giles

Department of Biology and Biotechnology
Worcester Polytechnic Institute
Worcester, Massachusetts 01609
USA

THE USE OF ARTIFICIAL LIPID VESICLES IN FUSION AND UPTAKE STUDIES

Studies on membrane fusion performed using lipid vesicles as model membrane systems have proven a powerful tool in the investigation of the mechanism of membrane interaction at the molecular level. Phospholipid vesicles, called liposomes, are stable structures with a limited permeability to small ions and large molecules and have been described extensively in literature (1, 2, 3). Since the construction of liposomes can be varied from single component systems to mixtures of various lipids differing both in their head group and hydrocarbon chain configuration as well as their charge, questions relating to head group specificity, bilayer fluidity, phase transitions and separations, and temperature dependent phase changes of the lipids can be studied. It is also possible to insert into the membrane other components such as cholesterol, glycolipids and proteins, as well as antibodies. Another advantage of these artificial vesicles is that known membrane-active drugs and molecules which are known to effect membrane fusion can be incorporated into the membrane and their mode of action studied in detail.

Lack of a rigorous distinction between fusion and uptake has caused disagreement about certain experimental observations. The evidence for fusion is based on observations such as the transfer of labelled lipid molecules between vesicles (4, 5), changes in the spectra of sonicated vseicles (6, 7) and increase in vesicle size (5, 7), mixing of lipid molecules from two vesicle populations (5, 8), transfer of fluorescent labels between vesicles and lipid films (9, 10), and the mixing and reactivity of two membrane proteins incorporated initially in different lipid vesicles (11). The difficulty arises from the interpretations placed on these observations in that probes or lipid molecules could equally well be exchanged by molecular exchange of diffusion of single molecules or even clusters of molecules rather than simply mixing by fusion.

Papahadjopoulos (5) has pointed out that the most incontrovertible evidence for fusion between vesicles would be a demonstration of stoichiometric mixing of vesicle membrane components as well as the mixing of the vesicle contents. Such stoichiometric mixing of lipid vesicles has been demonstrated for calcium induced fusion of acidic phospholipid vesicles (5, 12). The evidence, however, for the mixing of the water soluble vesicle contents has not been so easy to gather. This may well be due to the fact that vesicles can become leaky during the fusion process. The release of vesicle contents has been observed during calcium induced fusion of unilamellar phosphatidylserine vesicles (13).

Although it is possible that fusion events in biological membranes concern only calcium and phospholipids, it is probable that certain specific proteins may play important roles in either conferring specificity or organizing the biological membrane into certain specific microdomains which may either promote or inhibit fusion events. On this particular point the literature is confusing, since, although proteins of one type and another have been shown to play significant roles in fusion phenomena, the proteins used experimentally to this date have not been present in the matrix of natural membrane systems.

Concanavalin A (Con A) has been used widely in the characterization of cell surface phenomena because of its binding activity to small specific glycoproteins. Van der Bosch and MacConnell (14) report that with lecithin vesicles in the presence of Con A there is a large increase in the transfer rate of spin-labelled lipid molecules between vesicles. This effect was most pronounced at the phospholipid transition temperature, and it was concluded that Con A induced fusion between the vesicles, although again a rigorous exclusion of diffusion was not attempted.

Although it is still too early to really generalize about the effects of specific proteins on phospholipid membrane fusion-uptake phenomena, the existing data indicates that proteins and polypeptides enhance molecular mixing between vesicles. It is possible that this effect may be related to fusion. Nothing can yet be said concerning the mechanisms involved in the interactions.

COATED VESICLES AND THEIR OCCURENCE IN EUKARYOTIC CELL MEMBRANES

Coated vesicles seem to occur in most eukaryotic cells. They were first described by Rosenbluth and Wissig in 1963-4; in this case in the toad spinal ganglion. The vesicles are characterized by polygonal coats they have on the cytoplasmic surfaces of their lipid membranes. They vary in diameter from between 50 and 150 nm. Of their various functions one of the most striking is seen in developing chicken oocytes where they are concerned with the pinocytosis of yolk protein. In fibroblasts coated vesicles seemed to be responsible for the uptake of several macromolecules including low density lipoproteins and epidermal growth factors which ultimately reach lysosomes and are degraded. In other cell types the vesicles appear to be involved with secretions and associated with Golgi apparatus. Although much remains to be learned about the role of coated vesicles, they appear to be involved with the transfer of membrane and other selected macromolecules between different membranous sites within the cell in a defined direction. Clathrin, a polypeptide of molecular weight 180,000, is characteristic of coated vesicles. When disrupted in vitro, the vesicles give defined subunits containing several molecules of Clathrin which can under suitable conditions reassemble to form coat like particles. Electron micrographs of specimens show that their coats are made up of polyhedral lattices plus a few hexagons. Clathrin appears to form flexible lattices within these vesicles suited to the budding of their membranes.

Clathrin isolated from several different animal sources shows a high degree of conservation in its amino acid sequence. This is possibly as a result of Clathrin's many other protein associations which must be maintained within the cell and cell membrane network in order to insure its controlled activity during pinocytosis into the cell. The estimated half-life of these vesicles in fibroblasts is only a few minutes. They have, therefore, the potential for transferring considerable quantities of membrane components and specific protein contents around the cell.

Factors Affecting Liposome-Cell Interactions

Investigations of liposome-cell interaction have shown various factors to influence the delivery of macromolecules, such as liposome composition, incubation conditions, pH, and treatment with agents that enhance vesicle-cell interactions. Vesicles of a number of different lipid compositions have been reported to promote liposome-mediated delivery of macromolecules to mammalian cells in vitro (15, 16). Using a sensitive plaque assay for SV40 uptake, Fraley et al. (16) were able to quantitate the delivery of liposome-encapsulated SV40-DNA to cultured primate cells. These results indicate that negatively-charged vesicles composed of phosphatidylserine were at least ten-fold more effective than other phospholipids in promoting SV40-DNA uptake, even though the relative encapsulation efficiencies were identical in the various preparations. Therefore, differences in vesicle infectivity must reflect the extent of liposome-cell interactions, such as vesicle binding affinities or ability to deliver vesicle content to cells.

A variety of agents have been shown to promote the extent of liposome-mediated delivery of nucleic acids to cultured mammalian cells. The addition of RNAase to poliovirus RNA encapsulated in phosphatidylserine (PS) vesicles and primate cells has been demonstrated to enhance infectivity approximately ten-fold (17). Following incubation of vesicles with cells, treatments such as DMSO, PEG, or glycerol has increased vesicle encapsulated SV40 infectivity an additional ten- to twenty-fold (16).

Subsequent attempts to enhance vesicle-mediated delivery of encapsulated substances have been directed at increasing the affinity of liposomes for cells. One promising approach is the covalent coupling of IgG molecules on liposome surfaces which are direct against a specific cell type (18, 19). Antibody coated liposomes have been shown to dramatically increase liposome binding to cultured primate cells.

Liposome Interactions with Plant Protoplasts

Although liposomes have been widely employed for introducing a variety of macromolecules into mammalian cells, few reports have dealt with their possible appication in plant cell systems. Cassells (20) has shown that lecithin vesicles containing fluorescein diacetate transfer their contents to tomato protoplasts. However, this report fails to account for the possibility of liposome permeability and subsequent uptake of released fluorescein diacetate by the protoplast. *E. coli* RNA encapsulated in LUV liposomes was delivered to carrot protoplasts where subsequent degradation of the RNA occurred (21). pBR322 DNA sequestered in LUV liposomes was found tightly bound to cowpea protoplasts with subsequent transfer of pBR322 to the cell nuclei after incubation in the presence of polyethylene glycol (PEG) (22). In the same study it was shown that protoplast exudates possessing DNAase activity and pancreatic DNAase were ineffective in degrading pBR322 DNA encapsulated in liposome. Yet, in all the above studies, vesicles were composed of lecithin or a variation thereof. Results with mammalian systems demonstrating negatively-charged vesicles bind optimally to cells suggest that PS vesicles may enhance liposome interaction with protoplasts. In addition, lack of a sensitive assay for delivery, such as biological expression of donor nucleic acid requires a more critical evaluation of reported delivery in these systems.

With the technical problems and low efficiency of currently available techniques for the introduction of macromolecules into plant cells which have greatly restricted the scope of such investigations, a liposome-mediated delivery system may prove a valuable method for overcoming existing barriers. For example, most transformation attempts have employed the use of polycations, alkaline solutions, or some combination of the two, and resulted in increased membrane permeability, depressed host macromolecular synthesis and significant cytotoxicity. Phospholipid vesicles can be applied to most cells at high concentrations without significantly affecting subsequent cell growth and viability and present an efficient means for the introduction of nucleic acids into cells. Secondly, rapid degradation of donor DNA incubated with protoplasts can result from the release of high levels of deoxyribonuclease in the incubation medium (23, 24, 25). Attempts to overcome degradation of donor DNA by using a variety of polycation treatments suffer the main drawback of having deleterious effects on cell viability (26 - 29). Since encapsulation of RNA (17) and DNA (16) has been shown to protect these molecules from enzymatic degradation, it would appear that liposomes may offer means of delivering undigested DNA to protoplasts without drastically affecting protoplast viability. In addition, liposomes may enhance uptake of free DNA by overcoming the membrane restrictions of the uptake of large macromolecules, and delivery frequencies may be enhanced by varying lipid composition of the vesicles and post-incubation conditions with various agents which promote vesicle-cell interactions.

APPLICATION OF UPTAKE TECHNIQUE FOR CROP IMPROVEMENT

There is ample evidence in the literature to show that organelles, microorganisms, macromolecules, including DNA and RNA, can be taken up by plant protoplasts. The comments made in this paper suggest areas of research which might optimize those uptake phenomenon to a point where the efficiency of uptake can be realistically used for selection and transformation experiments. It might be realized that uptake alone is only the first step of many involved in the selection and possible incorporation of organelles and macromolecules into the host cell. The uptake of organelles into a preexisting population of organelles does not guarantee the survival or replication of that particular organelle in subsequent cell generations. In order to do this the efficiency of uptake must either be very high, such that a large number of introduced organelles are present in a given cell, or the selective advantage of the introduced organelles must be so carefully arranged that it is not only advantageous to but vital to the survival of the cell. It must be appreciated that for efficient incorporation of organelles or DNA into the cell this survival advantage given to the host cell must be expressed as soon as possible, preferably at the level of the first division of that protoplast such that selection procedures can pull out transformed cells or modified organelle populations within a cell as soon as possible. This obviates or partially eliminates the creation of chimeral or mixed cell populations.

In what particular areas then can the uptake of cell organelles be considered the usable method for crop improvement? There are obvious advantages to the forms of photosynthesis referred to as C_4 photosynthesis. However, many of the advantages accrued by this form of photosynthesis are dependent not only on the possession of chloroplasts of differential metabolic ability, but also on the structure and construction of the leaf tissue. Changes of the magnitude likely to be required for structural changes to the effective Krantz type anatomy of C_4 plants is unlikely to be brought about by the simple transfer of organelles, although some slight advantages in terms of PEP-carboxylase content might conceivably accrued by transfer of C_4 organelles into otherwise C_3 protoplasts. However, without strong selective advantage, it is questionable as to how long they might survive in the plastid population of the regenerating protoplast.

The uptake and exchange of mitochondria either intact or merely mitochondrial DNA and plastids into foreign protoplasts may have far reaching effects on the bioenergetics of the resulting cell. There is an increasing amount of evidence to suggest that ATP levels within cells are not only responsible for the energy required for metabolic activity, but that some basic threshold level of ATP is expended on the actual control mechanisms operating during that metabolism. If this is true, then the bioenergetic state of the cell may have extreme consequences with respect to its abilities to differentiate, continue growing, and possibly reproduce. These considerations become overwhelmingly important when one is considering the differences which exist between the juvenile and mature phases of many perennial species and the differences which can accrue during the senescence of plants, especially annual species. The potential conference of perenniality to annuals would have many advantages in a number of cropping situations. It would reduce the energy demands for plowing and resowing of seeds. It would prevent many of the weed problems from occurring by competing them out. It would in many countries have important soil erosion consequences in that minimum tillage methods oculd be employed. The possibility of this manipulation of the bioenergetic state by uptake of the mitochondria or mitochondrial DNA is a tantalizing objective.

The shreds of evidence which can be peiced together to formulate the hypothesis above stem from many animal and plant sources. It is a common observation that during the very early stages of senescence, mitochondria in both animal and plant cells begin to look distorted, balloon out, and finally collapse. The discovery and description of a plasmid-like DNA as part of the mitochondrial DNA of the fungus Podospora anserina by Esser and others (30) gives rise to the potential for an infectious senescence particle within the fungal mitochondria. The data published in this paper fit a previously published model in which during ageing plasmid DNA is excised from mitochondrial DNA and becomes autonomous for its replication and function. The plasmid DNA has been found to have a contour length of 0.75 microns, corresponding to a molecular weight of 2.4

kb (31, 32). The essential part of the model is the assumption that plasmid DNA originates from mitochondrial DNA. The plasmid DNA is integrated into mitochondrial DNA during the juvenile phase of growth but becomes liberated during the processes of ageing, when the mitochondria degenerate. After liberation, the plasmid DNA appears to replicate autonomously and becomes infective for juvenile hyphae, inducing their rapid senescence.

The observation that mitochondrial DNA is capable of losing part of its genetic information without immediate cell death resulting seems a general phenomenon. One of the most published examples is that involving the Petite mutants of _Saccharomyces cerevisiae_. In this yeast it was shown that certain types of Petite mutants (rho-) may survive despite having lost parts of their mitochondrial DNA. Deletions may comprise up to 99% of the mitochondrial DNA to the point in fact where no distinct mitochondrial DNA can be detected (33). Other examples may be quoted in this connection. The ragged mutants of _Aspergillus amstelodami_ show disturbed growth including hyphal degeneration but usually not leading to cellular death (34). Lazarus et al. (35) found that in ragged mutants a distinct part of the mitochondrial DNA was amplified and some mitochondrial DNA remained unchanged. In corn, mutations to male sterility involve alteration of the mitochondrial DNA. Male sterility in maize may be brought about by alterations in mitochondrial DNA, therefore. This involves the occurence of a plasmid DNA, the contour length of which varies in specific mutants from 1.42 to 2.35 kb. In wild strains the plasmid DNA seems to be a part of the mitochondrial DNA (36).

The potential opportunities for the expiration of the function of these mitochondrial plasmids and whole mitochondria and the maintenance of the bioenergetic state of the cell, its cell phase (in terms of juvenility and maturity, as well as senescence) appear to have interesting opportunities in terms of plant manipulation, modification and, hopefully, improvement. At this stage in the development of these techniques it would be foolhardy to claim that they hold the key to crop improvement. They do, however, hold promise of giving information which may be useful to plant breeders, plant biochemists, and molecular biologists who may be able to use this information for eventual improvement of existing crops. The basic understanding of plant development and differentiation as well as senescence and death may be of considerable interest and value to all fields of plant physiology and plant science.

REFERENCES

1. T.E. Thompson and F.A. Henn, Experimental phospholipid model membranes, _in_: "Membranes of Mitochondria and Chloroplasts," E. Racker, ed., Van Nostrand Reinhold Co., New York (1970).

2. D. Papahadjopoulos and H.K. Kimelberg, Phospholipid vesicles (liposomes) as models for biological membranes: their properties and interactions with cholesterol and proteins, in: "Progress in Surface Science," S.G. Davison, ed., Pergamon Press, Oxford, England (1973).
3. A.D. Bangham, M.W. Hill and N.G.A. Miller, Preparation and use of liposomes as models of biological membranes, in: "Methods in Membrane Biology," E.D. Korne, ed., Plenum Press, New York (1974).
4. T. Maeda and S.-I. Ohnishi, Membrane fusion transfer of phospholipid molecules between phospholipid bilayer membranes, Biochem. Biophys. Res. Commun. 60:1509-1516 (1974).
5. D. Papahadjopoulos, G. Poste, B.E. Schaeffer and W.J. Vail, Membrane fusion and molecular segregation in phospholipid vesicles, Biochim. Biophys. Acta 352:10-28 (1974).
6. J.H. Prestegard and B. Fellmeth, Fusion of dimyristoyl lecithin vesicles as studied by proton magnetic resonance spectroscopy, Biochemistry 13:1122-1126 (1974).
7. A.L.Y Lau and S.I. Chan, Alamethicin-mediated fusion of lecithin vesicles, Proc. Natl. Acad. Sci. USA 72:2170-2174 (1975).
8. F.J. Martin and R.C. MacDonald, Phospholipid exchange between bilayer membrane vesicles, Biochem. 15:321-327 (1976a).
9. G.W. Pohl, G. Stack and H.W. Trissl, Interaction of liposomes with black lipid membranes, Biochim. Biophys. Acta 318:478-481 (1973).
10. M.R. Moore, Fusion of liposomes containing conductance probes with black lipid films, Biochim. Biophys. Acta 426:765-771 (1976).
11. C. Miller and E. Racker, Fusion of phospholipid vesicles reconstituted with cytochrome C oxidase and mitochondrial hydrophobic protein, J. Mem. Biol. 26:319-333 (1976).
12. D. Papahadjopoulos, W.J. Vail, W.A. Pangborn and G. Poste, Studies on membrane fusion. II. Induction of fusion in pure phospholipid membranes by calcium and other divalent metals, Biochim. Biophys. Acta 448:265-283 (1976).
13. D. Papahadjopoulos, W.J. Vail, C. Newton, S. Nir, K. Jacobson, G. Poste and R. Lazo, Studies on membrane fusion. III. The role of calcium-induced phase changes, Biochim. Biophys. Acta 465:579-598 (1977).
14. J. Van der Bosch and H.M. McConnell, Fusion of dipalmitoyl phosphatidylcholine vesicle membranes induced by Concanavalin A, Proc. Natl. Acad. Sci. USA 72:4409-4413 (1975).
15. R.E. Pagano and J.N. Weinstein, Interactions of liposomes with mammalian cells, Ann. Rev. Biophys. Bioeng. 7:435-468 (1978).
16. R. Fraley, S. Subramani, P. Berg and D. Papahadjopoulos, Introduction of liposome-encapsulated SV40 DNA into cells, J. Biol. Chem. (in press) (1981).
17. T. Wilson, D. Papahadjopoulos and R. Taber, The introduction of poliovirus RNA into cells via lipid vesicles, Cell 17:77-

84 (1979).
18. T. Heath, R. Fraley and D. Papahadjopoulos, Antibody targeting of liposomes. Specific interaction of vesicles conjugated to anti-erythrocyte F $(ab')_2$, (Submitted) (1980).
19. T. Heath, B. Macher and D. Papahadjopoulos, Covalent attachment of proteins to liposomes via glycosphingolipids. I. Oxidation of vesicle lipids and aqueous contents, (Submitted) (1980).
20. A.C. Cassells, Uptake of charged lipid vesicles by isolated tomato protoplasts, Nature 275:760 (1978).
21. B. Matthews, D. Sheldon, J. Widholm and M. Ostro, Liposome mediated transfer of bacterial RNA into carrot protoplasts, Planta 145:37-44 (1979).
22. P.F. Lurquin, Entrapment of plasmid DNA by liposomes and their interactions with plant protoplasts, Nucleic Acid Res. (in press) (1980).
23. P.F. Lurquin and C.I. Kado, Escherichia coli plasmid pBR313 insertion into plant protoplasts and into their nuclei, Mol. Gen. Genet. 154:113-121 (1977).
24. B.G. Hughes, F.G. White and M.A. Smith, Fate of bacterial plasmid DNA during uptake by barley protoplasts, FEBS Lett. 79:80-84 (1977).
25. B.G. Hughes, F.G. White and M.A. Smith, Contribution of damaged protoplasts to DNA uptake by purified plant protoplasts, Plant Sci. Lett. 11:199-206 (1978).
26. M. Suzuki and I. Takebe, Uptake of single-stranded bacteriophage DNA by isolated tobacco protoplasts, Z. Pflanzenphysiol. 78:421-433 (1976).
27. S.M. Fernandez, P.F. Lurquin and C.I. Kado, Incorporation and maintenance of recombinant-DNA plasmid vehicles pBR313 and pCR1 in plant protoplasts, FEBS Lett. 87:277-282 (1978).
28. C.I. Kado and P.F. Lurquin, in: "Microbiology 1978," (1978).
29. P.F. Lurquin and L. Marton, DNA transfer experiments with plant protoplasts and bacterial plasmids, in: "Proc. Fifth Intl. Protoplast Symp.," (1980).
30. U. Kuck, U. Stahl and K. Esser, Curr. Gen. 3:151-156 (1981).
31. U. Stahl, U. Kuck, P. Tudzynski and K. Esser, Mol. Gen. Genet. 178:639-646 (1980).
32. D.H. Cummings, L. Belocur and C. Grandchamp, Mol. Gen. Genet. 171:239-250 (1979b).
33. N.W. Gillham, "Organelle Heredity," Raven Press, New York (1978).
34. C.E. Caten, J. Gen Microbiol. 72:221-229 (1972).
35. C.M. Lazarus, A.J. Earl, G. Turner and H. Kuntzel, Eur. J. Biochem. 106:633-641 (1980).
36. R.J. Kemble and J.R. Bedbrook, Nature 284:565-566 (1980).

INTERSPECIFIC SOMATIC PROTOPLAST FUSION PRODUCTS IN CULTIVATED JUTE SPECIES

P.M. Kumar, K. Das, R.R. Sinha, P. Mukherjee and S.K. Sen

Programme in Genetical Research
Department of Botany
Bose Institute
Calcutta 700009
INDIA

INTRODUCTION

Accounts of success on isolation and culture of plant protoplasts and various genetic manipulation studies including fusion of plant protoplasts from distantly related species have been reported in the past (1, 2, 3). However, records of success amongst major crop plants are not many. Jute is an important fiber crop plant extensively grown in the eastern part of India and Bangladesh. The two cultivated species (Corchorus capsularis and C. olitorius) complement the needs of each other. Production of sexual hybrids is prevented by barriers of sexual incompatibility between the two species. We reported recently (4) that interspecific somatic hybrids may be possible to evolve. On further improving our technical approach, we are convinced that somatic hybrid cell lines have been produced which, however, failed to differentiate into complete plants till date.

MATERIALS AND METHODS

Four strains, JRC 321, JRC 978, JRC 212 and Tripura capsularis of Corchorus capsularis, and JRO 3432, JRO 524, R-26 and white stem of C. olitorius were utilized for the study. Leaves from 3-4 week old apical shoot cultures of C. capsularis grown on SH (%) medium, modified with kinetin (0.2 mg/l) and NAA (0.1 mg/l) in addition to folic acid (0.5 mg/l) and biotin (0.002 mg/l), were the source of mesophyll protoplasts. Cultured cell suspensions of C. olitorius maintained on SH medium supplemented with kinetin (1.0 mg/l), 2,4-D (1.0 mg/l), pCPA (0.5 mg/l) and a mixture of amino acids (6) were the source of cultured cell protoplasts.

Isolation of Protoplasts

The one-step method (7) of enzyme treatment was employed for the isolation of protoplasts from both mesophyll and cultured cells. Young leaves from 3-4 week old apical shoot meristem cultures were pretreated in 0.3 M mannitol for 30 min. before chopping them fine in the same solution after removal of the lower epidermis. Minced leaf tissues were washed once in 35% sea water and subsequently incubated in enzyme mixture (1:10 w/v) containing Macerozyme R10 (0.7%), cellulase Onozuka R10 (1.5%) and Helicase (0.4%) in pH between 5.8 and 6.0 at an osmotic pressure of 710 mOs/kg.H_2O. Incubation was carried out at 28°C ± 1°C for 3 h in a roller shaker.

Cells from frequently subcultured suspension cultures were collected after 24-32 h of inoculation and incubated in the enzyme mixture (1:10 v/v) similar to that used for mesophyll cells, reducing the Helicase concentration to 0.3%. Osmotic pressure was adjusted to 605 mOs/kg.H_2O and incubated for 3 h at 28°C ± 1°C in a roller shaker.

Released protoplasts were purified following the procedure indicated earlier (4). Viability of the protoplasts was tested according to the procedure of Larkin (8).

Culture

Purified protoplasts of leaf mesophyll cells and cultured cells of all strains were subjected to different cultural conditions. Twelve established media (Table 1) with different combinations of phytohormones were tested for regeneration of the protoplasts using MDA (multiple droplet array) technique (9). Osmotic pressure of the media for mesophyll protoplasts was maintained at 710 mOs/kg.H_2O at cell density of 1.4-1.8 x 10^4 protoplasts/ml. Protoplasts from cultured cell suspension were maintained at 1.8-2.0 x 10^4 protoplasts/ml density and the medium had 605 mOs/kg.H_2O osmotic pressure. Petri plates sealed with parafilm were kept in transparent plastic boxes to provide diffused light and to maintain humidity.

Fusion

C. olitorius protoplasts from callus tissues (strains JRO 3432, JRO 524) and mesophyll protoplasts of *C. capsularis* (JRC 321 and JRC 978) responded favorably to protoplast cultural attempts with regard to regeneration. Protoplasts of the two different species were mixed together and about 150 µl of this mixed suspension were pipetted onto a cover glass (22 x 22 mm) placed on a drop of silicone 200

Table 1. List of Established Culture Media Tested.

Media	References
V47	10
SSH	13
MSH	4
SH	5
KM	1
NT	14
ER	15
F5	16
BL	17
MS	18
N6	19
B5	20

All media were modified with mannitol to give a final osmotic pressure of 710 mOs/kg.H_2O for leaf mesophyll protoplats and 605 mOs/kg H_2O for cultured suspension cells derived protoplasts.

fluid (100 cs) in a petri plats (60 mm). After 10 min., 450 µl of PEG fusion solution (4) was slowly added and left at room temperature for 50 min. 0.5 ml of culture medium was additionally put. Repeated washings were done with the same culture medium. Eight media - KM, NT, F5, BL, N6, ER, V47 and MS, all of which failed to respond while culturing parent protoplasts, were used for the culture of fusion products. Osmotic pressure of these were adjusted to 650 mOs/kg.H_2O.

Rationale for Screening the Fused Products

Initial screenings for obtaining the fused products and elimination of the growth of the parental protoplasts as well as selfed fusion products were carried out on the following two principles: 1) media which did not respond to parental protoplast growth are likely to permit only the hybrid cells to grow, if at all. Thus, evidence for growth in any of these media is likely to indicate for the hybrid cell lines, rather than the parental ones; and, 2) since culture of parental protoplasts required different osmotic pressures (605 mOs/kg.H_2O for cultured cell protoplasts and 710 mOs/kg.H_2O for mesophyll protoplasts), somatic hybrid cells as a result of fusion are likely to require intermediate osmotic pressure of the two. The intermediate range of the osmotic pressure of the culture medium is likely in turn not to suit any of the parental type of protopalsts, thereby eliminating them from any further growth.

Hybrid colonies obtained after following these screening procedures were further studied to confirm the nature of hybrid through: 1) chromosomal analysis - callus pieces were treated with colchicine (0.5%), 1 h, 12-16°C), fixed in acetic acid-ethanol (1:3) mixture for 4-20 h and the chromosomes were stained with orcein; and 2) isoenzyme studies - isoenzyme pattern study of peroxidase enzyme was carried out according to the method of Davis (10), using a slab gel.

RESULTS

Isolation of Protoplasts

Use of *in vitro* grown material as the source of protoplasts reduced the chance of contamination. Presoaking leaves in 0.3 M mannitol facilitated removal of the epidermis. Washing of the sliced leaf pieces with 35% sea water was found to remove mucilage to a great extent enabling the enzyme mixture to act faster. Yield and viability of mesophyll protoplasts of *C. capsularis* and cultured cell protoplasts of *C. olitorius* are presented in Table 2. It is evident from the table that release of viable protoplasts from leaf mesophyll cells of *C. capsularis* (Figure 1) and cultured cells of *C. olitorius* (Figure 2) were found to be genotype dependent.

Culture

Of the four cultivars taken for study, mesophyll protoplasts of only two cultivars (JRC 321 and JRC 212) responded favorably to our

Table 2. Release of Viable Protoplasts (in %) of Jute Species After the Enzyme Treatment.

Genetic lines	Protoplast release	Viable protoplasts
C. capsularis		
JRC 321	80-90	80-85
JRC 978	70-80	70-80
JRC 212	50-60	60-65
Tripura capsularis	40-45	40-50
C. olitorius		
JRO 3432	70-75	70-80
JRO 524	60-70	75-80
R-26-indigenous	50-60	60-70
White stem	60-70	70-80

Table 3. Response of Protoplasts of Various Genetic Lines of Jute Species to Different Culture Media.

Media	Strains of *C. capsularis* JRC 321	JRC 978	JRC 212	Tripura capsularis	Strains of *C. olitorius* JRO 3432	JRO 524	R-26-indigenous	White stem
SSH	++	-	++	-	++	-	-	-
MSH	+++	-	+++	-	+	-	-	-
SH	++	-	++	-	=	-	-	-
MS	-	-	=	-	-	-	-	-
B5	-	-	++	-	+	+	-	-
NT	=	-	-	=	-	=	-	-
KM	-	-	=	-	-	-	=	-
V47	-	-	-	-	=	-	-	-
ER	=	-	-	-	-	-	-	-
BL	-	-	-	-	-	-	-	-
N6	-	-	-	-	-	-	-	-
F5	-	=	-	-	-	-	-	-

- no response; = wall regeneration only; + less than 5% division; ++ 10-15% division; +++ 25-30% division.

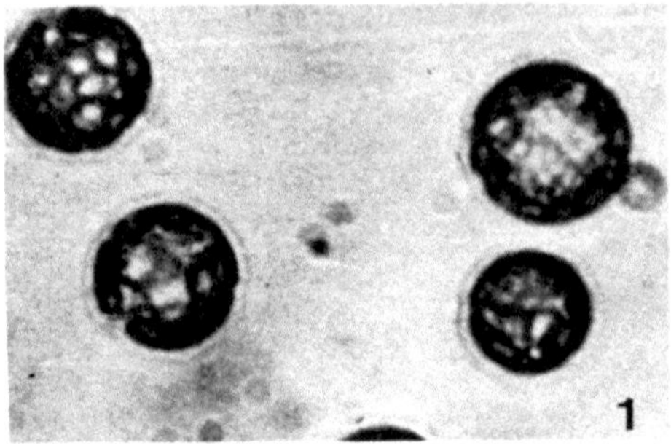

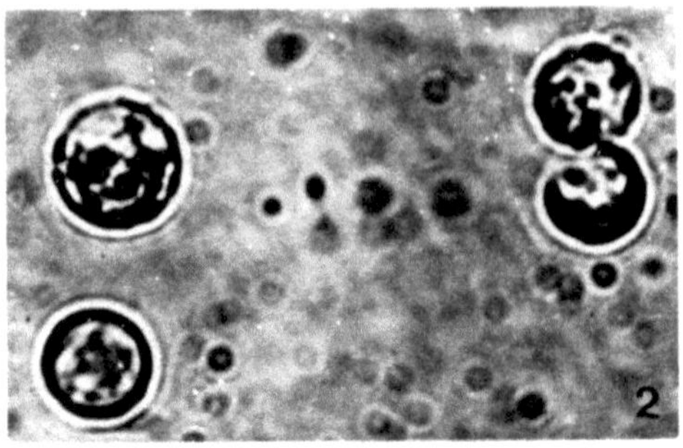

Figure 1. Freshly isolated mesophyll protoplasts under phase-contrast of C. capsularis.

Figure 2. Freshly isolated protoplasts under phase-contrast from cultured suspension cells of C. olitorius.

regeneration experiments (Table 3). Maximum frequency of division was observed in MSH medium (Table 3) containing 2,4-D and BAP at the level of 1.0 mg/l each. Change in shape of protoplast was observed after 48 h in culture followed by the first division. Repeated division was observed after 12 days when fresh medium with reduced osmoticum was added. Growth of small callus colonies were found to occur at the end of the third week.

Cultured cell protoplasts of cultivars JRO 3432 and JRO 524 responded to regeneration. Of these JRO 3432 was found to divide in SSH, MSH and B5 media, whereas JRO 524 showed division only in the B5 medium. Similar to leaf mesophyll protoplasts, cultured cell protoplasts divided only when these media were supplemented with BAP and 2,4-D at the level of 1.0 mg/l each. Even though protoplasts of JRO 3432 underwent few divisions in SH medium containing NAA and BAP, they failed to grow further. Change in shape followed by first division was observed after 52-54 h of culture. Small colonies could be found at the end of four weeks.

The plating efficiency of C. capsularis mesophyll and C. olitorius callus protoplasts ranged between 25-30% and 10-15% respectively.

Fusion

The technique used by us for fusion was found to be quite effective for causing high frequency fusion. Observations immediately after plating indicated that many mesophyll as well as callus protoplasts adhered to each other, though adherence between similar protoplasts was also seen. Occurrence for complete fusion (Figure 3) of

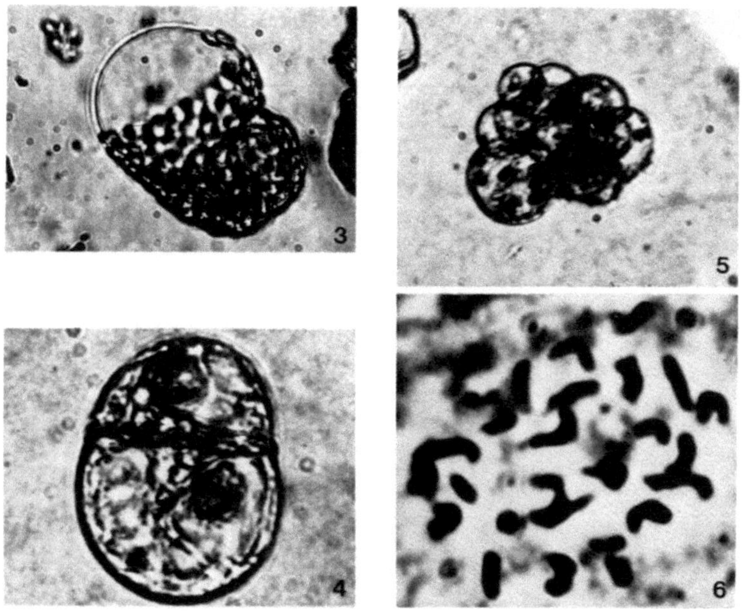

Figure 3. Mesophyll protoplasts and cultured cell suspension derived protoplasts in the process of complete fusion.

Figure 4. Division of the fused product.

Figure 5. Repeated divisions of the fused product.

Figure 6. Chromosome complements of the somatic hybrid cells.

two different types of protoplasts was recorded. Subsequently, within a couple of days we could observe few cells showing divisions (Figure 4) in V47 medium (11) with BAP and 2,4-D (both at the level of 1.0 mg/l). Only fusion products of JRC 321 x JRO 3432 regenerated their cell wall and showed division (Table 4). Repeated divisions (Figure 5) of some of these few cells were seen on addition of fresh medium and in about 32 days some calli could be observed. On addition of agar to the medium callus colonies grew further. Although about 1-2% of the cells could be seen to divide initially, only fourteen callus colonies could finally be recovered.

Cytological analysis of jute somatic chromosomes is a very difficult proposition. Even then at least five callus colonies (CLJ 79/4, 79/16, 79/18, 80/7, 81/11) of those fourteen callus lines cytologically proved (Figure 6) to contain some of the chromosome complements of both the species (Figures 7, 8). The isoenzyme pattern (Figure 9) also confirmed their hybrid nature. These cell lines were thus assigned by us to be true somatic hybrids. Except in two

Table 4. Response of Fused Protoplasts on Different Culture Media Tested.*

Media	JRC 321 + JRO 3432	JRC 321 + JRO 524	JRC 212 + JRO 3432	JRC 212 + JRO 524
MS	-	-	-	-
NT	=	=	-	-
KM	-	-	=	-
V47	+	-	=	-
ER	=	-	-	-
BL	-	-	-	-
N6	-	-	-	-
F5	-	-	-	-

* All media were modified with mannitol to give a final osmotic pressure of 650 mOs/kg.H_2O.

\- No response; = cell wall regeneration only; + 1-2% cells showing division.

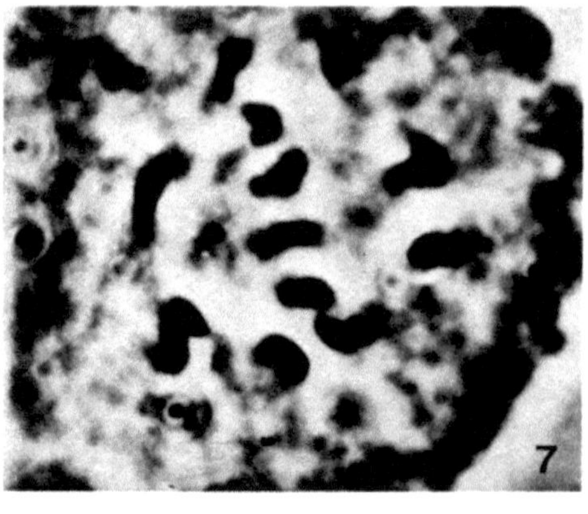

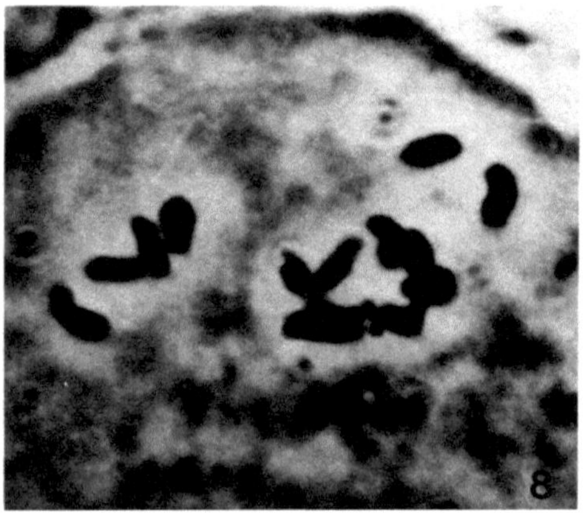

Figure 7. Chromosome complements of C. olitorius.

Figure 8. Chromosome complements of C. capsularis.

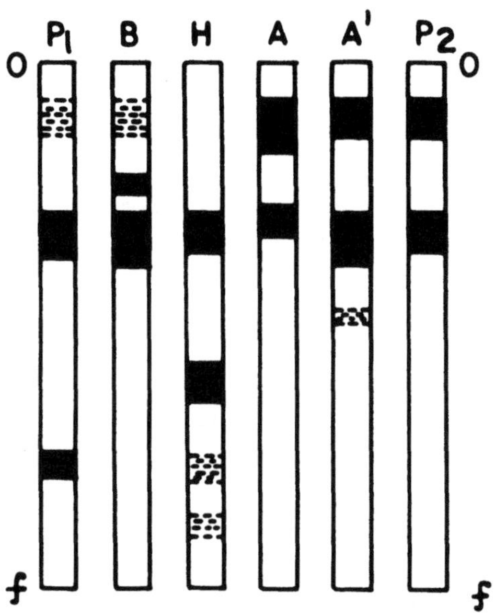

Figure 9. Peroxidase isoenzyme pattern of some of the callus and hybrid colonies. P_1: C. capsularis; P_2: C. olitorius; H: isoenzyme pattern of somatic hybrid colony CLJ 80/7; B: isoenzyme pattern of colony CLJ 80/16; A: isoenzyme pattern of colony CLJ 79/6; A': isoenzyme pattern of colony CLJ 80/2.

callus colonies (CLJ 79/20 and 80/10) where the chromosome number appeared to be at the diploid level (2n = 14), all other callus colonies contained chromosomes ranging from 22 to 28. Obviously, elimination of chromosomes appeared to be a frequent feature. The remaining seven callus colonies (CLJ 79/10, 79/6, 80/2, 80/16, 80/22, 81/5 and 81/8) having chromosome number 22 or more did not offer us a clear cut picture of their interspecific hybrid nature. The isoenzyme pattern of some of these lines did not also correspond to either the parental types or that of the lines which showed their hybrid nature in their chromosomal make up.

DISCUSSION

Genotype dependent response towards culture and fusion of protoplasts in the present case bore some similarity with the observation of Niizeki and Kita (12) in rice. The point which calls for special mention is the osmolarity of the culture medium. These were found to be quite critical in our studies. Mesophyll protoplasts required an osmotic pressure of 710 mOs/kg.H_2O, whereas cultured cell protoplasts required only 605 mOs/kg.H_2O. Any variation resulted in negative response of protoplast regeneration.

Screening principles for growth of somatic hybrids were considered to be apt. The strategies adopted in the present case bore some similarity with that of screening hybrid cells between *Arabidopsis thaliana* + *Brassica campestris* (3). The use of an intermediate osmotic pressure of the media is considered to be quite intriguing. The somatic hybrids were identified with the help of chromosome analysis and isoenzyme pattern as also adopted by earlier workers in a similar situation (3, 13). Work on the somatic hybrid calli lines are being pursued in order to cause differentiation.

ACKNOWLEDGEMENTS

Grants from the Department of Science and Technology of India to S.K. Sen and a fellowship from CSIR, New Delhi, to P.M. Kumar are gratefully acknowledged.

REFERENCES

1. K.N. Kao, F. Constable, M.R. Michayluk and O.L. Gamborg, Plant protoplast fusion and growth of intergeneric hybrid cells, *Planta* 120:215-227 (1974).
2. G. Melchers, M.D. Sacristan and A.A. Holder, Somatic hybrid plants of potato and tomato regenerated from fused protoplasts, *Carlsberg Res. Commun.* 43:203-218 (1978).
3. Y.Y. Gleba and F. Hoffmann, Hybrid cell lines of *Arabidopsis thaliana* + *Brassica campestris*. No evidence for specific chromosome elimination, *Molec. Gen. Genet.* 165:257-264 (1978).
4. K. Das, P.M. Kumar and S.K. Sen, Progress in application of protoplast technology in tackling genetical problems of cultivated jute species, *in*: "Proc. of International Workshop on Improvement of Tropical Crops Through Tissue Culture," A.S. Islam, ed., Dacca (1981).
5. R.U. Schenk and A.C. Hildebrandt, Medium and techniques for growth of monocotyledonous and dicotyledonous plant cell culture, *Can. J. Bot.* 50:199-201 (1972).
6. K. Das, Ph.D. Thesis, Calcutta University (1982).
7. J.B. Power and E.C. Cocking, Isolation of leaf protoplasts. Macromolecule uptake and growth substance response, *J. Exp. Bot.* 21:64-70 (1970).
8. P.J. Larkin, Purification and viability determinations of plant protoplasts, *Planta* 128:213-216 (1976).

9. I. Potrykus, C.T. Harms and H. Lorz, Callus formation from cell culture protoplasts of corn (Zea mays L.), Theor. Appl. Genet. 54:209-214 (1979).
10. B.J. Davis, Disc electrophoresis. II. Method and application to serum proteins, Ann. N.Y. Acad. Sci. 121:404-427 (1964).
11. H. Binding, Regeneration von haploiden und diploiden pflanzen aus protoplasten von Petunia hybrida L., Z. Pflanzenphysiol. 74:327-356 (1974).
12. M. Niizeki and F. Kita, Cell division of rice and soybean and their fused protoplasts, Japan J. Breed. 31:161-167 (1981).
13. P. Medgyesy, L. Menczel and P. Maliga, The use of cytoplasmic streptomycin resistance: chloroplast transfer from Nicotiana tabacum into Nicotiana sylvestris and isolation of their somatic hybrids, Mol. Gen. Genet. 179:693-698 (1980).
14. Nagata and Takebe, Planta 99:12 (1971).
15. Eriksson, Physiol. Plant. 18:976 (1965).
16. Frearson et al., Dev. Biol. 33:130 (1973).
17. D.F. Blaydes, Interaction of kinetin and various inhibitors in the growth of soybean tissue, Physiol. Plant. 19:738 (1966).
18. T. Murashige and F. Skoog, A revised medium for rapid growth and bioassays with tobacco tissue culture, Physiol. Plant. 15:473 (1962).
19. C.C. Chu, C.C. Wang, C.S. Sun, C. Hsu, K.C. Yin and C.Y. Chu, Establishment of an efficient medium for anther culture of rice through comparative experiments on the nitrogen sources, Scientia Sinica 18:659 (1975).
20. O.L. Gamborg, R.A. Miller and K. Ojima, Nutrient requirements of suspension cultures of soybean root cells, Exp. Cell Res. 50:151 (1968).

THE USE OF CHLOROPLAST PROTEINS IN CROP IMPROVEMENT

Birger Lindberg Møller, Gunilla Høyer-Hansen and Landis E.A. Henry

Department of Physiology
Carlsberg Laboratory
Gamle Carlsberg Vej 10
DK-2500 Copenhagen Valby
Denmark

INTRODUCTION

Detailed knowledge of the structure and function of the proteins involved in photosynthetic reactions can provide information to what extent these proteins can be modified in the course of crop improvement. This information also determines the usefulness of a given protein to serve as marker in cell cultures and to provide a basis for *in vitro* selection procedures. Our main interest has been directed towards an analysis of the proteins localized in the thylakoid membranes of the chloroplasts. The conversion of solar energy into chemical energy takes place in these thylakoid membranes and the organization of these thylakoids must thus allow so diversified processes as absorption of photons, charge separation, oxygen evolution and electron transport through the two photosystems resulting in the formation of the energy rich components ATP and NADPH. Using the electrophoretic properties of ribulose-1,5-bisphosphate carboxylase as phenotypic markers of the nuclear and chloroplast genomes, five somatic hybrids of potato and tomato obtained by protoplast fusion have been identified.

RESULTS AND DISCUSSION

The thylakoid membrane in higher plants and algae contains photosystems each with a specific reaction center plus antennae pigment system. Functional identification of the polypeptide components of these systems has been attempted using a variety of electrophoretic techniques often in combination with photosynthetic mutants.

Purified thylakoid membranes from barley can be resolved into ten chlorophyll containing bands plus free chlorophyll using SDS-polyacrylamide gel electrophoresis. The identity and probable function of eight of these bands were established by comparison of the wild-type pattern with those of isolated light-harvesting chlorophyll a/b-protein 2, photosystem I and II vesicles and a number of nuclear gene mutants of barley. It was found that the mutant chlorina-f2, which lacks chlorophyll b, contains only three chlorophyll a-proteins designated Chl_a-P1 (MW = 110,000), Chl_a-P2 (MW= 49,000) and Chl_a-P3 (MW = 41,000). Chl_a-P1 contains the reaction center of photosystem I (P700) and with the aid of the mutant viridis-m[29], Chl_a-P3 was deduced to be the most likely site of the reaction center (P680) of photosystem II. Two chlorophyll a/b-proteins were found, which differed in the a/b ratios and which are thought to play a role in light harvesting or light-focusing. These were designated $Chl_{a/b}$-P1 (MW = 32,000) and $Chl_{a/b}$-P2 (MW = 29,000), the latter being found also in multimeric complexes (1).

The five subunits of the chloroplast coupling factor (CF_1) have been identified in the total polypeptide pattern by comparison with the pattern obtained from purified CF_1 and by crossed immunoelectrophoresis (2). Isolation was also used to locate the position of the dicyclohexyl carbodiimide (DCCD) binding proton channel proteolipid which is a component of the CF_0 entity (3).

Isolation of photosystem I particles from barley was achieved by an initial digitonin fractionation of washed thylakoids. The photosystem I enriched pellet obtained at 144,000 g after differential centrifugation was treated for 20 min at 4°C with the zwitterionic detergent Empigen BB at a chlorophyll concentration of 0.5 mg/ml and a detergent concentration of 1%. The detergent concentration was diminished to 0.1% by addition of buffer and further purification was achieved either by ion exchange chromatography (Whatman DE52), by the use of a salt gradient or by gel filtration (Ultrogel AcA34). The fractions corresponding to the front of the main chlorophyll containing band eluted from both columns were found to be highly enriched in photosystem I and showed identical properties. The absorption maximum in the red region was 678 nm. The rate of photosystem I electron transport was 105 μmol oxygen/hr/mg chlorophyll when measured as the stimulation of oxygen consumption upon the addition of methyl viologen with ascorbate/DCIP present as the electron donor. The chemically determined chlorophyll to P700 ratio of the particles was 90, whereas the ratio as 110 when determined photochemically. The particles contained no cytochromes. When analyzed by SDS-PAGE, the photosystem I particles were found to be composed of three polypeptide bands with apparent molecular weights of 110,000, 18,300 and 15,200 (Figure 1). The band at 110,000 molecu-

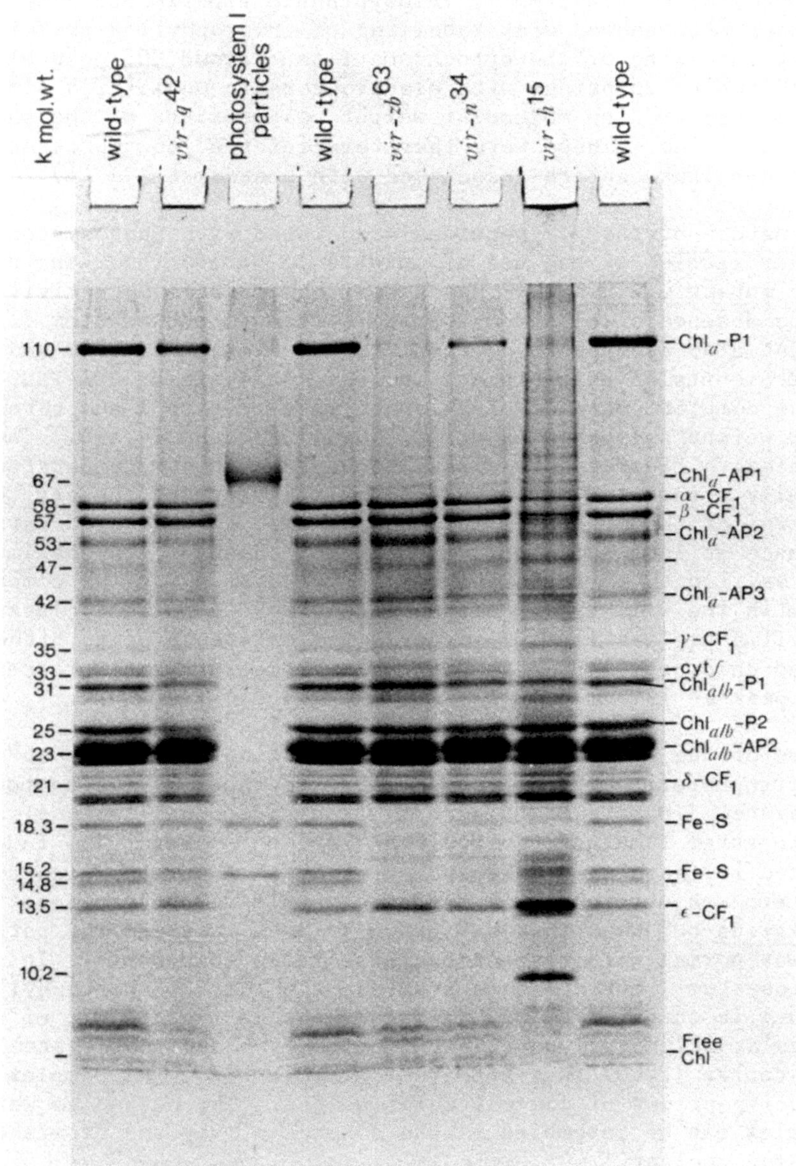

Figure 1. Thylakoid polypeptide pattern of wild-type and mutant barley compared with that of photosystem I particles. The same amount of chlorophyll was applied to each slot and the separation was carried out at 19°C on an 11-15% SDS-polyacrylamide gradient gel. The molecular weights assigned to the polypeptides in the low molecular weight region are inaccurate and only given as a guide.

lar weight is chlorophyll _a_-protein 1 and contains the reaction center (_P_700) of photosystem I. Biosynthetic experiments with the radioisotope ^{59}Fe showed weak labelling of chlorophyll _a_-protein 1 and strong labelling of the cytochrome _f_ band at 33,000 molecular weight and two polypeptides with electrophoretic mobilities similar to those of the two low molecular weight polypeptides of the photosystem I particles. These were therefore inferred to represent two of the iron-sulfur centers associated with photosystem I (4).

The nature of the polypeptides associated with photosystem I was further studied by the use of mutants in barley (5). The nuclear gene mutant _viridis-zb_^{63} has normal photosystem II activity but a complete absence of electron transport through photosystem I. This mutant also appears to be completely deficient in _P_700 and has decreased amounts of cytochrome _f_ and b_6. Analysis by SDS-PAGE showed the complete absence of chlorophyll _a_-protein 1 and three low molecular weight polypeptides at 18,300, 15,200 and 14,800. The long wavelength fluorescence emission observed at low temperatures is generally associated with the antennae pigments of photosystem I. The spectrum of the mutant _viridis-zb_^{63} showed an unaltered ratio of fluorescence at long and short wavelengths. However, the maximum of the long wavelength emission peak was shifted to 734 nm as compared to 740 nm in the wild-type. Since _viridis-zb_^{63} completely lacks chlorophyll _a_-protein 1, this suggests the existence of an hitherto undetected chlorophyll-protein with a low temperature fluorescence emission peak at 734 nm.

Three of the polypeptides which are missing in _viridis-zb_^{63} have electrophoretic mobilities identical to those of the components of photosystem I particles (Figure 1). An additional two nuclear gene photosystem I mutants, _viridis-n_^{34} and _-h_^{15}, were also found to be depleted in these same three polypeptides. A similar electrophoretic pattern was obtained from the cold sensitive nuclear gene mutant _tigrina-o_^{34} when this was grown at 20°C, whereas the pattern at 30°C was normal with respect to these three components. In all four photosystem I mutants here studied, the loss of chlorophyll _a_-protein 1 is accompanied by a similar loss in the content of three low molecular weight polypeptides. The chloroplast photosystem I reaction center (_P_700) thus appears to form an integral complex with several polypeptides of low molecular weight. The extent to which this complex can be assembled in the absence of any one of its components is yet unclear.

Photosystem II (PSII) vesicles were obtained using the two phase system of Albertsson (6) composed of 5.95% of dextran and polyethylene glycol. The PSII vesicles had a chlorophyll a/b ratio of 2.1. The low temperature fluorescence emission spectra of the different fractions obtained during the preparation of the PSII vesicles are shown on Figure 2. The PSII vesicles evolved 75 µmoles O_2/mg chlorophyll/hr with H_2O as electron donor and phenyl-_p_-benzo-

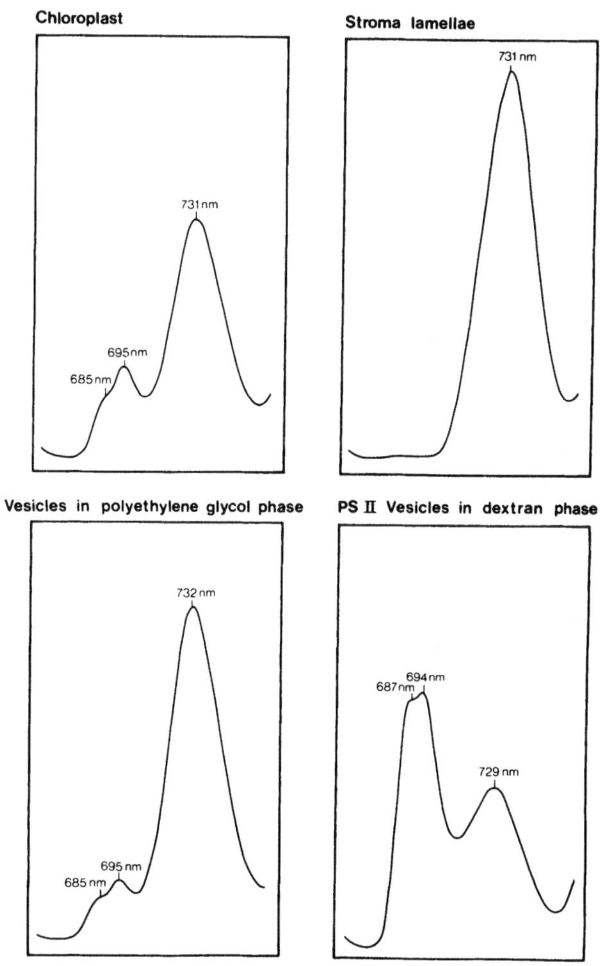

Figure 2. Low temperature fluorescence emission spectra of fractions obtained during preparation of the photosystem II vesicles.

quinone as acceptor. The PSII vesicle showed a net proton extrusion upon illumination corresponding to 6.9 nmoles H^+/mg chlorophyll in contrast to the chloroplasts which showed an alkalinization of the medium amounting to 46 nmoles H^+/mg chlorophyll. This indicates that the PSII vesicles are of opposite sidedness to the original thylakoid membrane. The PSII vesicles were enriched in cytochrome b-559HP and depleted of cytochromes b-563 and f. SDS-polyacrylamide gel electrophoresis revealed that the photosystem II vesicles contain four chlorophyll proteins, namely chlorophyll a-proteins 2 and 3, and chlorophyll a/b-proteins 1 and 2. Staining with Coomassie blue revealed an additional four polypeptides with molecular weights of 33,000, 22,000, 19,000 and 18,000 (7).

Reduced minus oxidized difference spectra showed the presence of high amounts of cytochrome b-559 in the PSII vesicle. To characterize the function of this cytochrome in photosystem II a method was developed for its isolation. The cytochrome was obtained from coupling factor depleted chloroplasts by extraction with 0.2% Triton and further purified by ion exchange chromatography on DEAE-Sepharose CL 6B. SDS-polyacrylamide gel electrophoresis revealed the presence of a low molecular weight component. When antibodies were raised either against purified cytochrome b-559 or against the low molecular weight band cut out from the gel, a cross reaction was observed in the 34,000 mol. wt. region at a position where no Coomassie blue stainable band was observed. Cytochrome b-559 isolated from etioplasts is much more labile than the protein isolated from chloroplasts. Thus, sonication or the presence of urea completely destroys the difference spectrum of cytochrome b-559 isolated from etioplasts whereas cytochrome b-559 isolated from chloroplasts is stable at these conditions (8).

The procedure of phase partitioning has previously been used to isolate inside-out vesicles from chloroplast grana but electrophoretic analysis showed that these vesicles contained polypeptides attributable to both photosystems (9). It is thought that the PSII vesicles obtained in this study are derived strictly from the appressed regions of the granal stacks due to the high pressures used in the French Press treatment and the conditions used strongly favoring stacking. The PSII vesicles obtained are sensitive to DCMU and must therefore contain the shielding protein which binds DCMU (10) and is reported to have a molecular weight of 32,000. Cytochrome b-559 has been reported to be composed of subunit polypeptides with molecular weights around 6,000 (11). No polypeptide band at this position was present in the vesicles. However, the isolation procedure for cytochrome b-559 here developed seems to indicate that this cytochrome has varying electrophoretic mobility depending on its degree of denaturation. The not completely denatured protein appears to have a very low staining affinity for Coomassie blue.

A comparison of the thylakoid polypeptide pattern of different

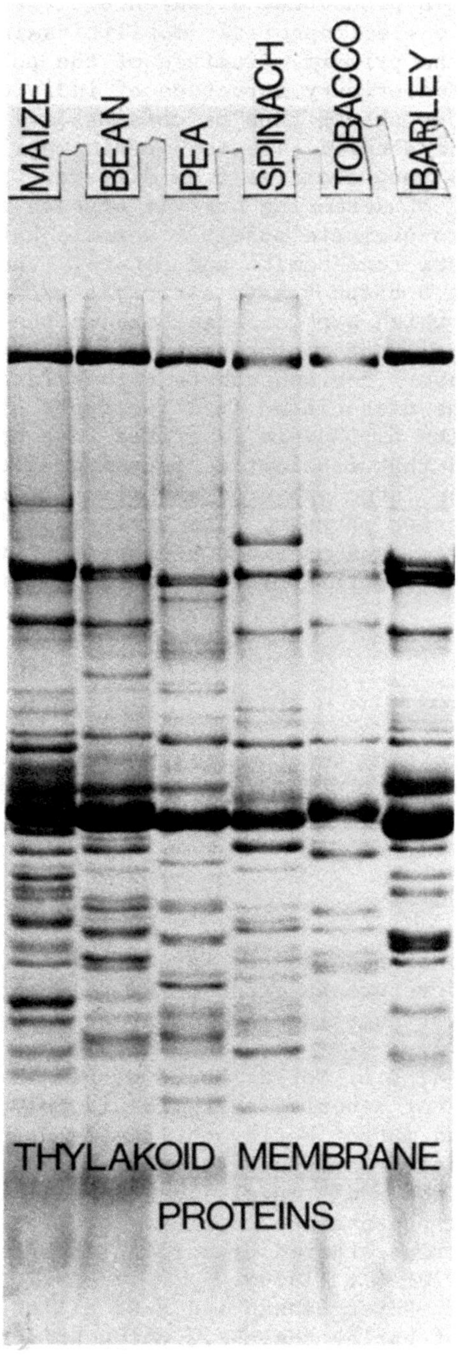

Figure 3. Thylakoid polypeptide pattern of different plant species as analyzed by SDS-polyacrylamide gel electrophoresis.

plant species reveals pronounced differences (Figure 3). These large alterations in electrophoretic mobilities indicate evolutionary divergence in the primary structure of the polypeptides. Induced alterations in the primary structure of individual photosynthetic membrane polypeptides should thus be obtainable in mutants of a single species. The altered electrophoretic mobilities observed for the same components when isolated from different plant species constitute an easy way of screening hybrids of such species. This approach was used to evaluate possible somatic hybrids obtained by fusion of protoplasts from tomato and potato. The marker enzyme used was ribulose-1,5-bisphosphate carboxylase/oxygenase, a soluble chloroplast enzyme which participates both in the photosynthetic carbon reduction cycle and in the photorespiratory carbon oxidation cycle (12). Ribulose-1,5-bisphosphate carboxylase isolated from higher plants can be dissociated into two types of subunits of which the large subunit (MW 55,000) is inherited in a Mendelian fashion (13). It was found that isoelectric focusing allows the separation of the small and the large subunits and also discriminates between the subunits of the two parents. The analysis of ribulose-1,5-bisphosphate carboxylase thus provided a convenient marker to demonstrate the hybrid nature of some of the plants obtained (14, 15).

ACKNOWLEDGEMENT

Financial support by the Commission of the European Communities within the solar energy program (Grant ESD-007-DK) is gratefully acknowledged.

REFERENCES

1. O. Machold, D. Simpson and G. Høyer-Hansen, Correlation between the freeze fracture appearance and polypeptide composition of thylakoid membranes in barley, Carlsberg Res. Commun. 42: 499-516 (1977).
2. G. Høyer-Hansen, B.L. Moller and L.C. Pan, Identification of coupling factor subunits in thylakoid polypeptide patterns of wild type and mutant barley thylakoids using crossed immunoelectrophoresis, Carlsberg Res. Commun. 44:337-351 (1979).
3. K. Sigrist-Nelson and A. Azzi, The proteolipid of chloroplast adenosine triphosphatase comples. Mobility, accessibility and interactions studied by a spin label technique, J. Biol. Chem. 254:4470-4474 (1979).
4. B.L. Møller, G. Høyer-Hansen and R.G. Hiller, Functional identification of barley thylakoid polypeptides resolved by SDS-polyacrylamide gel electrophoresis, in: "Proceedings of the Fifth International Congress on Photosynthesis," G. Akoyunoglou, ed., (in press) (1981).

5. R.G. Hiller, B.L. Møller and G. Høyer-Hansen, Characterization of six putative photosystem I mutants in barley, Carlsberg Res. Commun. 45:315-328 (1980).
6. P.A. Ålbertsson, Partition between polymer phases, J. Chromatogr. 159:11-122 (1978).
7. L.E.A. Henry and B.L. Møller, Polypeptide composition of an oxygen evolving photosystem II vesicle from spinach chloroplasts, Carlsberg Res. Commun. 46:227-242 (1981).
8. D. von Wettstein, B.L. Møller, F. Koenig, L.E.A. Henry and G. Høyer-Hansen, Polypeptides associated with photosystem II, in: "Proceedings Fifth Coordination Meeting," EC Solar Energy R & D Programme Series D, Cadarache, France, Vol. 1 (in press) (1981).
9. B. Andersson, H.E. Åkerlund and P.A. Ålbertsson, Light-induced reversible proton extrusion by spinach-chloroplast photosystem. II. Vesicles isolated by phase partitioning, FEBS Lett. 77:141-145 (1977).
10. A.K. Mattoo, U. Pick, H. Hoffman-Falk and M. Edelman, The rapidly metabolized 32,000-dalton polypeptide of the chloroplast is the "proteinaceous shield" regulating photosystem II electron transport and mediating diuron herbicide sensitivity, Proc. Natl. Acad. Sci USA 78:1572-1576 (1981).
11. H.S. Garewal and A.R. Wasserman, Triton X-100-4M urea as an extraction medium for membrane proteins. I. Purification of chloroplast cytochrome b-559, Biochemistry 13:4063-4079 (1974).
12. H.G. Lorimer, K.C. Woo, J.A. Berry and C.B. Osmond, The C_2 photorespiratory carbon oxidation cycle in leaves of higher plants: pathway and consequences, in: "Photosynthesis 1977," D.O. Hall, J. Coombs and T.W. Goodwin, eds., The Biochem. Society, London (1978).
13. K. Sakano, S.D. Kung and S.G. Wildman, Identification of several chloroplast DNA genes which code for the large subunit of Nicotiana Fraction 1 proteins, Molec. Gen. Genet. 130:91-97 (1974).
14. G. Melchers, M.D. Sacristán and A.A. Holder, Somatic hybrid plants of potato and tomato regenerated from fused protoplasts, Carlsberg Res. Commun. 43:203-218 (1978).
15. C. Poulsen, D. Porath, M.D. Sacristán and G. Melchers, Peptide mapping of the ribulose biphosphate carboxylase small subunit from the somatic hybrid of tomato and potato, Carlsberg Res. Commun. 45:249-267 (1980).

CURRENT STUDIES ON PLANT CELL TRANSFORMATION

D. Pental and E.C. Cocking

Plant Genetic Manipulation Group
Department of Botany
University of Nottingham
Nottingham NG7 2RD
United Kingdom

INTRODUCTION

As recently discussed (1) a generalized strategy for plant cell transformation requires the construction of a genetic vehicle which is capable either of integrating into the plant genome, or of independently replicating in the plant cell. The minimum attributes of such a vehicle would be the presence of a selectable plant marker and a site for integration of a passenger sequence.

THE USE OF THE TUMOR-INDUCING (TI) PLASMID OF *AGROBACTERIUM TUMEFACIENS* FOR GENETIC TRANSFORMATION OF PLANT CELLS

The use of the Ti plasmid as a natural vector for the genetic manipulation of higher plants has been suggested by several workers, and evidence has been presented that isolated Ti plasmid can transform plant protoplasts to phytohormone autotrophy, and that some of the resultant clones synthesize tumor-specific opines (1). However, unequivocal proof for transformation requires that the T-DNA is shown to be maintained in the transformed tissues derived from such protoplasts. Recently the presence has been demonstrated, by DNA-DNA hybridization, of DNA sequences homologous to the Ti plasmid in phytohormone-independent colonies derived from plasmid treated *Petunia* protoplasts (2).

Suspension cell protoplasts of *Petunia hybrida* were incubated with pTiACH5 using either glycine-NaOH buffer (pH 10.1) containing 20 mM $CaCl_2$ and 9% (w/v) mannitol, followed by mixing with 47% (w/v) polyethylene glycol 6000, or poly-L-ornithine (Mol. wt. 166,000) in 45 mM Na citrate buffer (pH 9.4) and 12% (w/v) mannitol. After nine months of subculturing, callus from eighteen hormone-independent clones from such plasmid treated protoplasts was harvested, and the DNA analyzed for the presence of Ti plasmid (pTiACH5) homologous sequences by DNA-DNA hybridization.

The transformants, analyzed by DNA-DNA hybridization, each contained pTiACH5 homologous sequences, but their hybridization patterns were not typical of that of authentic Petunia crown gall tissue induced by strain ACH5 of Agrobacterium. The results showed that a small fragment of pTiACH5, of approximately 2 Md, is stably maintained in each of the eighteen clones from plasmid treated protoplasts. It was suggested (2) that this DNA fragment could represent the smallest part of pTiACH5 necessary for protoplast transformation and the maintenance of phytohormone independent growth in derived colonies. There are, however, indications (3) from analysis of two calluses ensuing from Nicotiana tabacum leaf protoplasts incubated with octopine Ti-plasmid from the wild-type strain ACH5, in the presence of 40% w/v polyethylene glycol 6000 followed by a post-incubation with high Ca^{2+} concentration, that a larger part of the Ti-plasmid is integrated into the host genome than on infection with the whole bacterium. Whether these differences represent differing responses of different plant species must await further analysis, since only two calluses were analyzed in the tobacco protoplast experiments for the presence of Ti-plasmid DNA sequences (3).

Whilst it may be advantageous to utilize isolated Agrobacterium plasmids either from Agrobacterium tumefaciens or A. rhizogenes for the transformation of plant protoplasts, it may also be useful to use the normal infection processes of Agrobacteria to introduce Ti and Ri plasmids (4), into cells which have been formed from protoplasts regenerating a new cell wall (5) (Figure 1). The simplest way in which the infectivity of Agrobacteria can be used to introduce foreign DNA sequences into plants is to make tumors on intact plants, and then examine the tumor tissue for the presence and expression of foreign sequences. A method suggested for site-directed mutagenesis of nif genes of Rhizobium meliloti has been used for the insertion of Tn5, or other transposons, into different regions of the T-DNA of Agrobacterium (6, 7). Tumor cells produced by such Agrobacteria could then be screened for drug resistance and for the presence of Tn5 sequences. However, when intact plants are utilized it may be necessary to clone selected tissues at the single cell or protoplast level for adequate characterizations. A refined procedure for effective cloning of transformed cells is to infect protoplasts regenerating a cell wall with intact Agrobacterium. Cells regenerated from protoplasts can be transformed by Agrobacterium at the high frequency of 1-3% (8), and protoplast derived colonies can be screened for tumor characteristics and any foreign DNA inserted in the T-DNA.

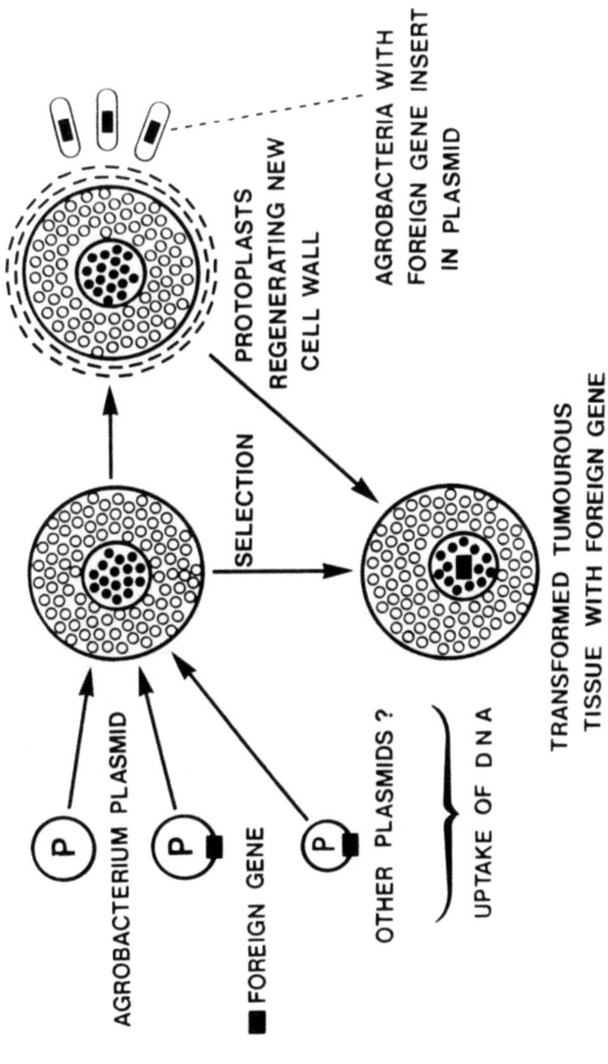

Figure 1. Strategy for Transformation using Protoplasts.

THE USE OF NITRATE REDUCTASE DEFICIENT MUTANTS FOR TRANSFORMATION ASSESSMENTS

There is a paucity of stable auxotrophs in higher plants, and one of the few such auxotrophs are nitrate reductase deficient lines of tobacco. These nia mutant lines, selected for chlorate resistance, lack nitrate reductase apoenzyme activity and cannot grow with nitrate as the sole source of nitrogen. Protoplasts can be readily isolated from the leaf mesophyll cells of nia-130 auxotrophs and show high plating efficiency, and cell colonies from the protoplasts regenerate into complete plants (9). Reconstruction experiments in which one wild type colony was introduced into 4×10^4 mutant colonies and recovered showed that selection for rare events of correction of nitrate reductase deficiency by back mutation, genetic transformation or cell fusion is possible. No nitrate utilizing revertants were observed when 1.8×10^7 nia-130 colonies were transferred from amino acid medium to selection medium containing nitrate as the sole source of nitrogen (9). It was suggested that this nia mutant protoplast system could be used to isolate genomic sequences of nitrate reductase of *Nicotiana*, or other higher plants, by functional complementation of mutant protoplasts by recombinant plasmids from a random genomic library of wild-type plant DNA.

One of the objectives of the fusion of protoplasts of widely divergent plant species is to attempt to incorporate into the recipient species some limited genetic attribute of a donor species. In the case of mammalian cells it is possible by x-ray irradiation of the parental cells, and other chromosome-destabilizing procedures, to bring about directional chromosome elimination after fusion; and there are indications that comparable elimination, with the retention of reconstructed chromosomes or a few genes, may also be possible when using protoplasts (1). When very high doses of irradiation are employed (20 to 100,000 rad), and the irradiated protoplasts are fused with normal protoplasts, the transfer of a few genes into a suitable recipient species may be possible (5). For adequate assessment suitable auxotrophic mutants capable of regeneration into plants are required; and this nitrate reductase deficient mutant of tobacco should be capable of detecting any such gene transfer from, say soybean into tobacco even if it occurs at a very low frequency.

FUSION-MEDIATED GENE TRANSFER BETWEEN PLANTS

Additional methods for transferring only part of the plant genome, both nuclear and cytoplasmic, between diverse species are highly desirable; and protoplast fusions provide novel opportunities for obtaining heterozygosity of extra chromosomal genes. As far as chloroplasts are concerned the characterization of chloroplast organelles indicate that the two types of parental chloroplasts in the somatic hybrid eventually sort out for one or the other, or in some cases only one of the parental types. The sorting out process of the parental chloroplasts in the somatic hybrid population is likely to be greatly influenced by many factors encountered during protoplast fusion and subsequent selection and regeneration of somatic hybrids. For instance, the analysis of the chloroplast DNA (cp DNA) by various restriction endonucleases of the three somatic hybrids previously produced by the fusion of protoplasts from wild-type mesophyll cells of *Petunia parodii* with protoplasts of *P. hybrida*, *P. inflata* and *P. parviflora* showed the presence of only cp DNA of *P. parodii* in the somatic hybrids (10). It was suggested that this unidirectional sorting out phenomenon, in favor of *P. parodii* chloroplasts, was most probably the result of the strong selections in favor of *P. parodii* used in the production of these somatic hybrids. It was not possible to determine the amount of chloroplast heterozygosity from the Fraction 1 profiles because the large subunit polypeptides (coded by chloroplast DNA) of the Fraction 1 protein are identical in these four *Petunia* species (11). No recombination between cp DNAs was detected in any of the three hybrids indicating that cp DNA recombinations must be an infrequent event, if it occurs at all. Opportunities for increasing cytoplasmic variability by protoplast fusions may be greater with mitochondria than with chloroplasts. Less detailed analysis has been carried out on mitochondrial DNA. However, there is some evidence for mitochondrial recombination in cytoplasmic hybrids of *N. tabacum* obtained by protoplast fusion (12). The heterogeneity of most mitochondrial DNAs isolated from *Nicotianas* is complex, and recent work suggests that those from some *Brassica* species are less complex, making *Brassicas* preferable for this type of analysis (13).

Microcell-mediated chromosome transfer has proved to be a promising approach for chromosome assignment of genes for mammalian cells. Nuclear material of cultured mammalian cells is fragmented by treatment of the cells with colchicine. The fragments consist of a limited amount of genetic material encapsulated in a cell membrane. Microcells are fused with recipient cells using routine somatic hybridization methods (14). Until recently there seemed little possibility that any comparable system could be developed for cultured plant cells apart from the production of subprotoplasts and cytoplasts (15). However, the discovery of plant microplast systems could lead to the development of such a microcell system, if fragmentation of the nuclear material can be induced with colchicine. Microplasts, surrounded by an inner membrane of the cell, most probably derived from the tonoplast, can be readily isolated by rupturing auxin-induced highly vacuolated thin-walled callus cells of several plant species (16). Even if fragmentation of nuclear material cannot readily be achieved, such enucleate subprotoplasts and microplasts could be used for the transfer of a range of cytoplasmic factors by fusion with nucleated protoplasts. Transfer of cytoplasmic based male sterility, and herbicide resistance, could readily be achieved between sexually incompatible species (5) (Figure 2).

FUSION MEDIATED GENE TRANSFER BETWEEN PLANT AND MICROBIAL SYSTEMS

The availability of plant protoplasts, and the development of procedures for their fusion, coupled with the finding that fusogens suitable for protoplast fusion are also suitable for animal cell and microbial protoplast fusions (17), has highlighted the possibilities of gene transfer by fusion between plant and microbial protoplasts and between animal cells and plant protoplasts. A significant recent development has been the introduction of Ti plasmid into *Vinca rosea* protoplasts by fusion and endocytosis with spheroplasts of *Agrobacterium tumefaciens* (18). These fusion studies have shown that, because of the Ti plasmid is probably introduced into the cytoplasm of the plant protoplasts in an intact state, the frequency of transformation is increased in comparison with the use of isolated plasmid. In this context it is particularly noteworthy that monkey cells were transformed by *Escherichia coli* harboring a recombinant plasmid of simian virus 40 (19). Infection frequency was greatly increased by fusion of lysozyme-treated bacteria with monkey cells in the presence of polyethylene glycol.

Parallel assessments will need to be undertaken utilizing *E. coli* spheroplasts harboring cloned Tn5 and other transposon recombinants at high copy number. These will need to be incubated with plant protoplasts in the presence of suitable fusogens. Selection for enhanced resistance to the appropriate antibiotic can then be undertaken.

As recently discussed (20) the limiting factor in transformation may be the extent of stabilization of the transforming DNA in the recipient cell. Protoplast fusions with bacterial protoplasts present large quantities of plasmid DNA to the recipient plant protoplasts. These workers have suggested that this excess of plasmid, or bacterial DNA, enhances the frequency of the transformation event. Whilst liposomes (21) have been effectively utilized for such plasmid-mediated transformations in protoplasts of microorganisms (22), protoplast fusions are themselves likely to prove to be a highly efficient method for effecting gene transfer for genetic engineering in higher plants. Moreover, there is no need to isolate and purify DNA.

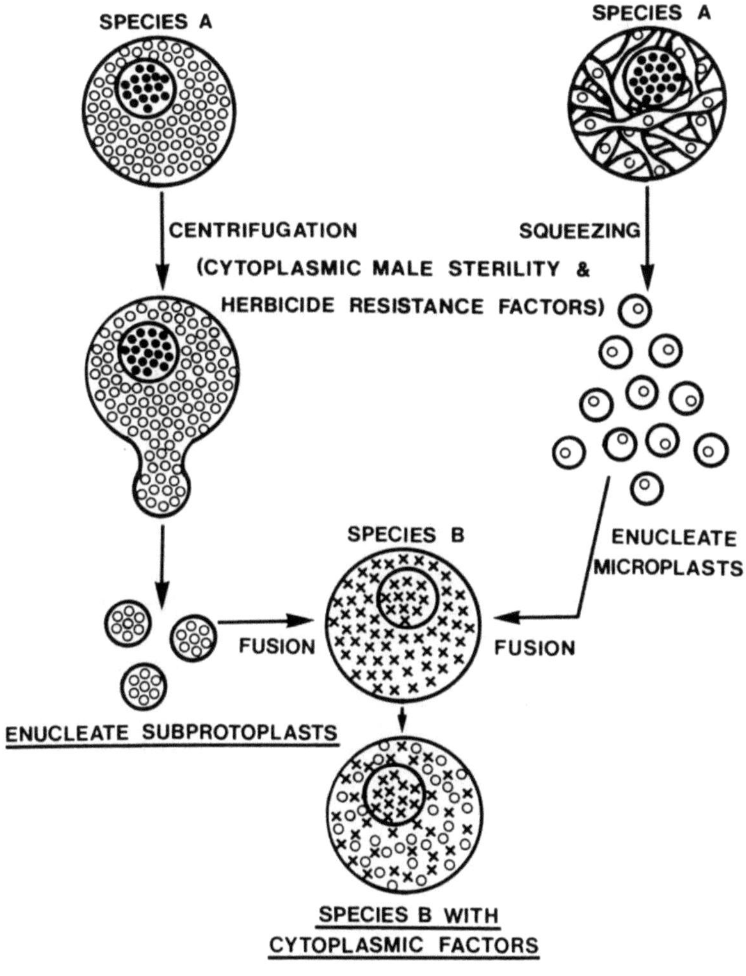

Figure 2. Transfer of Cytoplasmic Factors by Fusion.

GENERAL PROSPECTS

As has been emphasized it is likely that in the long term positive selection markers which can be added to a normal prototrophic genetic background will be preferred for the selection of transformants (1). An ideal genetic marker would be a drug or heavy metal resistance gene. Transformation of yeast protoplasts to a state of resistance to G418 has been achieved utilizing a plasmid carrying the transposable element Tn601, which renders bacteria resistant to the antibiotic (23). Transformation was unstable during mitosis on non-selective media because the DNA does not become integrated in the genome. It is particularly noteworthy that animal cells have been efficiently transformed to G418 resistance by part of the Tn5 transposon linked to a viral promoter (24). As far as plant cells are concerned the first requirement is to establish that comparable transformation can be achieved, and currently we are utilizing protoplasts isolated from leaves of *Nicotiana tabacum*, and from suspension cultures of *Salpiglossis sinuata*, for such assessments; and we now intend to extend these assessments to a range of crop species utilizing a range of protoplast systems isolated from leaf, cotyledon and root. Comparisons between meristematic type protoplasts [such as those from cotyledons (25) and roots (26)] and those from differentiated leaf cells are being undertaken in case there is a difference in competence for transformation. The availability of such positive selection markers would enable any co-transformation with two physically unlinked genes to be assessed in plants, comparable to the finding that co-transformed animal cells could be identified and isolated when one of the genes codes for a selectable marker such as thymidine kinase (27). The stability of any such transformations will also need to be determined, particularly whether the transformants are stable to meiotic stress in any plants regenerated from such genetically manipulated plant cell and protoplast systems. At present it is important to approach these transformation assessments on as broad a front as possible utilizing a wide range of plant species and a wide range of delivery systems.

REFERENCES

1. E.C. Cocking, M.R. Davey, D. Pental and J.B. Power, Aspects of plant genetic manipulation, *Nature* 293:265 (1981).
2. J. Draper, M.R. Davey, J.P. Freeman, E.C. Cocking and B.J. Cox, Ti plasmid homologous sequences present in tissues from *Agrobacterium* plasmid-transformed *Petunia* protoplasts, *Plant Cell Physiol*. 23(3):92 (1982).
3. F.A. Krens, L. Molendijk, G.J. Wullems and R.A. Schilperoort, *In vitro* transformation of plant protoplasts with Ti-plsamid DNA, *Nature* 298:72 (1982).
4. M-D. Chilton, D.A. Tepfer, A. Petit, C. David, F. Casse-Delbart and J. Tempe, *Agrobacterium rhizogenes* inserts T-DNA into the

genomes of the host plant root cells, Nature 295:432 (1982).
5. E.C. Cocking, Opportunities from the use of protoplasts, Phil. Trans. R. Soc. Lond. B 292:557 (1981).
6. D.J. Garfinkel, R.B. Simpson, L.W. Ream, F.F. White, M.P. Gordon and E.W. Nester, Genetic analysis of crown gall: fine structure map of the T-DNA by site-directed mutagenesis, Cell 27:143 (1981).
7. A.J.M. Matze and M-D. Chilton, Site-specific insertion of genes into T-DNA of the Agrobacterium tumor-inducing plasmid: an approach to genetic engineering of high plant cells, J. of Mol. and Appl. Genetics 1:39 (1981).
8. M.R. Davey, J.P. Freeman, J. Draper and E.C. Cocking, The use of protoplasts for transformation studies with Agrobacterium tumefaciens and its plasmids, in: "Tissue Culture Methods for Plant Pathologists," D.S. Ingram and J.P. Helgeson, eds., Blackwell, Oxford (1980).
9. D. Pental, S. Cooper-Bland, K. Harding, E.C. Cocking and A.J. Muller, Cultural studies on nitrate reductase deficient Nicotiana tabacum mutant protoplasts, Z. Pflanzenphysiol. 105:219 (1982).
10. A. Kumar, E.C. Cocking, W.A. Bovenberg and A.K. Kool, Restriction endonuclease analysis of chloroplast DNA in interspecies hybrids of Petunia, Theor. and Appl. Genet. (in press) (1982).
11. A. Kumar, D. Wilson and E.C. Cocking, Polypeptide composition of Fraction 1 protein of the somatic hybrid between P. parodii and P. parviflora, Biochem. Genetics 19:255 (1981).
12. G. Belliard, F. Vedel and G. Pelletier, Mitochondrial recombination in cytoplasmic hybrids of Nicotiana tabacum by protoplast fusion, Nature 281:401 (1979).
13. P. Lebacq and F. Vedel, Sal 1 restriction enzyme analysis of chloroplast and mitochondrial DNAs in the genus Brassica, Plant Sci. Letts. 23:1 (1981).
14. J.A. Wright, W.H. Lewis, C.L.J. Parfett, Somatic cell genetics: a review of drug resistance, lectin resistance and gene transfer in mammalian cells in culture, Can. J. Genet. and Cytol. 22:443 (1980).
15. H. Lorz, J. Paszhowski, C. Dierks-Ventling and I. Potrykus, Isolation and characterization of cytoplasts and miniprotoplasts derived from protoplasts of cultured cells, Physiologia Plantarum 53:386 (1981).
16. P.C. Bilkey, M.R. Davey and E.C. Cocking, Isolation, origin and properties of enucleate plant microplasts, Protoplasma 110:147 (1982).
17. L. Ferenczy, Microbial protoplast fusion, in: "Genetics as a Tool in Microbiology," S.W. Glover and D.A. Hopwood, eds., Cambridge University Press, Cambridge, 31st Symposium Society for Exp. Biol. (1981).
18. S. Hasezawar, T. Nagata and K. Syono, Transformation of Vinca protoplasts mediated by Agrobacterium spheroplasts, Mol. Gen.

Genet. 182:206 (1981).

19. W. Schaffner, Direct transfer of cloned genes from bacteria to mammalian cells, Proc. Natl. Acad. Sci. USA 77:2163 (1980).
20. R.M. Sandri-Goldin, A.L. Goldin, M. Levine and J.C. Glorioso, High-frequency transfer of cloned herpes simplex virus type 1 sequences to mammalian cells, Mole. and Cell Biology 1:743 (1981).
21. P. Nicholls, Liposomes - as artificial organelles, topochemical matrices, and therapeutic carrier systems, in: "International Review of Cytology," Suppl. 12:237 (1981).
22. J.F. Makins and G. Holt, Liposome mediated transformation of Streptomyces by chromosomal DNA, Nature 293:671 (1981).
23. A. Jiminez and J. Davies, Expression of a transposable antibiotic resistance element in Saccharomyces, Nature 287:869 (1980).
24. F. Coleber-Garapin, F. Horodniceanu, P. Kourilisky and A.-C. Garapin, A new dormant hybrid selection marker for higher embryotic cells, J. Molec. Biol. 150:1 (1981).
25. T.-Y. Lu, D. Pental and E.C. Cocking, Plant regeneration from cotyledon protoplasts, Z. Pflanzenphysiol. (in press) (1982).
26. Z.-H. Xu, M.R. Davey and E.C. Cocking, Isolation and sustained division of Phaseolus aureus (Mung bean) root protoplasts, Z. Pflanzenphysiol. 104:289 (1981).
27. M. Wigler, R. Sweet, G.K. Sim, B. Wold, A. Pellicer, E. Lacey, T. Maniatis, S. Silverstein and R. Axel, Transformation of mammalian cells with genes from prokaryotes and eukaryotes, Cell 16:177 (1979).

CROWN GALL: A MODEL SYSTEM FOR GENETIC MANIPULATION OF HIGHER PLANTS

G.J. Wullems, F.A. Krens, G. Ooms* and R.A. Schilperoort

Department of Biochemistry
State University of Leiden
Wassenaarseweg 64
2333 AL Leiden, THE NETHERLANDS

*Rothamsted Experimental Station
Harpenden,
Herts AL5 2JQ, UNITED KINGDOM

INTRODUCTION

The recent knowledge of the plant disease crown gall - caused by the soil bacterium *Agrobacterium tumefaciens* - has led to the recognition of this phenomenon as a model system for transformation of plants, e.g. to provide plants with foreign DNA.

The pathogenic *A. tumefaciens*, belonging to the family of *Rhizobiaceae*, induces the plant tumor crown gall on dicotyledonous plant species (1-6). The agentia, responsible for this tumor induction are plasmids, called tumor-inducing or Ti plasmids, present in the bacteria. Besides Ti plasmids being essential for tumor induction, some of them code for the production of tumor specific compounds. Amongst those there are amino acid derivatives (octopine type and nopaline type compounds). These compounds are synthesized by the plasmid coded enzymes lysopine dehydrogenase (LpDH) and nopaline dehydrogenase (NpDH) respectively. Other compounds are agropine and agrocinopines, which are sugar derivatives. The Ti plasmids code also for catabolic enzymes that convert tumor specific compounds into products that may serve as a carbon and nitrogen source for the bacterium. Therefore, the interaction between the bacterium and the plant cells results in a plant tumor which is for the benefit of the bacterium. The main properties of plant tumors are phytohormone independent growth (hormone autotrophy) of the tissue when brought in tissue culture; and the presence of opines (octopine, nopaline or agropine) in the tumor cells. These properties can be seen as phenotypic traits which are the result of the transformed nature of the tumor cells. This transformed nature is due to the integration of a segment of the Ti plasmid in the genome of the plant cells (7). This segment is called the T-region. From genetic analysis of mutated Ti plasmids, it is concluded that the T-region contains genes that, after transfer to plant cells, are involved in opine synthesis (8, 9) and in the regulation of cellular growth and differentiation (10, 11, 12). The detection of DNA sequences in tumor DNA that are homologous to T-region DNA came from Southern blotting analysis (7, 13, 14, 15). These tumor DNA sequences are called T-DNA. By preparing DNA from purified nuclei and from chloroplasts and mitochondria from tumor cells it was shown that the T-DNA is located only in the nucleus (16, 17). By comparing the organization of various T-DNA's (13, 14, 18) it has been found that crown gall cells usually contain a well defined T-DNA associated with different plant DNA sequences. T-DNA from nopaline tumors consists of a limited number of non-interrupted stretches of DNA (14), whereas T-DNA from octopine tumors might be composed of two separate stretches of T-DNA (13), which in the T-region of the Ti plasmid are linked together. These stretches are referred to as T_R- and T_L-DNA. In the octopine tumors, T_L-DNA is present as a "core" T-DNA in limited copies, whereas T_R-DNA, when present, exists in multiple copies and can vary in size.

The first evidence that the T-DNA is transcribed within the plant cell was obtained from hybridization experiments between total cellular RNA and T-region fragments (19, 20). More extended experiments involved the _in vitro_ transcription of cDNA prepared from crown gall derived nuclear RNA (21, 22, 23). The resulting RNA was then hybridized to T-DNA fragments. These studies have indicated that the T-DNA is transcribed over its entire length. That the transcription of T-DNA results in functional mRNA's was shown by the detection of a protein with the molecular weight of LpDH. This protein was specifically recognized by antibodies raised against purified LpDH (24).

Regarding these properties of the crown gall system, it is likely to consider Ti plasmids as natural vector systems which could be used, via cotransformation, for the promotion of the integration and expressions of foreign genes in plant cells. Therefore, the Ti plasmid should first be tamed in such a way that genes, involved in the neoplastic growth of the tissue, are suppressed. It is likely to speculate that the T-DNA is involved in the action of phytohormones that regulated growth and differentiation of plants. For reviews describing the physiological action of phytohormones see (25, 26, 27, 28). Concerning the involvement of T-DNA in these processes, most attention is focussed on the phytohormones belonging to the groups of auxins and cytokinins. Some physiological effects of auxins are the elongation of cells, the development of roots and inhibition of lateral shoot formation (apical dominance). It was shown already in 1928 by Went (29) that auxins are produced in shoot tips. Physiological effects of cytokinins can oppose those of auxins: inhibition of root development, stimulation of lateral shoot formation. Cytokinins are produced particularly in the root system. The addition to culture medium of relatively high concentrations of cytokinins versus auxins gives rise to shoot development on tobacco callus, whereas relatively high concentrations of auxins versus cytokinins result in root development. Balanced concentrations of both auxins and cytokinins result in the stimulation of undifferentiated callus growth. The relation between the action of these phytohormones and crown gall is based on two kinds of observations. Firstly, it is shown that crown galls maintained in tissue culture produce phytohormones (30, 31). Secondly, studies with mutated Ti plasmids have demonstrated that separated genes on T-DNA promote auxin- and cytokinin-like activities in tumor cells (32). *A. tumefaciens* mutant strain LBA 4060 harbors a plasmid with an insertion of an *A. tumefaciens* IS element, IS 60, in the left part of the T-region. This mutation is also found present in the T-DNA of tumors induced by this strain. Such tumors give rise to differentiation of shoots. A second Ti plasmid insertion mutant *A. tumefaciens* strain, LB 4210, with a Tn 904 transposon in the center of the T-region, induces tumors that exhibit root development on tobacco plants. Co-infection with equal amounts of both types of mutants has resulted in the formation of an undifferentiated wild type tumor. That the "abnormal" tumors can be correlated with the action of phytohormones is demonstrated by the observation that in the case the mutant LBA 4060, infection in the presence of exogenously supplied auxin results in a more or less normal tumor, whereas cytokinin supplied with infection by LBA 4210 has the same effect. The results described here suggest that it might be possible to mutate the Ti plasmid in such a way that the morphogenic potential of the transformed tissue can be restored.

TUMOR INDUCTION ON THE SINGLE CELL LEVEL

The properties of transformed crown gall cells, in particular the phytohormone independent growth, allow the selection for transformed cells present in a population of normal cells. With the development of methods to isolate single plant cells as protoplasts and to culture these protoplasts on culture media, resulting in the regeneration of plants that can be used in breeding programs, different strategies have been developed, based on the crown gall system, to obtain transformation of single cells. The phytohormone auxotrophy of plant cells might fill the gap that exists by the sparse occurence of other stable autotrophs in higher plants. On the other hand, mutant plant tissues like chlorophyll-deficient mutants (33), drug resistant mutants (34) and mutants carrying metabolic deficiencies (35) could be very useful in showing the applicability of transformation systems.

Transformation of Plant Cells by *Agrobacterium tumefaciens*

This approach of transforming isolated plant cells is based on the normal infective process of *A. tumefaciens* to introduce Ti-plasmid DNA into plant cells. Presumably based on the existence of a wound healing reaction to supply adequate conditions for infectivity, protoplasts, to be transformed through infection with *A. tumefaciens*, must have generated a new cell wall (36, 37) for the infection process to be effective. Regarding tobacco protoplasts, these conditions are present three days after the isolation. Mixed incubation of such protoplasts with virulent *A. tumefaciens* will result in aggregation of the protoplasts and the bacteria, presumably due to attachment of the bacteria to the partly synthesized new cell wall of the protoplasts. After growing cell colonies, derived from protoplasts treated in this way (in culture medium with phytohormones), a fraction of the cell colonies will continue its growth. Such colonies can be cultured continuously on hormone free medium. In this way we have induced hormone autotrophic calli with agrobacteria of the octopine- and the nopaline-type, and with agrobacteria harboring a cointegrate plasmid consisting of an octopine and nopaline Ti-plasmid. Tumorous tissues can be induced with a frequency of 1 to 0.1% of the original population of protoplasts. The characterization of these presumed transformants was based on the expression of specific crown gall tumor markers as hormone autotrophy (Aut) and opine synthesis (Ocs or Nos). In Table I the various phenotypes that have been characterized are summarized. These results are a clear demonstration of the segregation behavior of crown gall tumors induced *in vivo* on plant stems. Crown galls induced *in vivo* by a certain bacterial strain in general express the same phenotype. This difference with *in vitro* tumors might be explained by the fact that the *in vitro* tumors originate from one single transformed cell. This homogenous composition was shown by subcloning experiments via protoplast isolation. The *in vivo* tumors, however, might consist of a mixed population of different types of tumor cells, which all contribute to one resulting phenotype in which both the hormone autotrophy and the opine synthesis are expressed. Evidence for a mixed composition of *in vivo* tumors was shown by Ooms *et al*. (38).

Another interesting property of the *in vitro* transformants is the high morphogenic potential of some of those tumors. In general about 50% of the transformants induced by octopine bacteria showed regeneration of shoots, whereas regeneration of shoots has not been observed on *in vivo* octopine type tumors induced by the same bacteria on the same plants from which the protoplasts were isolated. The percentage of regeneration on nopaline type tumors was even higher (80%). Regenerants have been derived both from opine positive and opine negative transformants. In fact, the regenerants always have the same phenotype as the original callus tissue.

Although the presence of opines in phytohormone independent

Table 1. Summary of Phenotypes Derived Following Transformation of Tobacco Protoplasts by Different A. tumefaciens Strains.

Type of bacteria	Phenotype	Frequency (%)
octopine	Aut^+, Ocs^+	69
	Aut^+, Ocs^-	30
	Aut^-, Ocs^+	1
nopaline	Aut^+, Nos^+	77
	Aut^+, Nos^-	33
oc-nop cointegrate	Aut^+, Ocs^+, Nos^+	13
	Aut^+, Ocs^+, Nos^-	33
	Aut^+, Ocs^-, Nos^+	33
	Aut^+, Ocs^-, Nos^-	17
	Aut^-, Ocs^+, Nos^-	4

Aut: phytohormone autotrophy
Ocs: octopine synthesis
Nos: nopaline synthesis

growing tissues is a good criterion for genuine transformation, more direct evidence for this was the demonstration that T-DNA is present in the DNA of the tissues. These results were derived from hybridizations between cloned T-region fragments and Southern blots of DNA from protoplast derived transformants. The architecture of the T-DNA in octopine type transformants is similar with the T-DNA found in crown gall tissues induced on plant stems (in vivo tumors). It was found that, regarding the T-DNA, in vivo tumors are composed of a (limited) number of subpopulations, whereas in vitro derived transformants showed to be homogenous. The same conclusion could also be drawn from phenotypic analysis of subclones, derived via protoplast isolation. A detailed description of these experiments has been published elsewhere (37, 38).

In vitro Transformation of Plant Protoplasts by Ti-Plasmid DNA

Transformation of eukaryotic cells with foreign DNA has been increasingly successful in the past years. In particular in mammalian cells various prokaryotic and eukaryotic genes have been introduced into the genome where they are expressed (39, 40, 41). These developments have stimulated attempts to achieve also DNA mediated transformation of plant cells. With the availability of plant cells without a cell wall (protoplasts) it became possible to overcome a major obstruction that could hamper the uptake of added DNA. The most recent attempts to transform plant protoplasts by exogenous DNA are based on encapsulation of the DNA by liposomes (42) or on fusion like procedures (44). This last approach was based on a method to induce fusion of normal tobacco protoplasts and protoplasts from crown gall tissue with polyethyleneglycol (43). The transformation procedure included an incubation of tobacco protoplasts with octopine type Ti-plasmid DNA in the presence of PEG, followed by a post-incubation with high Ca^{++} concentration (44). After a pre-selective culture of treated cells in complete medium, selection of presumed transformants was achieved by culturing cell colonies in hormone free medium. In this way, we were able to isolate phytohormone independent calli in a frequency of approximately 0.01% of the original number of protoplasts. The selected calli of one such experiment were characterized concerning their phenotypic properties: hormone autotrophy (Aut) and octopine synthesis (Ocs). It was found that some lines were Aut^+, Ocs^+, others were Aut^+, Ocs^- whereas one line was Aut^-, Ocs^+. The distribution of these phenotypes was 50%, 41% and 9% respectively, but concerning the limited number of transformants analyzed so far, these numbers are rather premature. Nevertheless these data again show that segregation of these tumor markers occurs. In this respect, those pTi-transformants resemble the bacterial transformants described earlier.

Further evidence for the transformed nature of these pTi-transformants came from the demonstration that Ti-plasmid derived DNA sequences (T-DNA) were present in transformed callus tissue DNA (44). The analysis carried out so far for two transformants, both Aut^+, Ocs^+, has shown that the contiguous SmaI fragments 16a and 10c, composing the "core" T-DNA in octopine *in vivo* tumors, were present. In addition fragment SmaI - 17, located to the left side of SmaI - 16a, was also present. This fragment is not part of the "core" T-DNA. The presence of this fragment in these transformants might indicate that by transformation with pTi-DNA a larger part of the Ti-plasmid becomes integrated into the host genome than what is generally found after infection with the bacterium. Whether or not this is true has to be demonstrated by a more extended T-DNA analysis of other tissues. Nevertheless these data conclusively demonstrate that the Ti-plasmid is indeed the tumor-inducing principle, whose existence was proposed already in 1956 by Braun (45), and that *A. tumefaciens* itself is not a prerequisite to obtain transformation. This conclusion is important if one realizes the limited host range of *Agrobacterium*. As far as known the host range of *A. tumefaciens* is limited to dicotyledonous plants (46). The failure of *A. tumefaciens* to induce crown gall tumors, and so transformation, on monocotyledonous plants is thought to be due to the absence of adherence of the bacterium to the plant cell wall. If this is indeed the barrier to achieve transformation of monocots, it presumably might be overcome by the use of isolated Ti-plasmid DNA.

REGENERATION OF PLANTS FROM IN VITRO TRANSFORMED PLANT PROTOPLASTS

Differentiation of Shoots

If the aim of transformation experiments is the introduction of foreign genes in plants, it is of great importance that the neoplastic growth of the transformants is suppressed, resulting in a restoration of the morphogenic potential of the tissues. Until recently, spontaneous regeneration of shoots from which transformed plants could be generated seemed to be restricted to nopaline type crown gall tissues. Such regenerants have been described extensively by Braun and coworkers (47, 48, 49, 50). These regenerants, that were obtained from crown gall teratomata induced by *A. tumefaciens* strain T 37 on tobacco plants, had retained T-DNA sequences (51). However, haploid cells derived from anthers as well as diploid cells of the F1 generation have recovered from the tumorous state(49, 51). The potential use of the Ti-plasmid as a vector for introducing new genes into plants would be severely hampered if T-DNA sequences were lost in meiosis and so not transmitted to F1 progeny. As was described in a previous section, octopine type tumors can also regenerate shoots, if they were induced by certain mutant bacterial strains like LBA 4060 (32). The mutation in strain LBA 4060 has been located in the left part of the T-DNA (T_L-DNA). Because of the altered morphology of the tumors induced by this strain, the site of the mutation is supposed to be involved in cytokinin-like activities leading to shoot development. Comparable results have been observed by Schell and coworkers (52) with an *A. tumefaciens* strain carrying a mutation in the same region as the mutation in the mutant strain LBA 4060. With this mutant they were also able to obtain octopine synthesizing shoots. Although these data demonstrate that octopine tumors can also regenerate shoots, the frequency in which *in vivo* octopine tumors regenerate remains low in contrast with the demonstrated shooting potential of *in vitro* tumors.

Morphogenic tumor tissues, capable of differentiating into shoots have been obtained following three different approaches which are schematically represented in Fig. 1. These three approaches have the common feature that they all started on the single cell level: (1) transformation of three day old tobacco protoplast with *Agrobacterium tumefaciens* (36, 37, 53); (2) transformation of freshly isolated tobacco protoplasts with Ti-plasmid DNA (44); (3) fusion of normal tobacco protoplasts with protoplasts isolated from crown gall tissues (43).

The third approach involved PEG induced fusion of mesophyll protoplasts from a streptomycin resistant mutant tobacco plant (SR1) with crown gall protoplasts. From these experiments we were able to select somatic hybrid callus tissues in which tumor markers (Aut, Ocs, T-DNA) and markers specific for the SR1 (streptomycin resistance) were expressed. In general the behavior and properties of

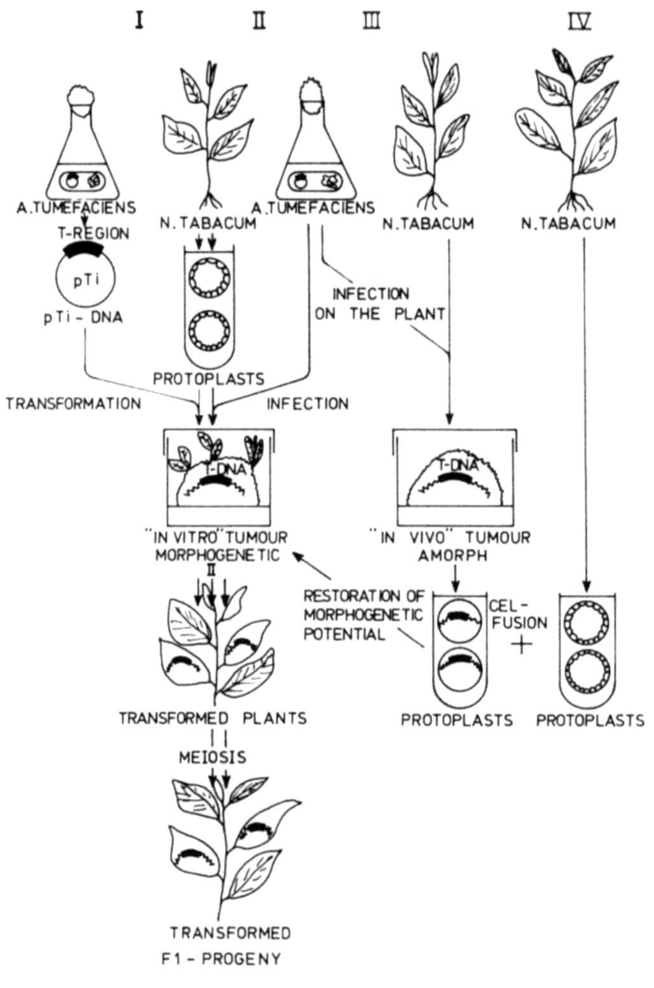

Figure 1. Different experimental systems for transformation of N. tabacum using Agrobacterium tumefaciens, its Ti-plasmid DNA and somatic hybridization. The systems I, II and III, based on transformation on the single cell level, result in the development of transformed plants. The transforming DNA is transmitted through meiosis in a Mendelian way.

these somatic hybrid tumor tissues was similar with the transformants (53). The shoots that appeared on the various tumors were cultured separately from the tumors on hormone free medium. These shoots could be distinguished from control tobacco plants by the absence of roots, their poor growth in size and their relatively thick stems and leaves. Additional markers were phytohormone autotrophy of shoot derived protoplasts when brought back into tissue culture, synthesis of opines (octopine or nopaline) and insusceptibility of the shoots for infection with A. tumefaciens. It was shown by protoplast isolation that individual shoots consisted of a homogeneous population of cells, regarding the markers mentioned above. In addition it was shown that the markers were stably present and were retained also under non-selective conditions. Southern blot hybridization experiments have demonstrated that in such tumor derived shoots, analyzed so far, T-DNA sequences were present. In octopine containing shoots from bacterial transformants and in octopine containing shoots of somatic hybrid origin, an internal fragment of the T-DNA was detected. This fragment, EcoRI fragment 7, is part of the "core" T-DNA which is found in most octopine crown gall tumors. T-DNA sequences located on the left side of this fragment 7 have not yet been detected in those shoots. If sequences from that position indeed are absent, it is likely to explain the regeneration of shoots in terms of auxin- and cytokinin-like activities, controlled by T-DNA genes. As we have seen from experiments with Ti-plasmids mutated in this area of the T-region, a cytokinin-like effect, resulting in shoot formation, was observed when a region on the left side of EcoRI fragment 7 was mutated. This hypothesis was supported by an observed deletion of the left part of the T-DNA in a bacterial transformant callus tissue and in a somatic hybrid callus tissue, both with a high shooting capacity (38).

Vegetative and Generative Propagation of Transformed Shoots

One of the properties of the transformed shoots is the failure to develop a root system. This might also be seen as a response of a high cytokinin-like activity in the shoot tissue, since we know that a high cytokinin versus auxin concentration can inhibit root formation. A consequence of the absence of roots is that the shoots have to be grafted for further development. Therefore, shoots from bacterial transformants of both octopine and nopaline type, shoots from pTi-DNA transformants (octopine type) and shoots from somatic hybrids (octopine type) were grafted on the stems of healthy tobacco plants. The response of the grafting was in all cases a development of the shoots into mature, flowering plants, with retention of the opine synthesis (Fig. 2). The flowers that appeared on the plants showed a remarkable difference with flowers on control tobacco plants. They were characterized by an outgrowth of the pistil (heterostyly) (Fig. 3). This developmental abnormality was observed on all transformed plants, whereas it has not been observed on grafted control plants, derived from control callus tissues. Whilst flowers on control plants formed seedbulbs as a result of self fertilization, what is natural for tobacco, the flowers on the transformants fell off without seed formation. Therefore, crosses were carried out by applying normal pollen on the stamen of the transformants and by applying pollen from transformants on stamina of control plants (Fig. 4). Only the first cross resulted in seed forma-

Figure 2. Representative transformed tobacco plant derived from morphogenic tumor tissue which was obtained after transformation of protoplasts by Ti-plasmid DNA. This grafted plant is male sterile and shows heterostyly.

tion. From such experiments which were carried out with bacterial transformants and with somatic hybrids it was concluded that those transformed plants were male sterile.

The seeds were germinated on culture medium at an efficiency of 90%. Initially, all seedlings developed a main root with root hairs and a hypocotyl. About two weeks after germination, two groups of seedlings could be distinguished. About 55% of the seedlings continued a normal development, whereas the other 45% remained smaller and the development of roots stopped. Finally, the main root degenerated into a callus. This group of seedlings developed further in a similar way to the parental transformants. Screening for opine syn-

Figure 3. (A) Flower of a normal tobacco plant. (B) Flower of a transformed tobacco plant, showing heterostyly. The pistil is growing outside the corolla.

thesis showed that the 55% "normal" seedlings were negative, whereas the 45% aberrant seedlings were positive with a few exceptions (Table 2). Based on the presence of opines, we concluded that these aberrant transformants must have retained T-DNA. If we assume that a low percentage of the seeds does not germinate due to the poor

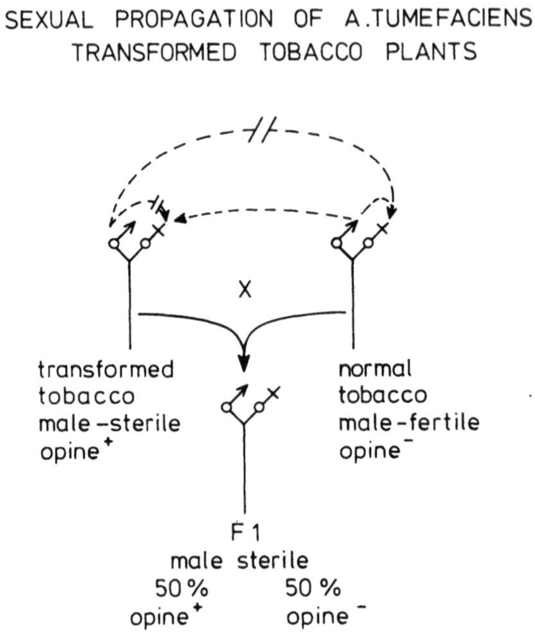

Figure 4. Schematic representation of the pollination of transformed tobacco plants. Due to the male sterility of the transformed plants, progeny was only obtained after cross-pollination.

root development, we would like to conclude that T-DNA linked genes (coding for LpDH or NpDH) are transmitted as a single Mendelian factor, and that the parental transformants are heterozygous for these traits.

Heterostyly and male sterility as observed on the parental transformants was also observed on the seedlings. A striking observation was that 100% of the seedlings expressed these traits. This implies that they were also observed on the "normal" opine negative seedlings. Since there is no segregation of the heterostyly and the male sterility, we think that they are inherited cytoplasmically. In fact we can conclude that these traits are present in all plants originated from T-DNA containing tumorous tissues. Therefore, we like to suggest that a positive relation exists between T-DNA and the presence of the male sterility. Such a suggestion has to be confirmed by the detection of T-DNA sequences, in particular in the "normal" seedlings. Therefore, DNA hybridization experiments with DNA from such "normal" seedlings are in progress.

Table 2. Transmission of Octopine and Nopaline Synthesis in Transformed Tobacco Plants.

Crosses	# of Seedlings Tested	Opine Positive	Opine Negative
SR1-4013-3 x SR1	243	113 (46.5%)	130 (53.5%)
B+S-42-3 x SR1	272	123 (45.2%)	149 (54.8%)
SR1-4058-38 x SR1	248	108 (43.5%)	140 (56.5%)
GWN1 x SR1	242	108 (44.6%)	134 (55.4%)

SR1-4013-3: octopine type bacterial transformant
B+S-42-3: octopine type somatic hybrid
SR1-4058-38: nopaline type bacterial transformant
GWN1: nopaline type F1 progeny plant from SR1-4058-38

CONCLUDING REMARKS

The aim of this article is not to present straightforward methods for the transfer at will of valuable foreign genes into plants. Too many aspects of the crown gall system are not yet understood. In particular the integration mechanism of Ti-plasmid sequences in the plant genome as T-DNA is unknown. In fact compared with the progress made in genetic manipulation with microorganisms and even with mammalian cells, the application of microbiological strategies to the plant kingdom has yet to be started. On the other hand, the present knowledge about the architecture of the T-DNA, its expression into RNA and proteins, and its inheritability justify the view that transformation systems based on the Ti-plasmid of *Agrobacterium* can be used as a starting point that will stimulate efforts to transfer foreign genes at will to plants. In this respect, the value of the crown gall system was demonstrated by T.C. Hall who was able to insert, via recombination of the Ti-plasmid, part of a gene for a storage protein from bean into sunflower crown gall tissue, where it was expressed (personal communication).

If an optimal result has to be reached with transformation experiments, the DNA of interest must be incubated with a large population of target cells, in order to obtain a few cells that stably incorporate the desired gene. In this respect, the methods described here, based on single plant protoplasts as target cells, are similar to those used in bacterial, yeast and mammalian cell genetics. Millions of single plant cells can be exposed to the transforming bacteria or DNA from which transformants can be selected. Integrated foreign DNA (T-DNA) is maintained and at least partly expressed (opine synthesis) in transformed tissues. With this, the similarity between plant transformation and the approaches used in the animal kingdom ends, and turns into a difference which might be in favor of the plant kingdom. Single, isolated plant cells have the potential to differentiate into new plants. The inserted T-DNA remains stably incorporated in the genome of such plants during development and is transmissable to Fl progeny plants in which it will be expressed. In addition, if transformed plant protoplasts might have lost their morphogenic potential, due to the transformation, it might be restored following cell fusion with morphogenic protoplasts derived from the same non-transformed species.

The development of protoplast transformation systems based on the Ti-plasmid has also reopened the search for transformation of monocotyledonous plants, a group to which all cereal crops belong. These plants are not susceptible to crown gall induction at the whole plant level. This failure might be due to the absence of a wound reaction which is a prerequisite for tumor induction on dicotyledonous plants. It should be of great interest to know whether or not this barrier could be bypassed through transformation of protoplasts by intact Agrobacteria or, more promising, by transformation of protoplasts by isolated Ti-plasmid DNA.

At present, the use of the crown gall system on the single cell level is highly dependent on the availability of isolated or synthesized plant genes of interest. Most progress has been made concerning genes for leghaemoglobin (54), the small subunit of ribulose-1,5-biphosphatecarboxylase (55), and a number of storage proteins including zein (56). Other genes of interest, including nitrogen fixing genes (nif genes), genes for most storage proteins or genes controlling disease resistancies are often clusters or collections of a number of separated genes and, hence, are rather complex regarding their availability, insertion, and expression in target cells. Although these technical problems are difficult to overcome, they, in time, will not prevent the application of microbiological and plant molecular biological methods in agriculture.

REFERENCES

1. A.C. Braun, Plant tumor research, in: "Progress in Experimental Tumor Research", F. Homburg, ed., S. Karger, Basel (1972).
2. R. Beiderbeck, Pflanzentumoren, in: "Ein Problem der Pflanzlichen Entwicklung", E. Ulmer, Stuttgart (1977).
3. R.A. Schilperoort, P.M. Klapsijk, G. Ooms and G.J. Wullems, Plant tumors caused by bacterial plasmids: Crown gall, in: "Genetic origin of Tumor Cells", G.J. Cleton and J.W.F.M. Simons, eds., M. Nijhoff, The Hague (1980).
4. E.W. Nester, Molecular studies on crown gall tumors - role of nucleic acids of Agrobacterium tumefaciens incorporated into the crown gall cell, in: "Neoplasma - Comparative Pathology

of Growth in Animals, Plants and Man", H.E. Kaiser, ed., Williams and Williams, Baltimore (1981).

5. J. Schell, M. Van Montagu, M. Holsters, J.P. Hernalsteens, J. Leemans, H. De Greve, Ch. Shaw, P. Zambryski, L. Willmitzer, L. Otten, J. Schröder and G. Schröder, Tumors as a result of gene transfer in plants, in: "Molecular Biology of Plant Development", H. Smith and D. Grierson, eds., Blackwell Scientific Publications, Oxford (in press).

6. G. Ooms, "Molecular biologic studies on Agrobacterium tumefaciens and crown gall plant tumors," Thesis, University of Leiden, The Netherlands (1981).

7. M.-D. Chilton, M.H. Drummond, D.J. Merlo, D. Sciaky, A.L. Montoya, M.P. Gordon and E.W. Nester, Stable incorporation of plasmid DNA into higher plant cells: the molecular basis of crown gall tumorigenesis, Cell 11:263-271 (1977).

8. M.F. Thomashow, C.G. Panagopoulos, M.P. Gordon and E.W. Nester, Host range of A. tumefaciens is determined by the Ti-plasmid, Nature 283:794-796 (1980).

9. B.P. Koekman, G. Ooms, P.M. Klapwijk and R.A. Schilperoort, Genetic map of an octopine Ti-plasmid, Plasmid 2:347-357. (1979).

10. P.M. Klapwijk, J. Van Breukelen, K. Korevaar, G. Ooms and R.A. Schilperoort, Transposition of Tn 904 encoding streptomycin resistance into the octopine Ti-plasmid of A. tumefaciens, J. Bacteriol. 141:129-136 (1980).

11. G. Ooms, P.M. Klapwijk, J.A. Poulis and R.A. Schilperoort, Characterization of Tn 904 insertions in octopine Ti-plasmid mutants of A. tumefaciens, J. Bacteriol. 144:82-91 (1980).

12. D.J. Garfinkel and E.W. Nester, A. tumefaciens mutants affected in crown gall tumorigenesis and octopine catabolism, J. Bacteriol. 144:732-743 (1980).

13. M.F. Thomashow, R. Nutter, A.L. Montoya, M.P. Gordon and E.W. Nester, Integration and organization of Ti-plasmid sequences in crown gall, Cell 19:729-739 (1980).

14. M. Lemmers, M. De Beukeleer, M. Holsters, P. Zambryski, A. Depicker, J.P. Hernalsteens, M. Van Montagu and J. Schell, Internal organization, boundaries and integration of Ti-plasmid DNA in nopaline crown gall tumors, J. Mol. Biol. 144: 353-376 (1980).

15. N.S. Yadav, K. Postle, R.V. Saiki, M.F. Thomashow and M.-D. Chilton, T-DNA of a crown gall teratoma is covalently joined to host plant DNA, Nature 287:458-461 (1980).

16. M.-D. Chilton, R.K. Saiki, N. Yadav, M.P. Gordon and F. Quetier, T-DNA from Agrobacterium Ti-plasmid is in the nuclear DNA fraction of crown gall tumor cells, Proc. Nat. Acad. Sci. (USA) 77:4060-4064 (1980).

17. L. Willmitzer, M. De Beuckeleer, M. Lemmers, M. Van Montagu and J. Schell, DNA from Ti-plasmid present in nucleus and absent from plastids of crown gall cells, Nature 287:359-361 (1980).

18. M. De Beuckeleer, M. Lemmers, G. De Vos, L. Willmitzer, M. Van Montagu and J. Schell, Further in sight on the transferred-DNA of octopine crown gall, Mol. Gen. Genet 183:283-288 (1981).
19. M.H. Drummond, M.P. Gordon, E.W. Nester and M.-D. Chilton, Foreign DNA of bacterial plasmid origin is transcribed in crown gall tumors, Nature 269:535-536 (1977).
20. W.B. Gurley, J.D. Kemp, M.J. Albert, D.W. Sutton and J. Callis, Transcription of Ti-plasmid derived sequences in three octopine type crown gall tumor lines, Proc. Nat. Acad. Sci. (USA) 76:2828-2832 (1979).
21. J.C. McPherson, E.W. Nester and M.P. Gordon, Proteins encoded by A. tumefaciens Ti-plasmid DNA (T-DNA) in crown gall tumors, Proc. Nat. Acad. Sci. (USA) 77:2666-2670 (1980).
22. L. Willmitzer, L. Otten, G. Simons, W. Schmalenbach, J. Schröder, G. Schröder, M. Van Montagu, G. De Vos and J. Schell, Nuclear and polysomal transcripts of T-DNA in octopine crown gall suspension and callus cultures, Mol. Gen. Genet., 182:255-262 (1981).
23. S.B. Gelvin, M.P. Gordon, E.W. Nester and A.I. Aronson, Transcription of the Agrobacterium Ti-plasmid in the bacterium and in crown gall tumors, Plasmid 6:17-29 (1981).
24. J. Schröder, G. Schröder, H. Huisman, R.A. Schilperoort and J. Schell, The mRNA for LpDH in plant tumor cells is complementary to a Ti-plasmid fragment, FEBS Letters 129:166-168 (1981).
25. D.S. Letham, P.B. Goodwin and T.J.V. Higgins, "Phytohormones and related compounds - A comprehensive treatise," Elsevier/North-Holland, Amsterdam (1978).
26. J.E. Varner and D. Ho Tuan-Hua, Hormone regulation of cell growth and development in vitro, in: "Frontiers of Plant Tissue Culture, T.A. Thorpe, ed., International Association for Plant Tissue Culture, Univ. of Calgary Press, Alberta, Canada (1978).
28. K.R. Libbenga, Hormone receptors in plants, in: "Frontiers of Plant Tissue Culture, T.A. Thorpe, ed., International Association for Plant Tissue Culture, Univ. of Calgary Press, Alberta, Canada (1978).
29. F.W. Went, Rec. Trav. Bot. Neerl. 25:1-116 (1928).
30. W. Pengelly and F. Meins, A specific radioimmuno-assay for nanogram quantities of the auxin, indole-3-acetic acid, Planta 136:173-180 (1977).
31. I.M. Scott, G. Browning and J. Eagles, Ribosylzeatin and zeatin in tobacco crown gall tissue, Planta 147:269-273 (1980).
32. G. Ooms, P.J.J. Hooykaas, G. Moolenaar and R.A. Schilperoort, Crown gall plant tumors of abnormal morphology induced by A. tumefaciens carrying mutated octopine Ti-plasmids; analysis of T-DNA functions, Gene 14:33-50 (1981).

33. O. Schieder, _in_: "Frontiers of Plant Tissue Culture, T.A. Thorpe, ed., International Association for Plant Tissue Culture, Univ. of Calgary Press, Alberta, Canada (1978).
34. P. Maliga, Resistance mutants and their use in genetic manipulation, _in_: "Frontiers in Plant Tissue Culture, T.A. Thorpe, ed., International Association for Plant Tissue Culture, Univ. of Calgary Press, Alberta, Canada (1978).
35. P. Maliga, Cell culture mutants and their uses, _in_: "Plant Improvement and Somatic Cell Genetics," I.K. Vasil, K.J. Frey and W.R. Scowcroft, eds., Academic Press, NY (1982).
36. L. Marton, G.J. Wullems, L. Molendijk and R.A. Schilperoort, _In vitro_ transformation of cultured cells from _N. tabacum_ by _A. tumefaciens_, _Nature_ 277:129-131 (1979).
37. G.J. Wullems, L. Molendijk, G. Ooms and R.A. Schilperoort, Differential expression of crown gall tumor markers in transformants obtained after _in vitro_ transformation of _N. tabacum_ protoplasts by _A. tumefaciens_, _Proc. Nat. Acad. Sci. (USA)_ 78:4344-4348 (1981).
38. G. Ooms, A. Bakker, L. Molendijk, G. Wullems, R.A. Schilperoort, M.P. Gordon and E.W. Nester, Organization of T-DNA and cellular composition of octopine crown gall tissues _in vivo_ and _in vitro_ by _A. tumefaciens_; impact of T-DNA on tumor phenotype, (in press).
39. A. Pellicer, D. Robins, B. Wold, R. Sweet, J. Jackson, I. Lowy, J.M. Roberts, G.K. Sim, S. Silverstein adn R. Axel, Altering genotype and phenotype by DNA-mediated gene transfer, _Science_ 209:1414-1422 (1980).
40. M. Wigler, R. Sweet, G.K. Sim, B. Wold, A. Pellicer, E. Lacy, T. Maniatis, S. Silverstein and R. Axel, Transformation of mammalian cells with genes from prokaryotes and eukaryotes, _Cell_ 16:777-785 (1979).
41. D.M. Robins, S. Ripley, A.S. Henderson, R. Axel, Transforming DNA integrates into the host chromosome, _Cell_ 23:29-39 (1981).
42. P.F. Lurquin, Entrapment of plasmid DNA by liposomes and their interactions with plant protoplasts, _Nucl. Acid. Res._ 6:3773-3783 (1979).
43. G.J. Wullems, L. Molendijk and R.A. Schilperoort, The expression of tumor markers in intra specific somatic hybrids of normal and crown gall cells from _N. tabacum_, _Theor. Appl. Genet._ 56:203-208 (1980).
44. F.A. Krens, L. Molendijk, G.J. Wullems and R.A. Schilperoort, _In vitro_ transformation of plant protoplasts with Ti-plasmid DNA, _Nature_ (in press).
45. A.C. Braun, The activation of two growth-substance systems accompanying the conversion of normal to tumor cells in crown gall, _Cancer Res._ 16:53-56 (1956).
46. M. De Cleene and J. De Ley, The host range of crown gall, _The Bot. Rev._ 42:389-466 (1976).

47. A.C. Braun, A demonstration of the recovery of the crown gall tumor cell with the use of complex tumors of single-cell origin, Proc. Nat. Acad. Sci. (USA) 45:932-938 (1959).
48. A.C. Braun and H.N. Wood, Suppression of the neoplastic state with the acquisition of specialized functions in cells, tissues and organs of crown gall teratomas of tobacco, Proc. Nat. Acad. Sci. (USA) 73:496-500 (1976).
49. R. Turgeon, H.N. Wood and A.C. Braun, Studies on the recovery of crown gall tumor cells, Proc. Nat. Acad. Sci. (USA) 73: 3562-3564 (1976).
50. H.N. Wood, A.N. Binns and A.C. Braun, Differential expression of oncogenicity and nopaline synthesis in intact leaves derived from crown gall teratomas of tobacco, Differentiation 11:175-180 (1978).
51. F. Yang, A.L. Montoya, D.J. Merlo, M.H. Drummond, M.-D. Chilton, E.W. Nester and M.P. Gordon, Foreign DNA sequences in crown gall teratomas and their fate during the loss of the tumorous traits, Mol. Gen Genet. 177:707-714 (1980).
52. H. De Greve, H. Decraemer, J. Seurinck, M. Van Montagu and J. Schell, The functional organization of the octopine A. tumefaciens plasmid pTi B6S3, Plasmid 6:235-248 (1981).
53. G.J. Wullems, L. Molendijk, G. Ooms and R.A. Schilperoort, Retention of tumor markers in Fl progeny plants from in vitro induced octopine and nopaline tumor tissues, Cell 24:719-727 (1981).
54.. D.P.S. Verma, in: "Genome Organization and Expresion in Plants," C.J. Leaver, ed., Plenum, NY (1979).
55. J.R. Bedbrook, S.S. Smith and J.R. Ellis, Molecular cloning and sequencing of cDNA encoding the precursor to the small subunit of chloroplast ribulose-1,5-biphosphate carboxylase, Nature 287:692-697 (1980).
56. B. Burr, in: "Genetic Engineering, Principles and Methods, Vol. 2," J.K. Setlow and A. Holaender, eds., Plenum, London (1980).

TI-PLASMIDS: GENETIC ENGINEERING OF PLANTS

J. Schröder[1], H. De Greve[2], J.-P. Hernalsteens[2],
J. Leemans[2], M. Van Montagu[2,3], L. Otten[1], G. Schröder[1],
L. Willmitzer[1] and J. Schell[1,3]

[1]Max-Planck-Institut für Züchtungsforschung
D-5000 Köln 30
FEDERAL REPUBLIC OF GERMANY

[2]Laboratorium Genetisch Virologie
Vrije Universiteit
Brussels
BELGIUM

[3]Laboratorium voor Genetika
Rijksuniversiteit
Gent
BELGIUM

INTRODUCTION

Crown gall is a neoplastic disease of most dicotyledonous plants and is caused by the soil bacterium *Agrobacterium tumefaciens*. A large extra-chromosomal plasmid in these bacteria was found to be responsible for its tumor-inducing capacity and was, therefore, called Ti-plasmid (1). Bacteria-free crown gall cells can be cultured in the absence of phytohormones and this hormone-independent growth defines tumor cells in plants (2). Sterile tumor tissues have been shown to contain a DNA segment (called T-DNA) which is homologous and colinear with a defined fragment of the Ti-plasmid, and it is covalently linked to plant DNA (3 - 9). The T-DNA has been localized in the nucleus (10, 11) and is directly responsible for the hormone-independent growth of the tumor cells. It is also responsible for the synthesis of low molecular weight compounds, called opines, which are not found in normal plant tissue. The opine produced defines crown galls as octopine, nopaline or agropine type tumors (12). Opines can be utilized by *A. tumefaciens* selectively as sources for carbon, nitrogen and energy, and, thus, the interaction between these bacteria and plants can be seen as a special parasitic relationship which benefits the bacteria (4).

This paper describes some recent advances in the analysis of this biological system: the expression of the T-DNA into RNA and proteins, and the use of Ti-plasmids as vectors for the genetic engineering of plants.

EXPRESSION OF T-DNA

Previous work had shown that the T-DNA is transcribed in transformed plant cells (13 - 16), but the number, sizes and location of the coding regions were not known. Recently these questions were investigated in some tumor cell lines (18). The cell suspension cultures from tobacco used in these experiments harbor the T-DNA from pTiA6, a Ti-plasmid which induces octopine synthesis in the transformed cells. The cells contain only the T_L-fragment of the T-region (Figure 1) (8, 17). Tumor-specific RNAs were detected and mapped by hybridization of ^{32}P-labelled Ti-plasmid fragments to polyadenylated RNA which had been separated on agarose gels and then transferred to DBM paper. The results, summarized in Figure 2, show that the cells contain a total of seven distinct transcripts which differ markedly in their relative abundance and in their sizes. They all bind to oligo(dT)-cellulose, indicating that they are polyadenylated. Thus, the T-DNA which was transferred from a prokaryotic organism must provide specific poly(A)-addition sites. The

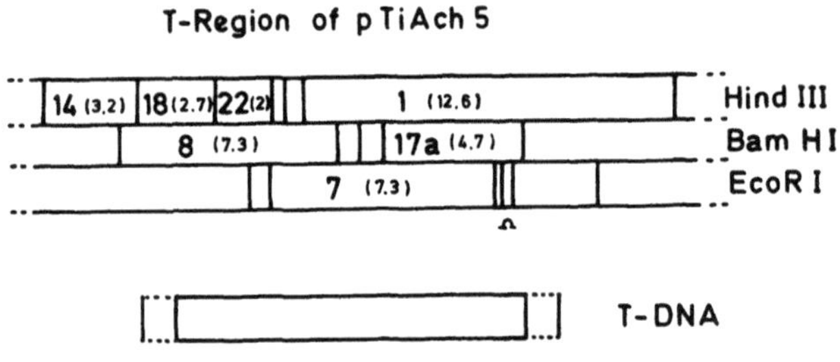

Figure 1. Restriction map of the T_L-region in octopine plasmid pTiAch5 which is identical to that of pTiA6 (38, 39), and the T_L-DNA in tobacco cell line A6-S1 (17). The dotted lines in the T-DNA indicate that the precise borders of the T-DNA have not yet been determined. The numbers in brackets refer to the size of the fragments in base pairs x 10^{-3}. Fragment EcoRI 32 (see last part of this article) is the small fragment left of EcoRI 7.

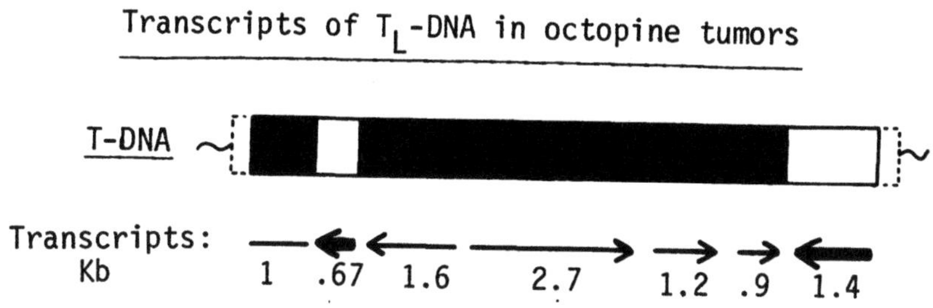

Figure 2. Transcripts in octopine tumor line A6-S1. The numbers refer to the size of the RNAs in bases x 10^{-3}. The thickness of the lines indicates the relative abundance of the RNAs, and the arrows describe the direction of transcription of the six transcripts where this has been determined.

direction of transcription was determined for six of the seven transcripts, and the location of the approximate 5'- and 3'-ends were mapped on the T_L-DNA.

All seven RNAs mapped within the T-DNA sequence. This and the observation that transcription is inhibited by low concentrations of -amanitin (19) seems to suggest that each transcript is determined by a specific promoter site on the T_L-DNA recognized by plant RNA polymerase II. The data presented here do not rule out that some T-DNA promoters serve for transcription of more than one RNA. Considering the groupwise orientation of several transcripts (Figure 2), the simplest model would assume one promoter site per group of transcripts. If so, one would expect that inactivation of a 5'-proximal gene of a group would also lead to disappearance of the transcripts from the 5'-distal genes. However, analysis of some cell lines containing the T-DNA of Ti-plasmid mutants indicates that groupwise inactivation of genes did not occur (20). The results available so far are consistent with the assumption that each gene of the T_L-DNA has its own signals for transcription in the eukaryotic plant cells.

To understand the mechanisms of T-DNA it is important to know whether these RNAs are translated into proteins and to analyze their functions. The hybridization studies and a previous report on T-DNA derived translatable RNAs (21) indicated that the concentrations of these RNAs are very low in transformed plant cells. It was, therefore, necessary to develop a hybridization selection procedure sufficiently sensitive and specific to detect mRNAs which represent about

0.0001% of the total mRNA activity in the plant cells. This procedure was used to enrich for T-DNA derived mRNAs by hybridization to Ti-plasmid fragments covalently bound to microcrystalline cellulose; the hybridized RNAs were eluted and translated in vitro in a cell-free system prepared from wheat germ.

The results obtained with this approach are summarized in Figure 3 (22). The octopine tumor cells contained at least three T-DNA derived mRNAs which are translated in vitro into distinct proteins, and the coding regions correlate with those of three transcripts. The protein encoded at the right end of the T_L-DNA (M_r 39,000) was of specific interest since previous genetic analysis indicated that this part is responsible for octopine synthesis (23 - 25) and since the size of the in vitro synthesized protein is identical with that of the octopine-synthesizing enzyme in octopine tumors. Immunological studies showed that this protein was recognized by antiserum against the tumor-specific enzyme (26). These results demonstrate that the structural gene for the octopine-synthesizing enzyme is on the Ti-plasmid. So far, this is the only protein product of the T-DNA with known enzymatic properties; the possible functions of the two smaller T-DNA derived proteins are not known. The

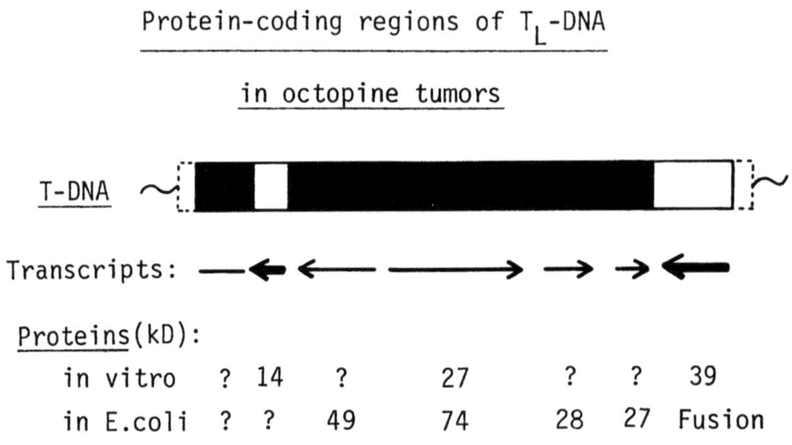

Figure 3. Correlation of transcripts with proteins encoded on T_L-DNA. Proteins were identified either by translation of hybrid-selected mRNA from tobacco cell line A6-S1 (in vitro) or by expression of coding regions in E. coli minicells (E. coli). The numbers refer to the size of the proteins in Daltons x 10^{-3}. Question marks indicate that it has not been possible so far to demonstrate proteins correlating with the transcripts. The protein with M_r 39,000 represents the octopine-synthesizing enzyme.

region coding for the octopine-synthesizing enzyme has recently been sequenced (De Greve et al., in preparation). According to these data, the sequences for initiation and termination of transcription reveal signals of eukaryotic rather than prokaryotic type, suggesting that this gene is designed for expression in eukaryotic cells.

However, this is not necessarily true for all genes of the T-region, since transcripts were also detected in Agrobacteria (16). It was therefore of interest whether the T-DNA derived mRNAs isolated from plant cells shared properties with typical prokaryotic or eukaryotic mRNAs. The fact that translation of all three mRNAs was inhibited by the cap analogue pm^7 G (Figure 4) suggests but does not prove that they contain a cap structure at the 5'-end. This would be typical for eukaryotic mRNA, since caps have not been described in prokaryotic RNA. All three mRNAs were found in polyadenylated as well as in nonpolyadenylated RNA fractions; this cannot be used for such a tentative classification, since both types of RNA have been described in eukaryotic as well as in prokaryotic cells.

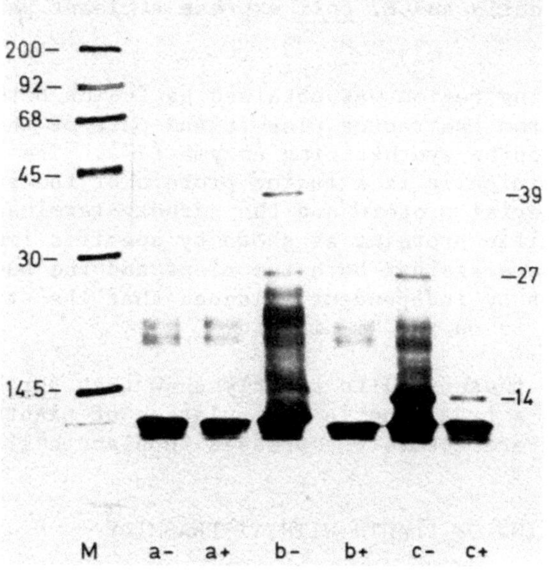

Figure 4. Inhibition of translation of T-DNA derived mRNAs *in vitro* by the cap analogue pm^7G. Affixes - and + indicate translations in absence and in presence (0.5 mM) of the drug. Lanes: (a) wheat germ extract without added RNA; (b) RNA annealed to Hind III 1; (c) RNA annealed to BamHI 8. The two restriction fragments together cover the entire length of the T-DNA present in tobacco cell line A6-S1 (see Figure 1).

The mRNAs for the three proteins each represent about 0.0001% of the total mRNA activity in polyribosomal RNA, and this appears to be the detection limit at present for translatable RNA. The other four transcripts detected by hybridization experiments are present at even lower concentrations, and, assuming that they possess mRNA activity, this is likely to be the reason why the corresponding proteins could not be identified by *in vitro* translation so far.

A different approach was therefore developed to search for coding regions on the T-DNA and their protein products. Fragments from the T-region were cloned into *E. coli* plasmids and analyzed for gene expression in *E. coli* minicells (27). Some of the results are summarized in Figure 3. There are at least four different coding regions within the T_L-DNA which can be expressed in minicells into distinct proteins from promoters which are active in the prokaryotic cells. The four regions expressed in *E. coli* correlate with four regions transcribed in plant cells into RNA. The plant transcripts are larger than the proteins in *E. coli*, and the regions expressed in minicells appear to lie within the regions transcribed in plant cells. Although other explanations are not ruled out, one may speculate that plant cells and *E. coli* express at least partly the same coding regions.

A fifth coding region was obtained as fusion between part of a bacterial gene from the vector plasmid and part of the structural gene for the octopine-synthesizing enzyme (27). The gene product of this hybrid in minicells is a fusion protein of the amino-terminal part of the bacterial protein and the carboxy-terminal part of the plant tumor-specific protein, as shown by specific immunoprecipitations with antisera against both the plant and the bacterial protein. This demonstrates by independent evidence that the structural gene for this protein is on the Ti-plasmid.

In summary, these results clearly show that *Agrobacterium tumefaciens* performs a true genetic manipulation of plants by transferring genes which are actively expressed in plant cells.

GENETIC ENGINEERING OF PLANTS WITH TI-PLASMIDS

There is considerable interest in using Ti-plasmids as vectors for the introduction of desired genes into plants, and it has been shown that foreign DNA can be inserted into the T-region of the plasmid and, thus, be cotransferred to the plant nucleus (28, 29). Since one of the goals of such experiments is the production of normal plants with interesting new properties, it is important to establish whether cells transformed with T-DNA can be regenerated into plants and whether the additional genetic information is maintained during sexual reproduction. These requirements are difficult to meet when using Ti-plasmids containing the genes for tumorous growth (30 - 36), probably because the activities of these genes cannot be entirely repressed. Since DNA transfer and tumorous growth are separate functions (20), an alternative approach is to alter Ti-plasmids genetically so as to inactivate the tumor genes in the T-region without interfering with the transfer of the T-DNA to the plant cells.

Such a pTi-mutant has been obtained by Tn7 mutagenesis (24); it contains the large transposon Tn7 inserted into the fragment EcoRI 32 of pTiB6S3 (see Figure 1), an octopine type plasmid. Tumors induced on tobacco with this plasmid are morphogenetically very active and produce numerous shoots.

The vast majority of these shoots can, upon separation from the tumor tissue, develop roots and grow into normal plants. These shoots do not contain any measurable T-DNA linked activity or function and are, therefore, presumably derived from either untransformed cells present in the initial, uncloned tumor tissue or from cells which have lost the T-DNA by deletion. By extensive screening of several hundred of these shoots, some were found to contain and express the gene for octopine synthase. These plants have been called rGV1-5 (De Greve et al., in preparation) and one of them, rGV-1, has been studied in great detail. It was found to regenerate into a morphologically normal tobacco plant which contained the additional enzyme activity in all cells of all the organs tested (37). Analysis of the T-DNA showed that it has an internal deletion which eliminated the Tn7 insertion and adjacent DNA sequences, but the octopine gene is present and also its transcripts, as would be expected from the presence of the enzyme activity (De Greve et al., in preparation). The experiments proved that the stable incorporation of a well defined additional gene into higher plant cells is possible without affecting the morphogenetic potential of the cells.

The plant flowered and set seeds quite normally, and it was investigated whether the progeny also contained the additional gene activity. The results, summarized in Table 1, show that the gene was transmitted as a stable, dominant Mendelian gene which is present once per haploid genome. Such experiments have been continued for several generations with the same result, indicating that the foreign gene in fact has become a stable part of the plant genome.

The results summarized here clearly indicate that a stable introduction of foreign genes via Ti-plasmids into plant cells is possible without affecting the morphogenetic potential of the cells. These are important prerequisites for successful genetic engineering of plants.

Table 1. Transmission of the Gene for Octopine Synthesis in Tobacco Plant rGV-1: The Gene is Transmitted to Progeny as a Dominant Mendelian Gene Which is Present Once per Haploid Genome.

Crosses	No. of Progeny Tested	Octopine Synthesis		
			+	-
rGV-1 ♂ x rGV-1 ♀	145		110 (76%)	35 (24%)
rGV-1 ♂ x rGV-1 ♀	200	++[1] 42 (21%)	+[1] 95 (48%)	63 (31%)
rGV-1 ♂ x Wildtype ♀	248		124 (50%)	124 (50%)
rGV-1 ♀ x Wildtype ♂	187		81 (43%)	106 (57%)
Plantlets derived from anther cultures (haploid)	102		47 (46%)	55 (54%)

[1] As estimated from a semiquantitative assay of enzyme activity.

REFERENCES

1. N. Van Larebeke, G. Engler, M. Holsters, S. Van den Elsacker, I. Zaenen, R.A. Schilperoort and J. Schell, Large plasmid in Agrobacterium tumefaciens essential for crown gall-inducing ability, Nature 252:169 (1974).
2. A.C. Braun, The activation of two growth substance systems accompanying the conversion of normal to tumor cells in crown gall, Cancer Res. 16:53 (1956).
3. M.-D. Chilton, H.J. Drummond, D.J. Merlo, D. Sciaky, A.L. Montoya, M.P. Gordon and E.W. Nester, Stable incorporation of plasmid DNA into higher plant cells: the molecular basis of crown gall tumorigenesis, Cell 11:263 (1977).
4. J. Schell, M. Van Montagu, M. De Beuckeleer, M. De Block, A. Depicker, M. De Wilde, G. Engler, C. Genetello, J.P.

Hernalsteens, M. Holsters, J. Seurinck, B. Silva, F. Van Vliet and R. Villarroel, Interactions and DNA transfer between Agrobacterium tumefaciens, the Ti-plasmid and the plant host, Proc. Roy. Soc. Lond. B 204:251 (1979).
5. M. Lemmers, M. De Beuckeleer, M. Holsters, P. Zambryski, A. Depicker, J.P. Hernalsteens, M. Van Montagu and J. Schell, Internal organization , boundaries and integration of Ti-plasmid DNA in nopaline crown gall tumors, J. Mol. Biol. 144:355 (1980).
6. P. Zambryski, M. Holsters, K. Kruger, A. Depicker, J. Schell M. Van Montagu and H.M. Goodman, Tumor DNA structure in plant cells transformed by A. tumefaciens, Science 209:1385 (1980).
7. N.S. Yadav, K. Postle, R.K. Saiki, M.F. Thomashow and M.-D. Chilton, T-DNA of crown gall teratoma is covalently joined to host plant DNA, Nature 287:458 (1980).
8. M.F. Thomashow, R. Nutter, A.L. Montoya, M.P. Gordon, E.W. Nester, Integration and organization of Ti-plasmid sequences in crown gall tumors, Cell 19:729 (1980a).
9. M.F. Thomashow, R. Nutter, K. Postle, M.-D. Chilton, F.R. Blattner, A. Powell, M.P. Gordon and E.W. Nester, Recombination between higher plant DNA and the Ti-plasmid of Agrobacterium tumefaciens, Proc. Natl. Acad. Sci. USA 77:6448 (1980b).
10. M.-D. Chilton, R.K. Saiki, N. Yadav, M.P. Gordon and F. Quetier, T-DNA from Agrobacterium Ti-plasmid is in the nuclear DNA fraction of crown gall tumor cells, Proc. Natl. Acad. Sci. USA 77:4060 (1980).
11. L. Willmitzer, M. De Beuckeleer, M. Lemmers, M. Van Montagu and J. Schell, The Ti-plasmid derived T-DNA is present in the nucleus and absent from plastids of plant crown gall cells, Nature 287:359 (1980).
12. P. Guyon, M.-D. Chilton, A. Petit and J. Tempe, Agropine in "null-type" crown gall tumors: evidence for the generality of the opine concept, Proc. Natl. Acad. Sci. USA 77:2693 (1980).
13. M.H. Drummond, M.P. Gordon, E.W. Nester and M.-D. Chilton, Foreign DNA of bacterial plasmid origin is transcribed in crown gall tumors, Nature 269:535 (1977).
14. W.B. Gurley, J.D. Kemp, M.J. Alber, D.W. Sutton and J. Gallis, Transcription of Ti-plasmid derived sequences in three octopine-type crown gall tumor lines, Proc. Natl. Acad. Sci. USA 76:2828 (1979).
15. L. Willmitzer, L. Otten, G. Simons, W. Schmalenbach, J. Schroder, G. Schroder, M. Van Montagu, G. De Vos and J. Schell, Nuclear and polysomal transcripts of T-DNA in octopine crown gall suspension and callus cultures, Mol. Gen. Genet. 182:255 (1981).
16. S.B. Gelvin, M.P. Gordon, E.W. Nester and A.I. Aronson, Transcription of the Agrobacterium Ti-plasmid in the bacterium and in crown gall tumors, Plasmid 6:17 (1981).

17. M. De Beuckeleer, M. Lemmers, G. De Vos, L. Willmitzer, M. Van Montagu and J. Schell, Further insight on the transferred-DNA of octopine crown gall, Mol. Gen. Genet. 193:283 (1981).
18. L. Willmitzer, G. Simons and J. Schell, the T_L-DNA in octopine crown gall tumors codes for seven well-defined polyadenylated transcripts, EMBO J. 1:139 (1982).
19. L. Willmitzer, W. Schmalenbach and J. Schell, Transcription of T-DNA in octopine and nopaline crown gall tumors is inhibited by low concentrations of -amanitin, Nucl. Acids Res. 9:4801 (1981).
20. J. Leemans, R. Deblaere, L. Willmitzer, H. De Greve, J.P. Hernalsteens, M. Van Montagu and J. Schell, Genetic identification of functions of T_L-DNA transcripts in octopine crown galls, EMBO J. 1:147 (1982).
21. J.C. McPherson, E.W. Nester and M.P. Gordon, Proteins encoded by Agrobacterium tumefaciens Ti-plasmid DNA (T-DNA) in crown gall tumors, Proc. Natl. Acad. Sci. USA 77:2666 (1980).
22. G. Schröder and J. Schröder, Hybridization selection and translation of T-DNA encoded mRNAs from octopine tumors, Mol. Gen. Genet. 185:51 (1982).
23. B.P. Koekman, G. Ooms, P.M. Klapwijk and R.A. Schilperorrt, Genetic map of an octopine Ti-plasmid, Plasmid 2:347 (1979).
24. H. De Greve, H. Decraemer, J. Seurinck, M. Van Montagu and J. Schell, The functional organization of the octopine Agrobacterium tumefaciens plasmid pTiB6S3, Plasmid 6:235 (1981).
25. D.J. Garfinkel, R.B. Simpson, L.W. Ream, F.F. White, M.P. Gordon and E.W. Nester, Genetic analysis of crown gall: fine structure map of the T-DNA by site-directed mutagenesis, Cell 27:143 (1981).
26. J. Schröder, G. Schröder, H. Huisman, R.A. Schilperoort and J. Schell, The mRNA for lysopine dehydrogenase in plant tumor cells is complementary to a Ti-plasmid fragment, FEBS Letters 129:166 (1981).
27. J. Schröder, A. Hillebrand, W. Klipp and A. Puhler, Expression of plant tumor-specific proteins in minicells of Escherichia coli: a fusion protein of lysopine dehydrogenase with chloramphenicol acetyltransferase, Nucl. Acids Res. 9:5187 (1981).
28. J.P. Hernalsteens, F. Van Vliet, M. De Beuckeleer, A. Depicker, G. Engler, M. Holsters, M. Van Montagu and J. Schell, The Agrobacterium tumefaciens Ti-plasmid as a host vector system for introducing foreign DNA in plant cells, Nature 287:654 (1980).
29. J. Leemans, Ch. Shaw, R. Deblaere, H. De Greve, J.P. Hernalsteens, M. Maes, M. Van Montagu and J. Schell, Site-specific mutagenesis of Agrobacterium Ti-plasmids and transfer of genes to plant cells, J. Mol. Appl. Genet. 1:149 (1981).
30. A.C. Braun, A demonstration of the recovery of the crown gall tumor cell with the use of complex tumors of single cell origin, Proc. Natl. Acad. Sci. USA 45:932 (1959).

31. A.C. Braun and H.N. Wood, Suppression of the neoplastic state with the acquisition of specialized functions in cells, tissues and organs of crown gall teratomas of tobacco, Proc. Natl. Acad. Sci. USA 73:496 (1976).
32. R. Turgeon, H.N. Wood and A.C. Braun, Studies on the recovery of crown gall tumor cells, Proc. Natl. Acad. Sci. USA 73:3562 (1976).
33. H.N. Wood, A.N. Binns and A.C. Braun, Differential expression of oncogenicity and nopaline synthesis in intact leaves derived from crown gall teratomas of tobacco, Differentiation 11:175 (1978).
34. F. Yang, A.L. Montoya, D.J. Merlo, M.H. Drummond, M.-D. Chilton, E.W. Nester and M.P. Gordon, Foreign DNA sequences in crown gall teratomas and their fate during loss of the tumorous traits, Mol. Gen. Genet. 177:707 (1980).
35. G.J. Wullems, L. Molendijk, G. Ooms and R.A. Schilperoort, Differential expression of crown gall tumor markers in transformants obtained after in vitro Agrobacterium tumefaciens-induced transformation of cell wall regenerating protoplasts derived from Nicotiana tabacum, Proc. Natl. Acad. Sci. USA 78:4344 (1981).
36. G.J. Wullems, L. Molendijk, G. Ooms and R.A. Schilperoort, Retention of tumor markers in F1 progeny plants from in vitro induced octopine and nopaline tumor tissues, Cell 24:719 (1981).
37. L. Otten, H. De Greve, J.P. Hernalsteens, M. Van Montagu, O. Schieder, J. Straub and J. Schell, Mendelian transmission of genes introduced into plants by the Ti-plasmids of Agrobacterium tumefaciens, Mol. Gen. Genet. 183:209 (1981).
38. G. Engler, M. Van Montagu, I. Zaenen and J. Schell, Homology between Ti-plasmids of Agrobacterium tumefaciens: hybridization studies using electron microscopy, Biochem. Soc. Trans. 5:930 (1977).
39. G. De Vos, M. De Beuckeleer, M. Van Montagu and J. Schell, Restriction endonuclease mapping of the octopine tumor-inducing plasmid pTiAch5 of Agrobacterium tumefaciens, Plasmid 6:249 (1981).

ESSENTIAL OIL PRODUCTION IN RELATION TO ORGANOGENESIS IN TISSUE CULTURES OF EUCALYPTUS CITRIODORA HOOK

P.K. Gupta and A.F. Mascarenhas

Biochemistry Division
National Chemical Laboratory
Poona,
INDIA

INTRODUCTION

Eucalyptus citriodora Hook is a fast growing forest tree valued for its wood, and its leaves contain an essential oil, the concentration of which varies in different trees. The concentration of the main components of the oil, citronellal and citronellol also vary in different trees. Citronellal is an important monoterpene aldehyde used in the synthesis of a number of high grade perfumery chemicals. Several reports are available on biosynthesis, biotransformation and, also, the production of secondary metabolites like alkaloids, steroids, glycosides, cardenolides, etc., using tissue cultures. Reports on the production and biosynthesis of monoterpenes by tissue culture are still rare (1 - 3). There are few reports on field data of clonally propagated mature trees.

The present paper describes: (1) the effect of organogenesis on the production and synthesis of the essential oil in callus and organized cultures of *E. citriodora*; and (2) results on the analysis of the oil content in clonally propagated plants.

MATERIALS AND METHODS

The media used in these studies are given in Table 1. Callus was isolated from leaves and stem segments of seedlings and mature trees on MS_1. Shoot regeneration was obtained on MS_2. Conditions for shoot and root regeneration from seedlings and mature trees have been described earlier (4, 5). Subcultures were made at intervals of 30 days.

Table 1. Composition of Media.

	Medium (mg/l)	Reference
Callus Initiation (MS_1)	MS + Kin (0.2) + BAP (0.3) + NAA (2)	Plant Sci. Lett. 20:195 (1981)
Callus differentiation and shoot formation (MS_2)	MS + Kin (0.2) + BAP (0.3) + Calcium pantothenate (0.1) + Biotin (0.1)	"
Root formation (W)	WB + NAA (2) (liquid) → WB (liquid) — MS (liquid)	"

Oil Analysis

Oil from field grown plants was extracted from fresh leaves by steam distillation, whereas cultured cells and leaves were extracted directly with ether at 5°C. Samples were made to a known volume and 1 µl used for analysis. Estimations were carried out using a Shimadzu gas liquid chromatography equipment (6 AMPF model) in a 2 meter length and 2 mm diameter stainless steel column packed with 10% carbowax, 20 M on chromosorb P. The column temperature of 175°C and the thermal conductivity detector (TCD) temperature at 200°C was maintained during analysis. The concentrations of oil, citronellal and citronellol have been calculated on the basis of dry weight. The GLC pattern of *E. citriodora* oil which is identical in tissue culture and normal plants is shown in Figure 1. Comparisons made in the present studies have been with the total oil and the main components, citronellal and citronellol.

RESULTS

Synthesis of Oil, Citronellal and Citronellol

Callus. Primary callus from leaf tissue contained the essential oil and its components at concentrations lower than the parent plant. These concentrations gradually diminished on subculture and were not detected by the tenth passage (Table 2). Attempts to obtain shoot differentiation from leaf callus using different media were unsuccessful. Table 3 shows the effect of organogenesis on oil content of callus obtained from stem of four month old seedlings.

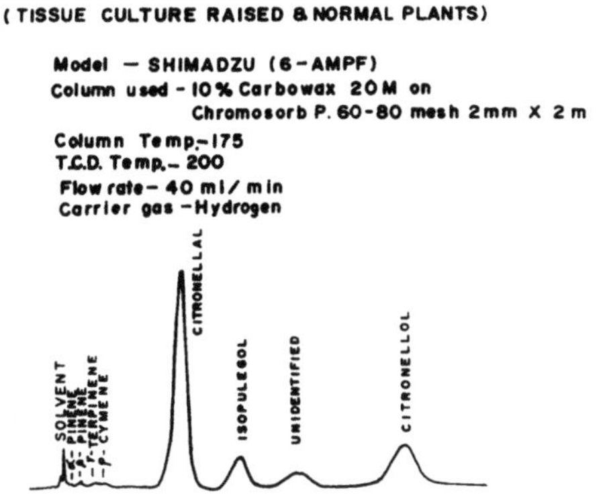

Figure 1. The GLC pattern of E. citriodora oil.

Table 2. Effect of Subculture on Oil Content in Callus.

	Parent plant %	Primary %	S_1 %	S_2 %	S_3 %	S_4 %	S_{10} %
Total oil (dry weight basis)	1.15	0.92	0.56	0.30	0.10	0.10	N.D.
Citronellal	70.20	42.50	25.00	12.10	8.20	7.00	N.D.
Citronellol	16.80	8.00	5.10	2.10	N.D.	N.D.	N.D.

N.D. - not detectable.
Initial inoculum - leaf callus from plant (1 year old).
Subculture period - 30 days.

Stem callus differentiated on medium MS_2 (5). In primary stem callus, and also just after shoot initiation, no oil was detected. With shoot elongation traces of oil and citronellal appeared. After root initiation, there was a sudden increase in the oil and citronellal concentration and also the appearance of citronellol. The optimum level was reached after the plants were grown in pots for four months.

Shoots. Shoots both from seedlings and mature trees could be subcultured repeatedly without any reduction in their organizing ability. Table 4 illustrates the effect of subculture on the oil content and its components from shoots isolated from four month old seedlings. There was a gradual reduction in the concentration of the oil and citronellal up to the third subculture. At this stage

Table 3. Effect of Organogenesis on Oil Content.

	Callus %	Shoot initiation (0.5-0.8 cm length) %	Shoot (2-3 cm length) %	Shoot (4-6 cm length) %	After root initiation %	Rooted plant in tube %	Plant in pot (4 months) %
Total oil (dry weight basis)	N.D.	N.D.	0.06	0.07	0.30	0.48	0.75
Citronellal	N.D.	N.D.	5.00	5.20	23.10	40.20	50.90
Citronellol	N.D.	N.D.	N.D.	N.D.	2.00	5.00	10.10

N.D. - not detectable.
Initial callus from seedling (4 months old) stem.
Period for shoot development - 60 days.
Period for root development - 30 days.

Table 4. Effect of Subculture on Oil Content of Shoots.

	Parent explant %	Primary %	S_1 %	S_2 %	S_3 %	S_4 %	S_{10} %
Total oil (dry weight basis)	1.06	0.08	0.38	0.15	0.12	0.12	0.10
Citronellal	62.60	43.80	20.20	11.00	10.80	10.30	10.20
Citronellol	13.50	6.20	2.40	1.00	N.D.	N.D.	N.D.

Initial inoculum - seedling (4 month old).
Subculture period - 30 days.

Table 5. Effect of Subculture on Oil Content of Shoots from Mature Tree.

	Parent tree %	Primary %	S_1 %	S_2 %	S_3 %	S_4 %	S_{10} %
Total oil (dry weight basis)	1.15	0.80	0.50	0.30	0.27	0.27	0.25
Citronellal	71.00	50.15	40.00	28.29	25.50	24.00	24.50
Citronellol	17.50	8.50	6.00	3.50	1.00	N.D.	N.D.

Subculture period - 30 days.
Initial inoculum - apical and axillary buds from mature tree (20 year old).

citronellol was not detected. The concentration of the total oil and citronellal was constant even after ten subcultures.

Mature Trees. In shoots from mature trees the effect of subculture on oil synthesis was more or less similar to the effect from seedling cultures, except that the concentration of the oil and citronellal at the end of ten subcultures was nearly double that of the seedlings (Table 5).

Rooting. The effect of rooting on the concentration of oil, citronellal and citronellol was studied both with tissues raised from seedlings and mature trees. This study was conducted on cultures which were in the fourth passage. In both cases the initiation of roots resulted in a sharp increase in the concentration of the oil and citronellal and was found to be nearly double the concentration in shoots alone. Citronellol was also detected. This level increased further with root elongation and in plants grown for four months in pots (Tables 6 and 7).

Table 8 summarizes the data on the oil, citronellal and citronellol contents in plants regenerated from callus, seedlings and mature trees. The percentage of oil present in shoots obtained from callus, seedlings and mature trees was 5, 10 and 20% respectively, when compared with the parents. The oil content of rooted plants from callus, seedlings and mature trees, however, was 40, 65 and 90% respectively, compared to the parents. More or less similar differences were also observed in the concentration of citronellal in shoots and rooted plants. Rooting had a striking effect on the synthesis of citronellol, which was not detected in shoots alone.

Analysis of Oil

In the second set of experiments, clonal plants grown in the field produced from vegetative buds of a mature tree were analyzed. The aim of this study was to determine the variation, if any, in the oil (citronellal and citronellol) concentrations in plants raised by tissue culture from the same parent tree. Table 9 documents the results of our analysis of plants at different ages. In this experiment only one plant was analyzed in each age group. Table 10 shows the results of four plants each, analyzed at 10 and 32 months respectively. In both these experiments, variation in the concentration of compounds in clonal plants compared to the parent tree was insignificant. These studies are preliminary and are being conducted using a higher number of replicates.

DISCUSSION

Our results demonstrate that: (1) the capacity for synthesis of significant quantities of monoterpenes, normally produced within

Table 6. Effect of Root Formation on Oil Content.

	Shoot %	After root initiation %	Rooted plant in tube %	Plant in pot (4 months) %
Total oil (dry weight basis)	0.12	0.32	0.65	0.89
Citronellal	10.70	25.60	49.60	56.70
Citronellol	N.D.	3.00	7.50	10.00

Shoots rooted at fourth subculture.
Initial inoculum from seedling (4 months).
N.D. - not detectable.

Table 7. Effect of Root Formation on Oil Content

	Parent tree (20 year) %	Shoot %	After root initiation %	Rooted plant in tube %	Plant in pot (3 months) %
Total oil (dry weight basis)	1.12	0.25	0.52	0.93	1.04
Citronellal	71.50	24.00	40.00	67.50	70.30
Citronellol	17.00	N.D.	4.38	11.25	15.50

N.D. - not detectable
Period for root development - 30 days.

Table 8. Comparative Data on Oil Content in Shoot and Rooted Plant Regenerated from Callus, Seedling and Mature Tree.

	Parent plant (4 month seedling) %	Parent tree (20 year old) %	Shoot (stem seedling callus) %	Shoot (seedling) %	Shoot (mature tree) %	Rooted plant in tube (callus) %	Rooted plant in tube (seedling) %	Rooted plant in tube (mature tree) %
Total oil (dry weight basis)	1.06	1.15	0.07	0.12	0.25	0.48	0.65	0.93
Citronellal	62.60	71.50	5.00	10.70	24.00	40.20	49.80	67.50
Citronellol	13.50	17.30	N.D.	N.D.	N.D.	5.00	7.50	11.20

Table 9. Oil Content in Plants Regenerated from Mature Trees in Field.

	Parent tree (20 year) %	Months				
		3 %	8 %	15 %	22 %	36 %
Total oil (dry weight basis)	1.15	1.06	1.10	1.11	1.09	1.13
Citronellal	71.50	70.80	71.02	70.91	70.25	70.56
Citronellol	17.30	15.90	16.28	16.86	16.90	17.00

Table 10. Oil Content of Tissue Culture Raised Plants.

S. No.	Months	Total oil (Dry weight basis) %	Citronellal %	Cirtronellol %
1	10	1.10	71.00	16.00
2	10	1.08	70.90	15.90
3	10	1.12	70.60	17.00
4	10	1.06	70.10	16.50
5	32	1.12	71.10	17.00
6	32	1.10	70.90	17.10
7	32	1.13	70.50	16.80
8	32	1.00	70.20	17.20
Parent tree	20 year old	1.15	71.50	17.30

or in association with specialized cells, is lost during callus growth. This finding confirms earlier reports (1, 2); (2) organogenesis is necessary, and roots play an important role in oil biosynthesis. This is evident from the fact that there is marked increase in oil content with the appearance of citronellol on root initiation; (3) plants regenerated from mature tree explants contain higher concentrations of oil, citronellal and citronellol than plants regenerated from seedlings or callus cultures; and (5) the concentration of the total oil, citronellal and citronellol in clonally produced plants were identical to the 20 year old parent tree. This preliminary experiment conducted on a small number of plants also demonstrates the vast potential of this technique for clonally multiplying high oil containing "elite" Eucalyptus trees.

ACKNOWLEDGEMENT

The authors thank Dr. V. Jagannathan for his interest during the course of this study.

REFERENCES

1. A.D. Krikorian and F.C. Steward, Biochemical differentiation: The biosynthetic potentialities of growing and quiescent tissue, in: "Plant Physiology, Volume VB," F.C. Steward, ed., Academic Press, New York (1969).
2. T. Suga, T. Hirata and Y. Yomamoto, Lipid constituent of callus tissues of Mentha spicata, Agric. Biol. Chem. 44:1817-1820 (1980).
3. H. Sugisawa and Y. Ohnishi, Isolation and identification of monoterpenes from cultured cells of Perilla, Plant Agric. Biol. Chem. 40:231-232 (1975).
4. P.K. Gupta, R.A. Nadgauda, R.R. Hendre, A.F. Mascarenhas and V. Jagannathan, Tissue culture of forest trees. 1. Multiplication of Eucalyptus citriodora Hook from isolated seedling segments, in: "Proc. All India Symp. and Third Conf. Plant Tissue Culture," Baroda (1978).
5. P.K. Gupta, A.F. Mascarenhas and V. Jagannathan, Tissue culture of forest trees - clonal propagation of mature trees of Eucalyptus citriodora Hook by tissue culture, Plant Sci. Lett. 22:195-201 (1981).
6. T. Murashige and F. Skoog, A revised medium for rapid growth and bioassays with tobacco tissue culture, Physiol. Plant. 15:473-497 (1962).
7. P.R. White, "The Cultivation of Animal and Plant Cells," The Ronald Press Co., New York (1963).

IN VITRO PROPAGATION OF APPLE

G.S. Cheema and D.P. Sharma

Regional Fruit Research Station
Mashobra,
Simla 171007
INDIA

INTRODUCTION

The apple industry is undergoing change throughout the world. The old established cultivars like Delicious are giving way to new compact spur types that hold the advantage of producing more fruit of better quality per unit land area per unit time in high density planting systems (1). Breeding programs also are producing new cultivars which are more attractive, high yielding, insect and pest resistant, and are of better storage quality (2 - 7). Such cultivars, however, are available in limited quantity and it is just not possible to multiply trees at a faster rate, since most of scion cultivars are difficult to root from shoot cuttings. The alternative method of producing shoots on root cuttings has also proved quite unsuitable (8). In nursery, apple scion cultivars are propagated by budding or grafting onto seedlings from seeds or vegetatively propagated root stocks which themselves are raised by stooling or layering; a process that usually takes three years.

Propagation through *in vitro* cultures of apple cultivars has been attempted in the past (9 - 20) with varying success, utilizing seedling and mature tree explants. The objective of the present study was to focus attention on the problem areas which impede commercial tissue culture development and devise reliable, reproducible and rapid methods for mass clonal propagation of apple cultivars using explants strictly from *in vivo* growing mature trees. This paper describes techniques for extra-rapid production of self-rooted trees, which can supply 1,000,000 plants in six months from a single shoot-tip.

MATERIALS AND METHODS

Choice of Material

The material chosen for study comprised shoot-tips taken from *in vivo* growing mature seedling trees over 25 years old so that the methods developed can be easily applied to other cultivars of apple (*Malus domestica* Borkh.).

Culture Media

A number of culture media, MS (21), LS (22), J (23), and Ch (12), with various modifications with respect to quantities of minerals and growth substances, were tried to select a basal medium. These were supplemented with indoleacetic acid (IAA), α-naphthalene-acetic acid (NAA), 6-benzylaminopurine (BAP), gibberellic acid (GA_3) (0.01-4.0 mg/l), casein hydrolysate (CH) (200-2000 mg/l), phloroglucinol (PG) (162 mg/l), activated charcoal (Ac) (1-2%), singly and in combination. All these media were invariably supplemented with 3% sucrose unless mentioned otherwise. A pH of 5.5-5.8 was adjusted with 1 N NaOH or 1 N HCl before adding agar (0.8-1.0%). The media were steam sterilized at 1.1 kg/cm^2 pressure for 15-20 min.

Cultural Conditions

All the cultures were kept in a temperature controlled room at relative humidity of 55-60%. The temperature of the room was maintained at 25°C ± 3°C. The source of light consisted of two four foot white fluorescent tubes (40W) and an incandescent lamp (25W, 100W). The intensity of light at the level of cultures was 3000-3500 lux. The light regime consisted of 16 h light/3 h dark. The subculture (passage) of cultures by transfer onto fresh media was generally performed after 4 weeks, unless desired otherwise and mentioned.

RESULTS AND DISCUSSION

Establishment of Sterile Cultures

Three to four days after inoculation of explants, it was observed that the explants exuded yellowish-brown exudates into the medium, which not only inhibited further growth of explants but also adversely affected their survival. A regular transfer to fresh media of the same composition after every 7 days was a must. The use of liquid media did not help much in this problem because of very poor survival of shoot-tips/axillary buds after this treatment. The horizontal positioning and embedding the explants half of their thickness into the medium was very effective. The inclusion of phloroglucinol in media was particularly effective in the early establishment of explants and retardation of microbial growth as observed by Jones et al. (16). Since these authors showed their cultures infected with latent microbes and could not establish sterile cultures without PG in the medium, a severe screening for sterile cultures was effected.

Screening for Sterile Explants

Each of the explants established was tested for latent infection of bacteria/fungi by growing them on media further supplemented with casein hydrolysate (2000 mg/l) and incubating the cultures at 30°C in the dark for four weeks. This procedure showed that infected cultures show profuse contaminant growth within a week, covering the whole explant. Such cultures were discarded. Only on the negative test were the cultures utilized further for study. Figure 1 shows an apple shoot-tip on basal MS medium after 4 weeks growth. The establishment of sterile explant culture was considered complete when no exudation in subsequent subcultures was evident.

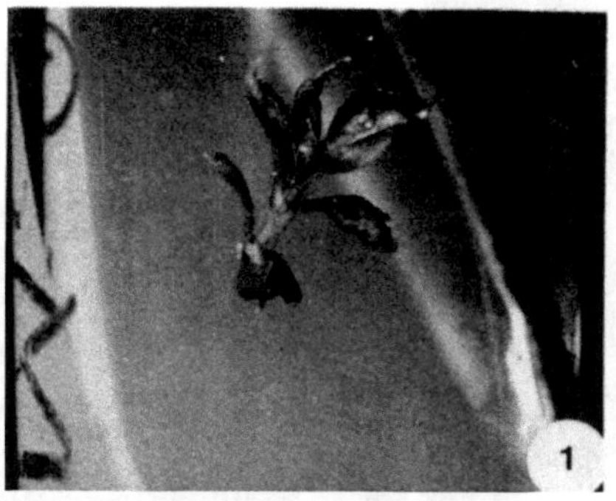

Figure 1. 4 week old apple shoot-tip on basal medium.

Proliferation of Axillary Buds

The different media tested had profound effect with the same growth regulator content, which revealed that the concentration of minerals and the organic constituents were perhaps the major factors causing difference in response. The various parameters tested such as axillary bud break, axillary growth, number of leaves/axillary bud, characteristics of leaves and shoots etc., led us to adopt MS. The reduction of concentration of minerals to half that of MS (as in Ch medium) produced highly hydrated shoots, which were very sensitive to injury while handling and initiated much basal callusing.

On MS + BAP (2.0 mg/l) + NAA (0.1 mg/l) + Ch (2000 mg/l) (MBNC), there was prolific growth of axillary shoots and many adventitious shoots developed from the basal internodes and its callus (Figure 2). On modified MS + BAP (1.0 mg/l) + IBA (0.1 mg/l) + GA_3 (0.5 mg/l), there was prolific growth of slender shoot axes accompanied with a good amount of basal callus, but leaves remained relatively small and shoots again were sensitive to injury while handling (Figure 3).

The best medium so far devised was MS + BAP (1.0 mg/l) + IBA (0.2 mg/l) (MBI), which supported very healthy, dark green, prolifically growing axillary shoots ans well developed leaves (Figure 4). A multiplication ratio of 1:8-14 shoots per four week interval has been obtained. The shoot axes remained relatively smaller in MBI than in MBNC, but leaves were better developed in the former. It was interesting to note that in none of our media with or without Ch was there any browning of basal portions of shoot-tips even after prolonged culture, i.e., 10-12 weeks, on the same medium. This was in contrast to phloroglucinol supplemented media where this was a

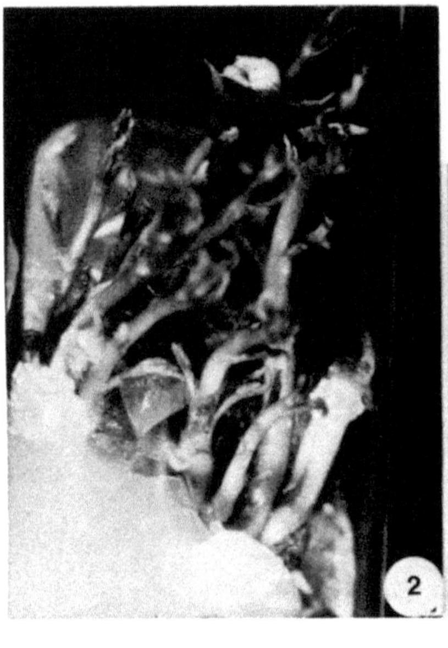

Figure 2. 8 week old cultures of a single shoot-tip showing prolific growth of axillaries and adventitious buds from basal internodes and its callus on MS + BAP (2.0 mg/l) + NAA (0.1 mg/l) + Ch (2000 mg/l).

Figure 3. 8 week old culture of a single shoot-tip showing prolific growth of shoot axes and basal callus on modified MS + BAP (1.0 mg/l) + IBA (0.1 mg/l) + GA_3 (0.5 mg/l).

Figure 4. One week old culture of shoots from medium (Fig. 3) in MS + BAP (1.0 mg/l) + IBA (0.1 mg/l) + GA_3 (0.5 mg/l) showing excellent shoot and leaf growth.

very characteristic feature of ageing cultures as observed while working with apple and other woody plants.

Since inclusion of PG in the medium was consistently giving reduced growth, comparison of growth and proliferation of shoot-tips was made in MBI medium supplemented with PG and AC (2%), the results of which are presented in Table 1 (Figure 5). As is evident, in the case of charcoal medium, the proliferaton of shoot-tip was scarce, but good shoot and leaf growth was attained. The baneful PG effect evident in the present study is consistent with the observation of Zimmerman and Broome (19) in apple cultivars, Jones and Hopgood (24) in mazzard, Hammerschlag (25) in peach, Zimmerman and Broome (26) in thornless blackberries, and Singha (27) in pear and sweet cherry, apple rootstocks, mazzard propagation (Cheema, unpublished). Therefore, it is not longer included in the multiplication medium. However, this shall have to be tested with other apple cultivars also.

Organogenesis/Adventitious Shoot Formation

In the multiplication media, MBNC and MBI, in a good number of cultures adventitious shoot development from the basal internodes and/or its callus has been noted quite often (Figures 2, 5). In still other cases, internodes and leaves, on touching the medium, produced green compact nodules which again organized into slow growing shoots. This system is being further strengthened for routine regeneration/differentiation and is being utilized for histological study of shoot induction.

Table 1. Comparison of Growth Rate, Multiplication Rate and Mode of Proliferation of Apple Shoots in MBI Medium Supplemented with Phloroglucinol and Activated Charcoal. (Data based on minimum of four replicates.)

Medium	Growth value*	Multiplication rate	Remarks
MBI	247.18	1:8-14	Prolific multiplication of shoots by axillaries, adventitious shoots in callus and internodes. (Figure 5A)
MBI + PG (162 mg/l)	226.50	1:5	Axillary shoot growth only, marked browning of shoot bases, adventitious shoots scarce. (Figure 5B)
MBI + AC (2%)	83.83	1:1-2	Axillary and adventitious shoot growth very scarce; original shoot/explant and leaf growth. (Figure 5C)

* Growth Value (After 4 wks) $= \frac{\text{Increase in Fresh Weight}}{\text{Initial Fresh Weight}} \times 100$

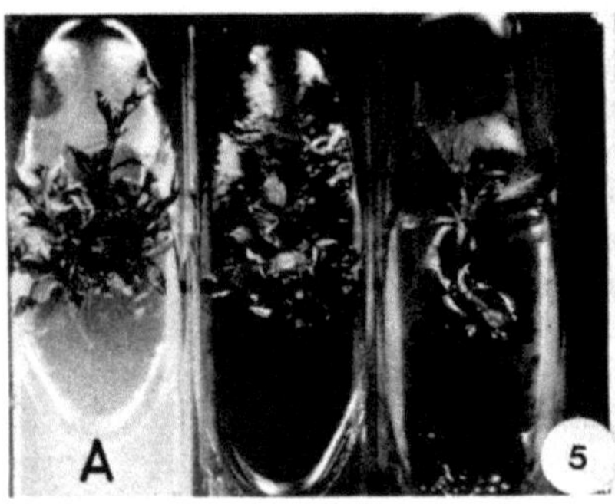

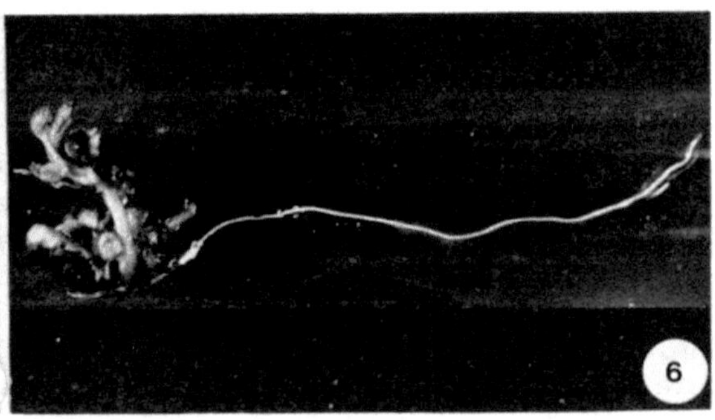

Figure 5. Comparison of shoot-tip proliferation and growth on: A = MS + BAP (1.0 mg/l) + IBA (0.2 mg/l); B = A + PG (162 mg/l); C = A + AC (2%). Also note prolific adventitious shoot formation from basal portions on A, all after four weeks.

Figure 6. A shoot rooted on modified MS + IBA (5.0 mg/l) showing five week growth on modified basal MS.

Induction of Rooting

Green healthy shoots (0.5-1.0 cm) whether resulting from callus or axillary/adventitious shoots on multiplication medium are excised and transferred to modified MS (half strength macroelements, 2% sucrose, 100 mg/l mesoinositol, and 0.5 mg/l thiamine HCl, while minor- and micro-elements remain the same) (MMS) + IBS (3-5 mg/l). Within a period of 4-5 weeks, 33% of the cultures showed 2-3 strongly developed roots on higher concentration of IBA. The plantlets showed active growth of root and shoot when transferred to the same medium but devoid of any growth substances (Figure 6). The inclusion of AC (2%) in the rooting medium has been found to increase the percentage of rooting and reduce basal callusing.

Acclimatization of Tissue Cultured Plantlets in Soil

To date very little is known on this very important transition stage. It has been the most difficult of all stages to us. We have made efforts to overcome this problem by taking into consideration the environmental condition of the different potting mixture substrates and atmospheric conditions necessary for proper hardening of plantlets. A number of substrates, viz. sand only, sand:soil: compost (1:1:1), moss only, agar gelled-water medium of different concentrations, were tried, but plants have succumbed ot everything once out of culture vessel. It has been observed that plants during this stage died more due to high moisture (which damaged roots) and high temperatures (25°C-28°C), than of moisture stress in roots at a reasonable temperature of 20°C ± 2°C. The experience of apple rootstocks experiments have shown that before transferring the rooted plantlets to field conditions, a 2-5 min. dip in 0.2% urea solution with IBA (1.0 mg/l), yielded good success in transplantation even in direct transfer to ordinary garden soil at above conditions. By this time, we have successfully acclimatized apple rootstock plantlets in pots, however, the apple plantlets are being acclimatized on similar lines so that they can be successfully transferred to nursery or field.

The procedures outlined in this study will be particularly helpful in:

1. Year-round large scale multiplication of commercial cultivars of new varieties for meeting the demands of growers;
2. Obtaining disease-free plantlets by meristem culture;
3. Incompatibility studies under controlled conditions in short time should lead to easy assessment of the nature of rootstock-scion interaction.
4. Shoot-tip grafting *in vitro* for raising and rearing hybrid embryos from distant crosses and early release of new varieties;

5. Study of infection mechanism, establishment and perpetuation of causal pathogen;
6. Study of the physiology of phase change and dormancy;
7. Mutation breeding;
8. Creation of germplasm bank by low temperature storage and/or cryopreservation for conservation, maintenance, and/or exchange of various cultivars.

REFERENCES

1. C.J. Seelay, Apple tree photosynthesis: an introduction, Hort. Sci. 13:640 (1978).
2. A.G. Brown, Apple, in: "Advances in Fruit Breeding," J. Janick and J.N. Moore, eds., Purdue University Press, Indiana (1975).
3. M.B. Crane, R.M. Greenslade and R.M. Massee, Studies on the persistence and immunity of appels to wooly aphid (Eriosoma lanigerum Hausm.), J. Pomology 14:137-163 (1936).
4. R.L. Knight, J.B. Brigs, A.M. Massee and H.M. Tydeman, The inheritance of resistance to wolld aphid, Eriosoma lanigerum Hausm. in apple, J. Hort. Sci. 37:207-217 (1962).
5. R.C. Lamb, H.S. Aldwincle, R.D. Way and D.E. Terry, "Liberty" apple, Hort. Sci. 14:757-758 (1979).
6. D.F. Dayton, J.B. Mowry, E.B. Williams, J. Janick, F.U. Emerson, L.F. Hough and C. Bailey, "Jonafree" apple, Hort. Sci. 14:551-552 (1979).
7. C. Stushnoff, S. Munson, L.B. Hertz, W. Gray and D.K. Wildung, "State Fair", "Sweet Sixteen", and "Keepsake" apple, Hort. Sci. 15:542-543 (1980).
8. J.C. Robinson and W.W. Schwabe, Studies on the regeneration of apple cultivars from root cuttings. I. Propagations aspects, J. Hort. Sci. 52:205-220 (1977).
9. R.F. Elliott, Axenic culture of shoot apices of apple, NZ J. Bot. 10:254-258 (1972).
10. D.G.A. Walkey, Production of apple plantlets form axillary bud meristems, Can. J. Plant Sci. 52:1085-1087 (1972).
11. A.J. Abbott and E. Whiteley, Culture of Malus tissues in vitro. I. Multiplication of apple plants from isolated shoot apices, Scientia Hortic. 4:183-189 (1976).
12. T.Y. Cheng, Clonal propagation of woody species through tissue culture techniques, Proc. Plant Prop. Soc. 28:139-155 (1978).
13. W. Huth, Kulture von apfelpflanzen aus apikalen meristemen, Gartenbauw. 43:163-166 (1978).
14. W. Lane, Regeneration of apple plants from shoot meristem tips, Plant Sci. Lett. 13:281-286 (1978).
15. S.F. Liu, W.L. Chen, H.H. Wang and S.Y. Yang, Shoot-tip culture of apple rootstocks and apple seedlings in vitro, in: "Proc. Symp. Plant Tissue Culture," Science Press, Peking (1978).

16. O.P. Jones, C.A. Pontikis and M.E. Hopgood, Propagation *in vitro* of five apple scion cultivars, *J. Hort. Sci.* 54:155-158 (1979).
17. S. Morini, Revista della ortoflorofruitcoltura (Preliminary studies on the micropropagation of apple), *Italiana* 64:41-51 (1980).
18. C. lundergan and J. Janick, Regulation of apple shoot proliferation and growth *in vitro*, *Hort. Res.* 20:19-24 (1980).
19. R.H. Zimmerman and O.C. Broome, Apple cultivar micropropagation, *in*: "Proc. Conf. Nursery Production of Fruit Plants Through Tissue Culture - Applications and Feasibility," USDA, Beltsville, MD (1980b).
20. S. Sriskandarajah and M.G. Mullins, Micropropagation of "Granny Smith" apple: factors affecting root formation *in vitro*, *J. Hort. Sci.* 56:71-76 (1981).
21. T. Murashige and F. Skoog, A revised medium for rapid growth and bioassays with tobacco tissue cultures, *Physiol. Plant.* 15:473-497 (1962).
22. E.M. Linsmaier and F. Skoog, Organic factor requirements of tobacco tissue cultures, *Physiol. Plant.* 18:100-127 (1965).
23. O.P. Jones, M.E. Hopgood and D. O'Farrell, Propagation *in vitro* of M.26 apple rootstock, *J. Hort. Sci.* 52:235-238 (1977).
24. O.P. Jones and M.E. Hopgood, The successful propagation *in vitro* of two rootstocks of *Prunus*: the plum rootstock "Pixy" (*P. institia*) and the cherry rootstock F12/1 (*P. avium*), *J. Hort. Sci.* 54:63-66 (1979).
25. F. Hammerschlag, Peach micropropagation, *in*: "Proc. Conf. Nursery Production of Fruit Plants Through Tissue Culture: Application and Feasibility," USDA, Beltsville, MD (1980).
26. R.H. Zimmerman and O.C. Broome, Micropropagation of thornless black berries, *in*: "Proc. Conf. Nursery Production of Fruit Plants through Tissue Culture: Applications and Feasibility," USDA, Beltsville, MD (1980a).
27. S. Singha, *In vitro* propagation of "Seckel" pear, *in*: "Proc. Conf. Nursery Production of Fruit Plants Through Tissue Culture: Applications and Feasibility," USDA, Beltsville, MD (1980).

ASEXUAL MASS PROPAGATION OF ORCHIDS AND ITS COMMERCIALIZATION: A REVIEW OF THE PRESENT STATUS

C.J. Goh

Botany Department
National University of Singapore
Singapore 0511

INTRODUCTION

Orchid is the first horticultural crop successfully mass propagated through tissue culture. It was first shown by Morel (1) in his attempt to produce virus-free Cymbidium from diseased plants. By culturing shoot apices on Knudson III (C) agar medium in small test-tubes, Morel observed that the colorless explant became green and enlarged slowly as a small flat bulblet looking exactly like the protocorm which develops from an embryo. These protocorm-like bodies further differentiated into clumps of protocorms, each developing into a new plant. Later, Morel (2) discovered that the protocorms, when cut into small pieces and subcultured, could regenerate more protocorms, and he estimated it was possible to obtain more than 4,000,000 plants in a year from a single bud. This clonal propagation technique created intense interest among orchid growers and had a tremendous impact on the orchid industry. Many leading orchid growers and nurseries around the world include tissue culture technique as a routine in their operations.

Since the early work of Morel with Cymbidium, many orchid genera have been successfully cloned by this tissue culture technique. Murashige (3) enumerated 22 genera, while Arditti (4) listed 35 including some intergeneric hybrids. Further successes in other species or hybrids reported recently included Aranthera (5), Ascocenda (6), Holttumara (7), and Renantanda (8).

SOURCE OF EXPLANTS

Since Morel's work in 1960, shoot tips remained the most commonly used explants. It should be noted that while Morel (1) used very small explants (0.1 mm thick) consisting of shoot apical meristem together with a few primordial leaves, most workers used much larger pieces of about 3-5 mm. Thus, shoot tips of Cymbidium hybrids were used as explants by Wimber (9), Wilfret (10), Steward and Mapes (11), Wreckmeister (12), and Fonnesbech (13). For Cattleya, Morel (2) cultured the shoot tips but found that it was relatively difficult to propagate compared to cymbidiums. Scully (14), however, successfully cultured shoot tips of Cattleya and obtained protocorms. Shoot tips of Dendrobium were also cultured by Kim et al. (15) who obtained protocorms after 1 month. Intuwong and Sagawa (16) and Singh (17) reported that shoot tips from side shoots of Dendrobium were also able to grow easily in vitro. Shoot tips of many other sympodial orchids had also been used as explants (4).

For monopodial orchids, success was obtained in Rhynchostylis (18), Vanda (19, 20, 21), Phalaenopsis (22), Aranda (23, 24), as well as Aranthera (5).

Wilfret (10) demonstrated that formation of protocorm-like bodies was not restricted to the apical dome of Cymbidium. Subapical tissues, 8 mm from the tip, also proliferated to produce protocorm-like bodies. Wreckmeister (12) observed that axillary buds of Cymbidium, when cultured on agar medium, proliferated and embryo-like bodies were formed. Axillary buds of Dendrobium were also successfully cultured which gave rise to plantlets (15, 17, 25). Kim et al. (15) while working on Dendrobium, also showed the relative efficiency of protocorm production from shoot tip and axillary buds. From most of the shoot tips, protocorms developed in one month old cultures, but most of the axillary buds grew directly into shoots without giving rise to any proliferating tissues. Almost similar observations were made by Sagawa and Shoji (26). Mosich et al. (25) reasoned that the lower percentage of protocorm production in cultured Dendrobium buds was due to auxin inhibition produced by apical meristem. Axillary buds in many monopodial orchids such as Vanda and Aranda are usually poorly developed. However, they had also been cultured successfully (24, 27), although a decrease in the generative capacity was observed (28).

Other plant parts like roots, leaves, inflorescence axis, and flower bud primordia had also been used, but success is limited to a few specific examples. Root was used as an explant in Neottia nidus-avis (29). Recently, plantlet roots of Epidendrum were shown to develop callus when cultured and in one case gave rise to one plantlet (30).

More successful results were obtained from propagation of

Phalaenopsis through buds excised from floral stalks (31, 32, 33). Inflorescence of three different Sarcanthus orchid hybrids cultured in vitro successfully developed proliferating tissues which later differentiated into plantlets (34).

Leaf tissues of orchids appeared to be a potential source of explants. In Malaxis paludosa, foliar embryos developed spontaneously (35). In the case of Cymbidium, Wimber (36) reported dedifferentiation of tissues at leaf tips leading to plantlet formation in vitro. The growth potential of seedling leaves was again demonstrated by Champagnet et al. (37). The meristem at detached leaf base gave rise to callus which in turn differentiated into many protocorms. In Laeliocattleya and Epidendrum, leaf tips cultured in vitro produced proliferating bodies (38, 39). Growth of callus tissues obtained from leaf tips of Epidendrum was found to be influenced by orthochlorophenoxyacetic acid (40). Juvenile leaves were found to be very suitable material for propagation and production of additional plants in Aranda Nancy (41). Similar results were obtained in Phalaenopsis (42, 43), Cattleya and Dendrobium (6).

COMPOSITION OF CULTURE MEDIA

Nutritional requirements of plant tissues cultured in vitro were discussed by White (44) and Street (45). Useful reviews were published by Gautheret (46), Butenko (47) and Murshige (3).

Marston and Voraurai (48) and Morel (49) have also published reviews associated with composition of nutrient media particularly suitable for orchid tissues. Most of the media used for orchid tissues cultured in vitro are those which have been used for many years for the germination of orchid seeds (50, 51). Detailed composition of each medium is available in the manual compiled by Arditti (4). Three of these (namely, Vacin and Went, Knudson C, and Murashige and Skoog), which are used in our laboratory, are presented in Table 1 for easy reference.

Coconut water is commonly used as additive to basal media for orchid tissues. The usefulness of coconut water for in vitro culture of plant tissues and embryos was proposed by Van Overbeek et al. (52). Coconut water was found to induce cell division in excised carrot phloem (53), tobacco pith (54) and many other plant tissues. In orhcids it has long been known that inclusion of coconut water into basal media improved seed germination and seedling growth (52, 55, 56, 57).

The effect of coconut water on growth of meristem tissues of Cattleya, Dendrobium or Vanda was very striking. It promoted the division of epidermal cells, leading to the formation of protocorms (20). It was added to the basal medium in the range of 10 to 25%

Table 1. Chemical Composition of Vacin and Went's Medium (VS), Murashige and Skoog's Medium (MS) and Knudson C Medium (KC); expressed in mg/l.

Chemicals	VW	MS	KC
A. Major Elements			
$Ca_3(PO_4)_2$	200		
$CaCl_2.2H_2O$		440	
$Ca(NO_3)_2.4H_2O$			1,000
Ferric Citrate	28		
$FeSO_4.7H_2O$		27.8	25
KNO_3	525	1,900	
KH_2PO_4	250	170	250
$MgSO_4.7H_2O$	250	370	250
$(NH_4)_2SO_4$	500		250
NH_4NO_3		1,650	
Na_2-EDTA		37.3	
B. Minor Elements			
$CoCl_2.6H_2O$		0.025	
$CuSO_4.5H_2O$		0.025	
H_3BO_3		6.2	
KI		0.83	
$MnSO_4$	7	22.3	7.5
$Na_2MoO_4.2H_2O$		0.25	
$ZnSO_4$		8.6	
C. Organic Constituents			
Glycine		2.0	
Myo-inositol		100	
Nicotinic Acid		0.5	
Pyridosin-C+HCl		0.5	
Thiamin-HCl		0.1	

(11, 14, 15, 19, 21, 24, 26, 34, 58, 59). In our laboratory, we have routinely used 20% (v/v) of coconut water in our cultures of explants. The amount may be reduced to 10% after differentiation occurs on subcultures (28).

The beneficial effect of coconut water could be related to its growth regulator content. The most important ones are perhaps the cytokinins (60, 61). Instead of coconut water, other workers have added cytokinins and/or auxins in the media (11, 13, 14, 59). The pH of culture medium is usually adjusted to within the range of 5.2-5.4.

ISOLATION OF SHOOT TIPS AND AXILLARY BUDS (ARANDA HYBRIDS)

Shoots with 10-15 leaves are excised from plants and all the leaves are removed to expose the axillary buds and the shoot tips. The stem is first washed with running tap water and then cut into sections with 2-3 nodes each. These sections are placed in a beaker of diluted soap water with a few drops of Tween 80 as wetting agent for 3 minutes with constant shaking. They are again washed with a soft brush or non-absorbent cotton wool to remove any remaining dirt or sooty molds on the stem surface. After rinsing thoroughly with distilled water, the stem sections are surface sterilized in "Clorox" (commercial bleach - active ingredient: 5.25% sodium hypochlorite) solution.

Shoot tips are sterilized in 200 ml of 10% "Clorox" solution for 15 minutes. Longer exposure of shoot tips to "Clorox" is lethal and the tissues usually turn brown. 15% "Clorox" solution is used for stem sections with axillary buds. They are treated with two changes of solution each of 10 minutes duration. After sterilization, they are throoughly washed 2-3 times in sterilized water before isolation.

Isolation and inoculation of shoot tips and axillary buds are done under aseptic conditions. The stem sections are transferred to sterilized petri dishes. With the help of a pair of small forceps, leaf bases near the tip are removed and the apex (about 5 mm in length) is excised for inoculation.

For axillary buds, two incisions are made at right angles to each other to remove the bud from the stem. The shoot tips or axillary buds are sterilized again by dipping in 5% "Clorox" for one or two minutes with agitation followed by rinsing with sterilized distilled water. These explants are then inoculated into the flasks with nutrient medium.

CULTURE CONDITIONS

Liquid medium is preferred for the initiation of protocorm or callus proliferation of the explant materials. However, for the continued development of the protocorm to plantlet and seedling stages, agar medium can be used. When liquid medium is employed, culture flasks are usually placed on a shaker, either gyratory or reciprocal (lateral) movements are satisfactory. This agitation of medium serves several functions: 1) aeration of the medium, 2) dispersal and dilution of toxic chemicals produced by wounded or lysing cells, 3) easy availability of nutrients to explants, and 4) preventing the establishment of polarity. The same objectives can be achieved by bubbling filtered sterile air into the medium.

Light conditions are not critical in the initial establishment of the cultures. Our cultures are placed on a reciprocal shaker along the side of the laboratory with natural light (about 1,000 lux) filtered through glass windows. In culture rooms, "Gro-lux" fluorescent tubes (12 hr/day) at about 28°C provide satisfactory conditions for growth and development of cultures.

GROWTH AND DEVELOPMENT OF *ARANDA* EXPLANTS

Our work with *Aranda* shoot tip and axillary bud explants had been reported elsewhere (23, 24, 27). Most of these were conducted with Vacin and Went medium supplemented with 2% sucrose and 20% coconut water. In *Aranda* Hilda Galistan, explants of shoot tips began to enlarge in about three weeks. In many explants, the leaf primordia in the original explant continued to develop to from young leaves. However, the growth of these leaves was slowed down when protocorms began to develop from the explant tissues nearer the leaf bases (Figure 1a). In others, the growth of leaves was not observed and protocorms developed in about 2 months after inoculation. The protocorms proliferated very quickly and the whole explant was subsequently covered with protocorms (Figure 1b). These protocorms developed into plantlets. Yellowish callus tissues were also produced either from the explant or the differentiating protocorms. Callus masses were frequently detached from the clumps of differentiating protocorms due to the agitated condition of the medium. When these callus tissues were subcultured on agar medium, further proliferation of callus masses occurred rapidly, followed by differentiation. In some cases, protocorm-like bodies were observed while in others the tissues differentiated directly into leaf-like structures (Figure 1c). In both cases they continued to develop and normal plantlets with well formed leaves and roots were obtained in about a six month period (Figures 1d and e).

With axillary bud explants, enlargement of the explants was evident in about three weeks, resulting in a dome shaped structure

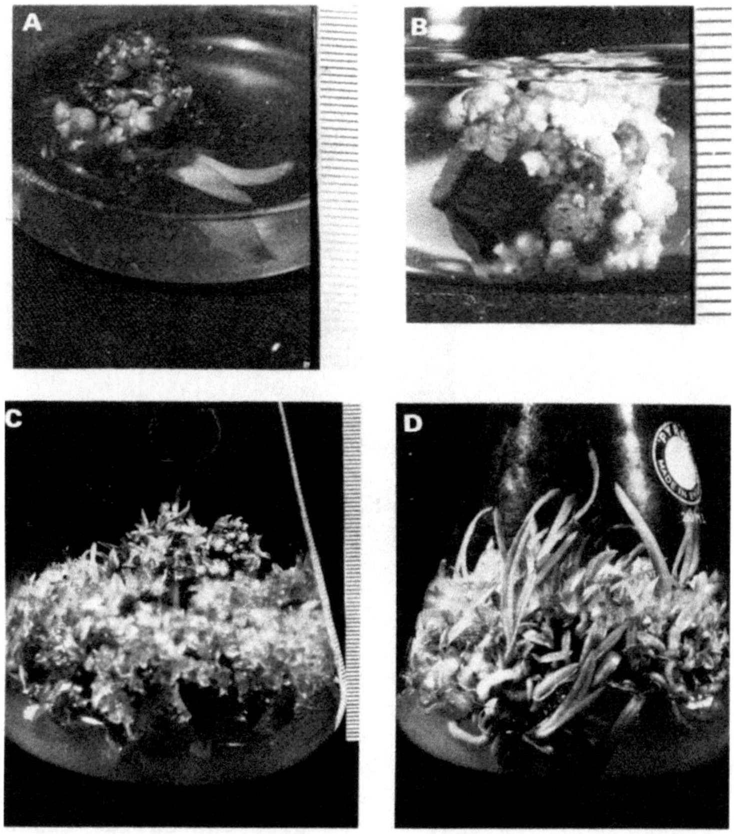

Figure 1. Tissue culture of Aranda Hilda Galistan (scale in mm).
A) (top left) Shoot tip explant with leaves, 4 months in culture.
B) (top right) Shoot tip explant without leaves, 4 months in culture.
C) (bottom left) Callus tissues differentiated into plantlets, 7 months in culture.
D) (bottom right) Plantlets, 8 month old culture.

in about two months. Protocorms began to differentiate from the explant in about three to four months. Very frequently, two groups of protocorms were produced diagonally from the basal portion of the dome shaped explant (Figure 1f). Subsequently, more and more protocorms developed and the whole explant was covered with them. These protocorms generally continued to differentiate into shoots while still attached to each other. Later, some roots were produced and plantlets were thus formed. As in the case of shoot tip cultures, callus tissues were also produced, and they differentiated into plantlets when subcultured on agar medium.

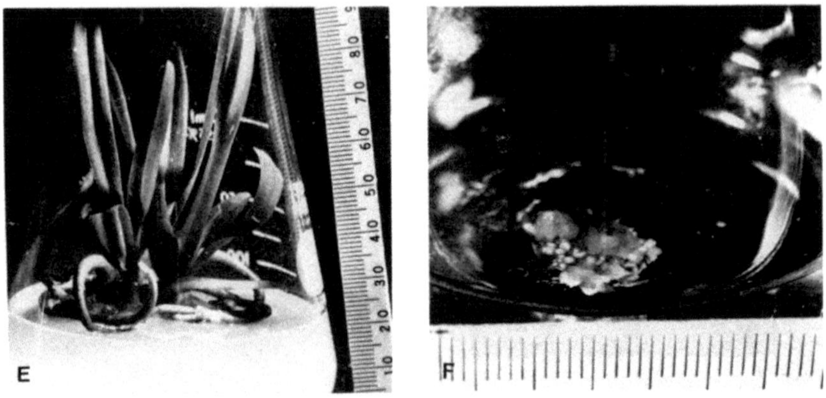

Figure 1. E) (left) Seedlings complete with leaves and roots, 12 months in culture.
F) (right) Protocorm production in axillary bud explant, 4 months in culture.

With Aranca Lucy Laycock, the growth of shoot tip explants was slower than that of A. Hilda Galistan. The developmental sequence was, however, similar and many protocorms developed in about five months after isolation (Figure 2a). These protocorms normally differentiated very fast into shoots while still attached to the original explants. Some protocorms detached from the clumps due to the agitated condition of the medium and produced callus from the wound surfaces (Figure 2b). On agar medium also the callus tissues proliferated and differentiated into protocorms (Figure 2c). Plantlets of about 3-4 cm in length were obtained in about 12 months (Figure 2d).

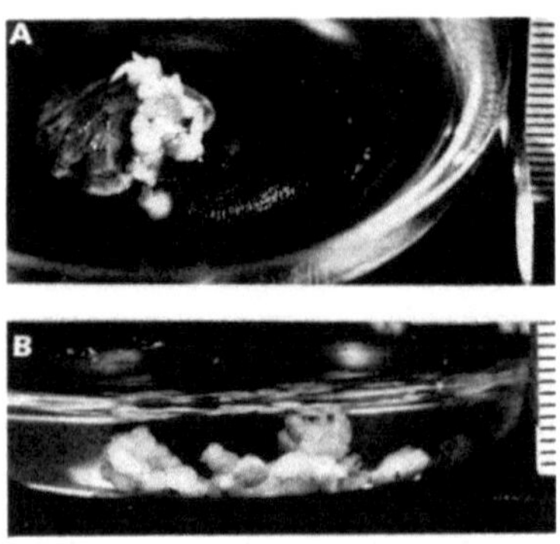

Figure 2. Tissue culture of Aranda Lucy Laycock (scale in mm).
A) (left) Shoot tip explant, 5 months in culture.
B) (right) Callus tissues produced from differentiating protocorms.

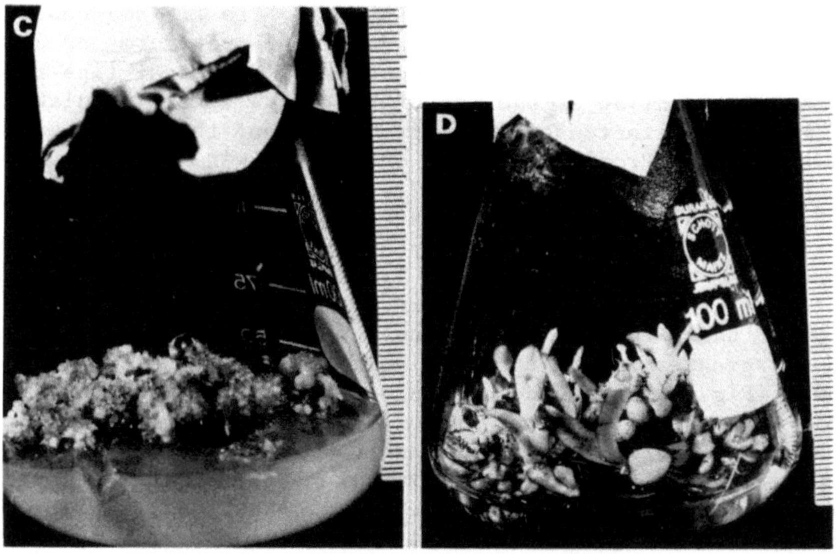

Figure 2. C) (left) Callus tissues differentiating into protocorms on agar medium.
D) (right) Plantlets of 12 month old cultures.

The axillary bud explants did not show any increase in size during the first two months of culture. Subsequently, the bud enlargement occurred and two to three small leaves of 0.5 cm long developed. Numerous protocorms then appeared form the basal part of the dome shaped explant and they developed into plantlets (Figure 2e). A twelve month period of culture was necessary to obtain plantlets of about 3 cm in length.

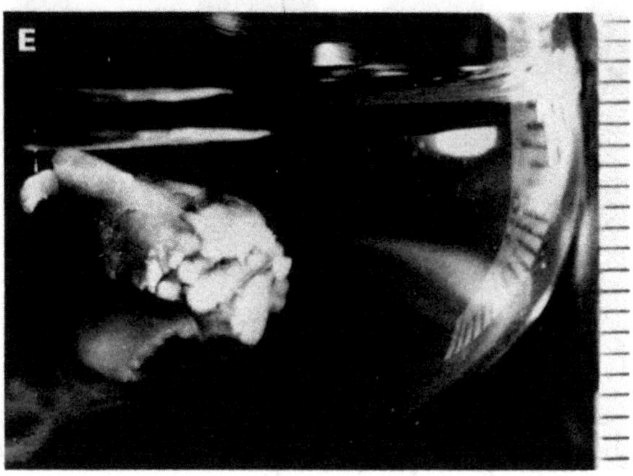

Figure 2. E) Axillary bud explant showing leaf and protocorm production, 5 months in culture.

The leaves of young plantlets (about 0.5 to 1.0 cm) could be used as explants for further proliferation of protocorms or callus tissues (41). With cultures of Aranda Nancy, the leaf tips remained green and were swollen at the end of two weeks after inoculation. Few of the leaves increased in their length and their tips were curved slightly. Prominent callus outgrowths were visible at the cut ends of the leaves and many callus lobes were prominent in 4-6 weeks time (Figures 3a and b). The cream colored callus showed smooth dome shaped outline, made up of homogeneous parenchyma, and the units were distinctly separate from each other. Such callus bodies were light green in color and proliferated in one or two weeks after their initial appearance. The volume of callus tissue formed was greater on the upper epidermis side of the leaf than on the lower (Figures 3c and d). Each unit of callus mass grew further and in 8-10 weeks after first inoculation gave rise to one or two protocorm-like bodies (Figures 3c and d). The leaf portion used as explant, callus as well as protocorm-like bodies that developed on it remained green throughout the period of experiment. Some of the protocorms were elongated, while others were globose or spherical. In 12-14 week old cultures, a clump of protocorms, each measuring about 1-2 cm, were present and these were compactly arranged. From 12 weeks onwards such protocorms were detached from the explant and they grew independently in the liquid medium (Figure 3e). On further development these developed into regular seedlings (Figure 3f).

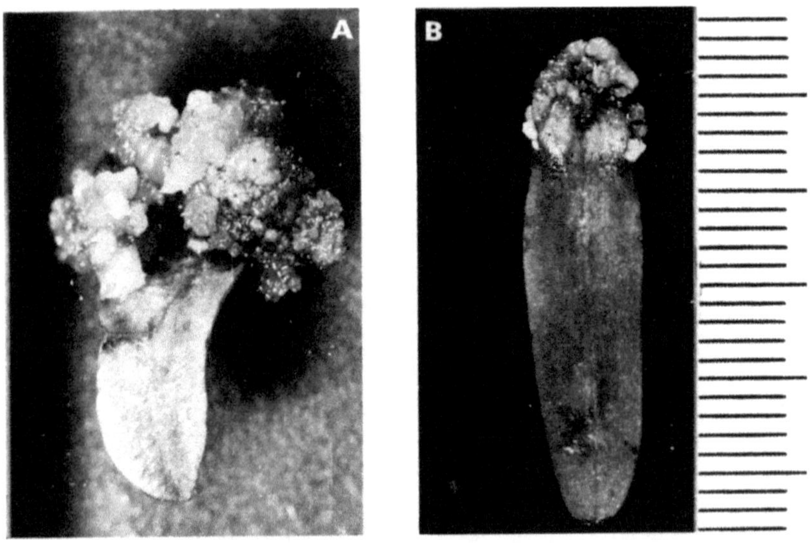

Figure 3. Leaf culture of Aranda Nancy (scale in mm).
A) (left) Callus production on cut end of leaf tip explant, 8 weeks in culture.
B) (right) Callus production of whole leaf explant, 4 weeks in culture.

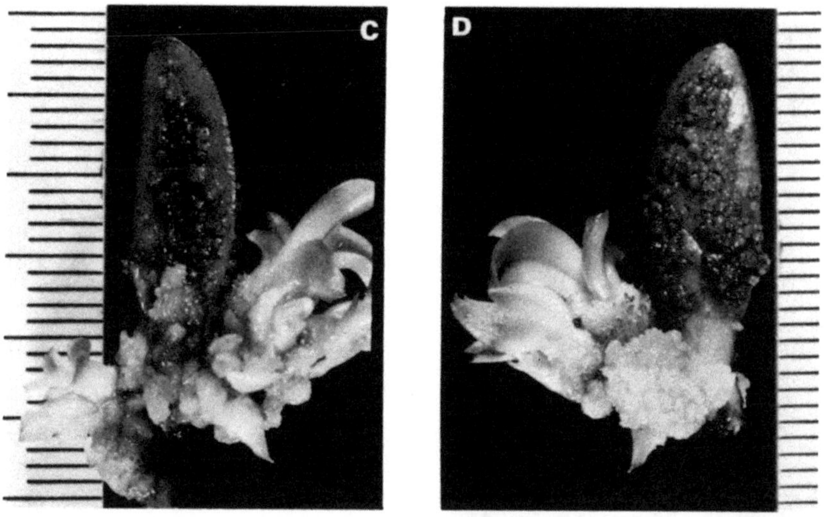

Figure 3. C) (left) Callus and protocorm production of whole leaf explant, 4 months in culture; view from upper epidermis.
D) (right) Same as (C), lower epidermis.

When the plantlets are about 10 to 15 cm tall, they are removed from culture flasks and planted into community pots using charcoal chips as the potting medium. They have to be sheltered from direct sun until new roots are produced and the seedlings are well established. Further transfer of the seedlings into bigger pots or eventually into soil (beds) depends on the rate of growth of the seedlings. The procedure and care for these seedlings are the same as those devised for seedlings from seeds.

MASS PROPAGATION

From the above account, it is seen that practically all parts of the orchid plant could be used as the initial explant materials as long as meristematic tissues are present. In our experiments, seedling leaves are most prolific in terms of protocorm production (41). Indeed, all parts of young *Phalaenopsis* seedlings were shown to produce more protocorms (43). Morel also used protocorms or young leaves for further proliferation of protocorms (2, 49). Thus, it is suggested that the apical shoot-tip, axillary buds, as well as young leaves of the mother plant are all to be used as initial explants. The protocorms generated from these explants are allowed to develop leaves, and these young leaves are to be excised for further protocorm production, thereby increasing the number of cultures to effect mass propagation in the shortest time possible.

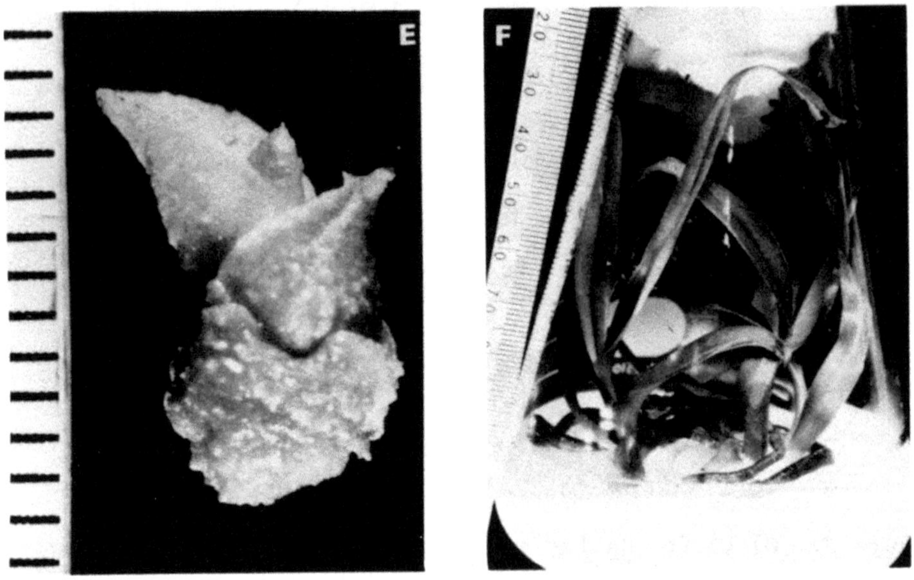

Figure 3. E) (left) A plantlet with 3 leaves but no roots, 3 1/2 months in culture.
F) (right) Plantlets complete with roots, 6 months in culture.

COMMERCIAL APPLICATIONS

The commercial aspects of the orchid industry have recently been summarized by Rao (62). Orchid nurseries are basically involved in two different aspects of operations: the sale of plants and the sale of blooms. Actually most nurseries do engage in both types of activities when business opportunities arise. However, they normally concentrate on only one of the two. This tendency is clearly advertised by the nurseries in orchid journals.

With the success in clonal propagation through tissue cultures, commercial nurseries were quick to employ the technique for mass propagation of selected hybrids or varieties. It is well known that organizations like Vacherot and Lecoufle in France were the first to adopt the techniques (62). Indeed, mericlones have been available from leading nurseries in western countries since the '60s. These mericlones are offered to orchid enthusiasts for home growing, as part of their orchid collection to be admired. Thus, rare plants or prized hybrids/varieties, which command lofty prices, are made available at a very much reduced rate. A majority of orchid nurseries in the western countries are engaged in this aspect of operation.

On the other hand, most of the orchid nurseries in South East Asia are dealing with cut flower trade. The orchid blooms are exported world-wide, but mainly to the European and American markets. The volume of orchid cut flower trade has more than doubled since the report by Rao (62). For example, the official value of Singapore's orchid export was $ 8.3 million (U.S. $ 4.0 million) in 1976, but increased to $ 16.9 million (U.S. $ 8.2 million) in 1980. This vast increase in the sale of orchid blooms indicates the increasing number of orchid plants required in the nurseries. As the conventional vegetative propagation by shoot cutting is a slow process, many nurseries are now resorting to tissue culture to supply the plants needed for flower production.

It has to be noted, however, that when there is already a good stock of an established clone, the conventional shoot cutting method of vegetative propagation is not only more economical, but also more reliable for multiplying the stock. With monopodial orchids, each plant can easily produce 4 to 10 side-shoots (keikis) in a year. Some nurserymen can obtain as many as 18 plants from one single mother plant in a year. The sympodials can also be caused to produce three or four new leads per plant per year. However, when one wants to mass propagate a new award-winning hybrid or variety (usually from one single plant or at most a few plants) within a short time period, tissue culture is the only method currently available.

Due to the very nature of the orchid industry - depending on the likes and dislikes of human idiosyncrasies, many new hybrids and varieties are regularly produced. Thus, whether a nursery is involved in the sale of plants or the sale of flowers, it has to offer new hybrids and varieties from time to time to satisfy public fancy. In this respect, perhaps, the cut flower trade is even more vulnerable to the pressure for change. Apart from the human desire for new flowers, they are also subject to keen competition amongst themselves: when blooms of a particular hybrid are in demand, more nurseries will grow that hybrid, leading eventually to over-production. Since flowers cannot be stored for long periods, competition for sale is keen and, consequently, the price is bound to decline and the returns for orchid growers become unprofitable. Whereas new varieties may fetch one to two (Singapore) dollars per inflorescence on the export market, the old ones may be down to 20 cents. The option for growers is obvious, bearing in mind that the same amount of care and labor is required to tend these orchid plants.

In the Singapore orchid cult flower trade, *Aranda* Wendy Scott is a good example, once a top export flower, it is being replaced by varieties of *Aranda* Christine. Thus, leading orchid growers are always on the lookout for new and better hybrids or varieties. The new find is then mass propagated to supply the blooms.

One other reason for the need of mass propagation of established varieties by tissue culture is the very reason which started Morel's experiments: to produce virus-free plants. Diseased plants are much less floriferous and the inflorescences are ususally distorted and become unacceptable for export. It is, therefore, necessary to produce disease-free clones.

While mericlones have been produced in many leading orchid nurseries in the West since the '60s, commercial tissue culture laboratories are recent establishments in South East Asian countries. The most active work is progressing in Thailand and numerous mericlones have been produced, currently there are six commercial laboratories producing about five million plantlets a year. In Singapore, though research work started early in the '60s and satisfactory results were obtained from various explant materials, commercial scale mericloning started only five or six years ago. Similarly, commercial mericloning in Malaysia was also rather recent. There are two laboratories in the Philippines which offer mericlones for sale. In Indonesia, tissue culture work was initiated recently.

CONCLUDING REMARKS

Orchids are the first horticultural crop successfully cloned through tissue culture. In fact, the current wave of tissue culture activity with other crops is an outgrowth of the success with orchids. While the intention of clonal propagation through tissue culture is to mass produce selected plants of desired quality or characters, it should be pointed out that the early claim that "mericlone" plants are all true to type (63) needs some qualifications. Variations of the progenies in varying degrees have been reported, and the causes for these slight variations are discussed by Vajrabhaya (64). Nevertheless, the percentage of plants which are apparently true to type is high, ranging from 78 - 100%, and many of the variations may indeed be beneficial, such as chromosome doubling or larger flower types (65). Thus, tissue culture method of clonal propagation becomes a standard in the orchid industry.

REFERENCES

1. G. Morel, Producing virus-free _Cymbidiums_, _Amer. Orchid Soc. Bull._ 29:495-497 (1960).
2. G. Morel, Tissue culture - a new means of clonal propagation of orchids, _Amer. Orchid Soc. Bull._ 33:473-478 (1964).
3. T. Murashige, Plant propagation through tissue culture, _Ann. Rev. Plant Physiol._ 25:135-166 (1974).

4. J. Arditti, Clonal propagation of orchids by means of tissue culture - a manual, in: "Orchid Biology - Reviews and Perspective 1," J. Arditti, ed., Cornell University Press, Ithaca (1977).
5. S.S. Irawati, S. Harjadi, H. Suseno and S. Idris, Tissue culture of Aranthera James Storie, Orchid Rev. 85:138-142 (1977).
6. F.M.L. Fu, Studies on the tissue culture of orchids. 2. Clonal propagation of Aranda, Ascocenda, Cattleya by leaf tissue culture, Orchid Rev. 87:343-346 (1979).
7. K.H. Teo and C.H. Wong, Effects of sucrose on the growth of protocorms of Holttumara Loke Tuck Yip, Orchid Rev. 86:285-289 (1978).
8. C.J. Goh and H. Tan, Clonal propagation from leaf explants in an orchid hybrid Renantanda Ammani, Plant Physiol. Suppl. 63: 116 (1979).
9. D.E. Wimber, Clonal multiplication of Cymbidium through tissue culture of the shoot meristem, Amer. Orchid Soc. Bull. 32:105-107 (1963).
10. G.J. Wilfret, Formation of protocorm-like bodies on excised Cymbidium shoot-tips, Amer. Orchid Soc. Bull. 35:823-827 (1966).
11. F.C. Steward and M.O. Mapes, Morphogenesis in aseptic cell cultures of Cymbidium, Bot. Gaz. 132:65-70 (1971).
12. P. Wreckmeister, Light induction of geotropism, and the control of proliferation and growth of Cymbidium in tissue culture, Bot. Gaz. 132:346-350 (1971).
13. M. Fonnesbech, Growth hormones and propagation of Cymbidium in vitro, Physiol. plant. 27:310-316 (1972).
14. R.M. Scully, Aspects of meristem culture of Cattleya alliance, Amer. Orchid Soc. Bull. 36:103-108 (1967).
15. K.K. Kim, J.T. Kunisaki and Y. Sagawa, Shoot-tip culture of Dendrobiums, Amer. Orchid Soc. Bull. 39:1071-1080 (1970).
16. O. Intuwong and Y. Sagawa, Clonal propagation of Dendrobium golden wave and other noble types, Amer. Orchid Soc. Bull. 44:319-323 (1975).
17. H. Singh, Meristem tissue culture of Dendrobium Ng Eng Cheow, Gardens' Bull. Singapore 28:259-267 (1976).
18. M Vajrabhaya and T. Vajrabhaya, Tissue culture of Rhynchostylis gigantea, a monopodial orchid, Amer. Orchid Soc. Bull. 39:907-910 (1970).
19. J.T. Kunisaki, K.K. Kim and Y. Sagawa, Shoot-tip culture of Vanda, Amer. Orchid Soc. Bull. 41:435-439 (1972).
20. G. Morel, Morphogenesis of stem apical meristem cultivated in vitro; application to clonal propagation, Phytomorphology 22:265-277 (1972).
21. K H. Teo, T. Kunisaki and Y. Sagawa, Clonal propagation of strap-leafed Vanda by shoot-tip culture, Amer. Orchid Soc. Bull. 42:402-405 (1973).
22. O. Intuwong and Y. Sagawa, Clonal propagation of Phalaenopsis

by shoot-tip culture, Amer. Orchid Soc. Bull. 43:893-895 (1974).

23. C.J. Goh, Meristem culture of Aranda Deborah, Malayan Orchid Rev. 11:10-15 (1973).

24. C.J. Goh, C.S. Loh and A.N. Rao, Clonal propagation of Aranda hybrids through shoot meristem culture, in: "Proceedings of Seminar Biology in Society," Singapore Natl. Acad. Science and Singapore Institute of Biology, Singapore (1975).

25. S.K. Mosich, E.A. Ball and J. Arditti, Clonal propagation of Dendrobium by means of node culture, Amer. Orchid Soc. Bull. 43:1055-1061 (1974).

26. Y. Sagawa and T. Shoji, Clonal propagation of Dendrobiums through shoot meristem culture, Amer. Orchid Soc. Bull. 36:856-859 (1967).

27. C.J. Goh and C.S. Loh, Further observations on meristem culture of Aranda Deborah, Malayan Orchid Rev. 12:10-13 (1975).

28. C.S. Loh, C.J. Goh and A.N. Rao, Some factors affecting morphogenesis of Aranda orchid tissue culture, in: "Proceedings of the Symposium or Orchidology," E.S. Teoh, A.N. Rao and C.J. Goh, eds., Orchid Society of South-East Asia, Singapore (1978).

29. M. Champagnat, Recherches sur la multiplication vegetative de Neottia nidus-avis Rich., Ann. Sci. Nat. Bot. Biol. Veg. 12:209-247 (1971).

30. J. Stewart and J. Button, Development of callus and plantlets from Epidendrum root tips cultured in vitro, Amer. Orchid Soc.Bull. 47:607-612 (1978).

31. R.M. Scully, Stem propagation of Phalaenopsis, Amer. Orchid Soc. Bull. 35:40-52 (1966).

32. A.T.Y. Tse, R.J. Smith and W.P. Hackett, Adventitious shoot formation on Phalaenopsis nodes, Amer. Orchid Soc. Bull. 40:807-810 (1971).

33. O. Intuwong, J.T. Kunisaki and Y. Sagawa, Vegetative propagation of Phalaenopsis by flower stalk cuttings, Na Okika O Hawaii - Hawaii Orchid Journal 1:13-18 (1972).

34. O. Intuwong and Y. Sagawa, Clonal propagation of Sarcanthus orchids by aseptic culture of inflorescences, Amer. Orchid Soc. Bull. 42:209-215 (1973).

35. R.L. Taylor, The foliar embryos of Malaxis paludosa, Can. J. Bot. 45:1553-1556 (1967).

36. D.E. Wimber, Additional observations on clonal multiplication of Cymbidium through culture of shoot meristems, Cym. Soc. News 20:7-10 (1965).

37. M. Champagnat, G. Morel and B. Monnetou, La multiplication vege tative des Cattleya a partir de jeunes feuilles cultiveęs aseptiqueruent in vitro, Ann. Sci. Nat. Bot. Biol. Veg. 11:87-114 (1970).

38. M.E. Churchill, J. Arditti and E.A. Ball, Clonal propagation of orchids from leaf tips, Amer. Orchid Soc. Bull. 40:109-113 (1970).

39. M.E. Churchill, E.A. Ball and J. Arditti, Tissue culture of orchids. I. Methods for leaf tips, New Phytol. 72:161-166 (1973).
40. J.M. Rudolph, E.A. Ball and J. Arditti, Tissue culture of orchids. III. Does orthochlorophenoxy acetic acid select for or induce anthocyanin production?, Amer. Orchid Soc. bull. 41:1074-1078 (1972).
41. C.S. Loh, A.N. Rao and C.J. Goh, Clonal propagation from leaves in the orchid Aranda, Jour. Singapore Natl. Acad. Sci. 4:97-99 (1975).
42. M. Tanaka, A. Hasegawa and M. Goi, Studies on the clonal propagation of monopodial orchids by tissue culture, J. Jap. Soc. Sci. (cited by M. Tanaka and A. Hasegawa, 1976) Amer. Orchid Soc. Bull. 45:1022-1024 (1974).
43. K Zimmer and W. Pieper, Clonal propagation of Phalaenopsis by excised buds, Orchid Rev. 86:223-227 (1978).
44. P.R. White, "A Handbook of Plant Tissue Culture," Jacques Catell Press, Lancaster, Pennsylvania, USA (1943).
45. H.E. Street, "Plant Tissue and Cell culture," Blackwell Scientific Publications, Oxford, England (1973).
46. R.J. Gautheret, The nutrition of plant tissue cultures, Ann. Rev. Plant Physiol. 6:433-484 (1955).
47. R.G. Butenko, "Plant Tissue Culture and Plant Morphogenesis," Israel Program for Scientific Translations, Jerusalem, Israel (1964).
48. M.E. Marston and P. Voraurai, "Multiplication of orchid clones by shoot meristem culture - a review of the literature," Dept. of Horticulture, University of Nottingham, England, Misc. Publ. #17 (1967).
49. G. Morel, Clonal multiplication of orchids, in: "The Orchids - A Scientific Study," C.L. Withner, ed., John Wiley and Sons, London (1974).
50. C.L. Withner, "The Orhcids: A Scientific Survey," The Ronald Press, New York (1959).
51. J Arditti, Factors affecting the germination of orchid seeds, Bot. Rev. 33:1-97 (1967).
52. J. Van Overbeek, M.E. Concklin and A.F. Blackeslee, Cultivation in vitro of small Datura embryos, Amer. J. Bot. 29:472-477 (1942).
53. S.M. Caplin, F.C. Steward, Effect of coconut milk on the growth of explants from carrot root, Science 108:655-657 (1948).
54. J.R. Jablonski and F. Skoog, Cell enlargement and cell division in excised tobacco pith tissues, Physiol. Plant. 7:16-24 (1954).
55. C P. Hegarty, Observations on the germination of orchid seed, Amer. Orchid Soc. Bull. 24:457-464 (1955).
56. D. Lawrence and J. Arditti, A new medium for the germination of orchid seeds, Amer. Orchid Soc. Bull. 33:766-768 (1964).
57. D.K. McIntyre, G.J. Veitch and J.W. Wrigley, Australian terrestrial orchids from seeds. II. Improvement in techniques and

further success, Amer. Orchid Soc. Bull. 43:52-53 (1974).
58. C.J. Goh, Tissue culture of Vanda Miss Joaquim, J. Singapore Natl. Acad. Sci. 2:31-33 (1970).
59. E.G.P. Lindemann, J.E. Gunckel and O.W. Davidson, Meristem culture of Cattleya, Amer. Orchid Soc. Bull. 39:1002-1004 (1970).
60. D.S. Letham, Regulators of cell division in plant tissues. XX. The cytokinins of coconut milk, Physiol. Plant. 32:66-70 (1974).
61. J. Van Standen and S.E. Drewes, Identification of zeatin and zeatin-riboside in coconut milk, Physiol. Plant. 34:106-109 (1975).
62. A N. Rao, Tissue culture in the orchid industry, in: "Applied and Fundamental Aspects of Plant Cell, tissue and Organ Culture," J. Reinert and Y.P.S. Bajaj, eds., Springer-Verlag, Berlin, Heidelberg (1977).
63. M. Lecoufle, A mass blooming of mericlones, Amer. Orchid Soc. Bull. 36:579-581 (1967).
64. T. Vajrabhaya, Variation in clonal propagation, in: "Orchid Biology: Reviews and Perspectives I.," J. Arditti, ed., Cornell University Press, Ithaca, New York, USA (1977).
65. U. Charanasri and J. Wanichkul, Qualitative and quantitative aspects of variation in mericlones of Dendrobium cultivars, in: "Proceedings of the Symposium on Orhcidology," E.S. Teoh, A.N. Rao and C.J. Goh, eds., Orchid Society of South-East Asia, Singapore (1978).

NEW STRATEGY TO TACKLE BREEDING PROBLEMS OF POTATO

G. Wenzel[1], V.A. Bapat[2] and H. Uhrig

Max-Planck-Institut fur Züchtungsforschung
Erwin Baur Institut
D-5000 Köln 30 (Vogelsang)
Federal Republic of Germany

[1]Biologische Bundesanstalt fur Land-und-Forstwirtschaft
Institut fur Resistenzgenetik
D-8059 Grünbach
Federal Republic of Germany

[2]Bhabha Atomic Research Centre
Bio-Organic Division
Bombay, India

INTRODUCTION

Indian and German potato growing have one common possession: the growing area for potatoes of about 500,000 ha is identical. For a small country like Germany this is, however, a relatively large area, and it demonstrates the importance of potato as one of the central German crops. It ranks on the fourth position after wheat, barley and sugar beet. On the other hand, in Germany the importance of potato is decreasing, while in India - like in most countries with subtropical climatic areas, cultivation of potato is increasing. Consequently, it should be useful to discuss possibilities and approaches to the present situation of how the breeding of new potato varieties may be enhanced.

To allow a judgement on the efficiency of new breeding strategies, one should know how a new variety is produced today. Potato is presently exclusively propagated vegetatively, and, therefore, its breeding is very easy. Each promising F_1 seedling arisen from any cross can be propagated to yield a new variety. In sharp contrast to other plants, one needs no further breeding for consistency and variety production. The autotetraploid nature of potato demands, however, large seedling numbers to ensure the production of a new variety. This counterbalances all other advantages of the classical breeding technique. The present calculation reckons 500,000 to 2,000,000 F_1 seedlings for selection of a new variety. The possible reduction of this number by using semiclassical or unconventional breeding procedures will be discussed in the following paragraphs.

MATERIALS AND METHODS

Parthenogenesis

Dihaploids with 2n = 2x =24 chromosomes were extracted from tetraploid breeding clones or varieties by pollination with the *S. phureja* clones IvP 35 and IvP 48 (1). Fertilization was performed in a greenhouse (temperature 22 $\pm$ 2°C) on cut plant stems kept in nutritional solutions.

Androgenesis

For androgenesis, we routinely use the technique of microspore culture within the anther (2). To summarize the most important steps: the anther donor plants are grafted on tomato stocks to prolong flowering by preventing tuberization and are illuminated during the short day season with 10,000 lux to give a 16h photoperiod at temperatures of 18 $\pm$ 2°C. From such plants flower buds of 4-6 mm length were harvested and precultured in dry test tubes for two days at 6°C in the dark. The anthers were dissected and plated on a Murashige and Skoog medium (MS) (3) with a total of 6% sucrose, 0.5% active charcoal, and 1 mgl^{-1} 6-BAP. The macroscopic structures which developed were transferred to MS medium with 10% coconut milk, 0.3 mgl^{-1} zeatin and 3% sucrose. After morphogenesis the tissue could be transferred to MS medium with 0.5 mgl^{-1} 6-BAP instead of zeatin. Functional shoots were transplanted as cuttings to the greenhouse.

Protoplast Culture and Fusion

Protoplasts were isolated from leaves of shoot cultures of dihaploid *S. tuberosum* clones. In Kao Michayluk medium (KM) (4) sustained cell divisions were obtained. After about 10 days the microcolonies were transferred to MS medium identical to the microspore regeneration medium (5). The axeneity of the shoot tip cultures has to be carefully monitored by parallel samples in KM medium.

Fusion was performed by both methods: high Ca^{++} and high pH (6) or with PEG (7). With high PEG (25%) the relative rate of fusion was better, but the survival rate of fusion products decreased. As our selection system is not too essential, and, consequently, many non-fused protoplasts have to be regenerated, we use the more stringent PEG method predominantly (8).

RESULTS

Ploidy Level Reduction from 4x to 2x

It was proposed as early as 1963 by Chase (9) to convert potato breeding from the tetraploid to the dihaploid level. After the discovery of superior pollinators within the primitive variety *S. phureja*, the number of dihaploids could be steadily increased. It is not necessary to go into the details of this technique, which is published (1); however, there is one point to stress: it is worthwhile to use fraternal markers so that all hybrids are directly visible and can be discarded. In the case of potato this is the nodal band marker visible already in the seeds as embryo spot. This marker was also used in protoplast fusion experiments. By parthenogenesis we produced 5,000 different primary dihaploid clones from 70 different tetraploid cultivars during the last three years.

Besides the parthenogenetic way, we also tested the chances to reduce the ploidy level via androgenesis. Up to now, there was limited success in regenerating haploids via microspore culture from tetraploid potatoes (10). However, using exactly the same technique as described for monohaploid production, we could produce several dihaploids from microspores cultured within the anthers of tetraploids. Their percentage is rather low compared to the percentage of parthenogenetic dihaploid production (Table 1). Taking into consideration the higher expenses of this procedure, the parthenogenetic way is superior for 4x to 2x ploidy level reduction. Although the number of plants produced is small, we can state that there is no positive or negative correlation between the success rate of partheno- and androgenesis. The strongest restriction for both procedures is poor flowering, which in some genotypes could not be enhanced by grafting.

Ploidy Level Reduction from 2x to 1x

Figure 1 demonstrates that by using dihaploids instead of tetraploids in the breeding procedure, the chance to find a desired genotype is drastically increased, but there is good reason to go one step further, to pass to the monohaploid state with $2n = 1x = 12$ chromosomes.

Again both possibilities, the parthenogenetic and the androgenetic technique may be used for ploidy level reduction. Comparing the results of both methods, we find a clear advantage of androgenesis (1,067 haploids from 13,575 seeds in comparison to 5 monohaploids

Table 1. Comparison of Parthenogenetic and Androgenetic Ploidy Level Reduction from 4x to 2x.

anther donor/ mother clone	number of anthers plated	number of plants regenerated	andro- genesis %	number of fertilized flowers	number of plants regenerated	partheno- genesis %
W1	300	0	0	101	1	1
Renema	300	0	0	457	26	5.7
632.07	300	4	1.3	500	50+	10.0
77/194	300	0	0	245	29	11.8
Kero	300	0	0	387	73	18.9
77/7	300	8	2.7	48	10+	20.8
22/71	300	4	1.3	1292	324	25.1
Wega	300	0	0	67	23	34.3
Granola	300	0	0	75	44	58.7

+ calculated from the number of berries.

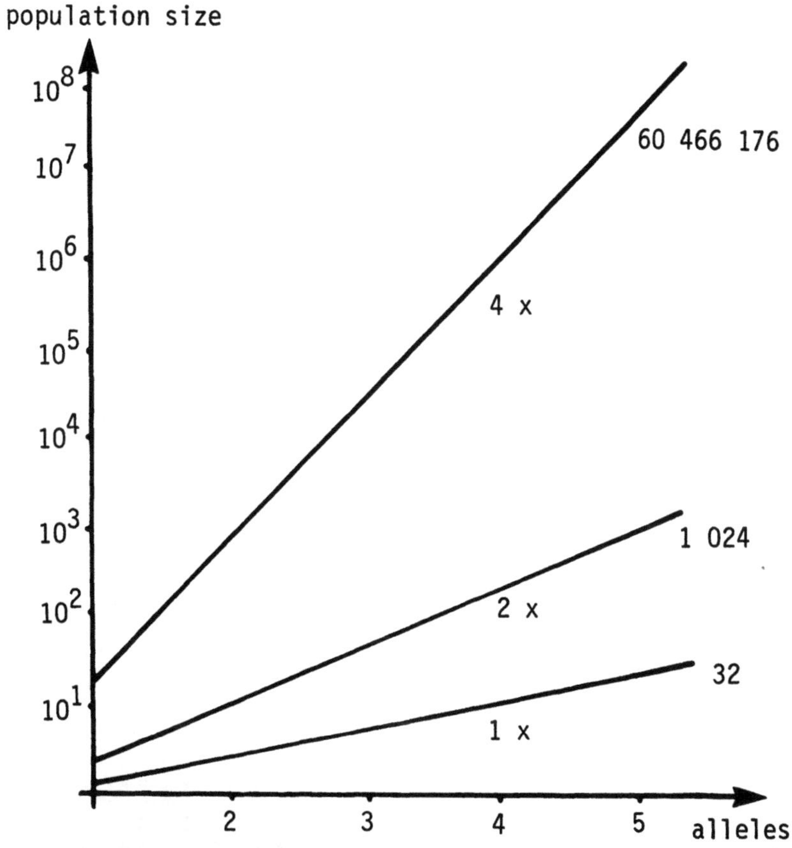

Figure 1. Statistical minimum population size to find a specific genotype in the homozygous condition in dependence of the number of alleles responsible for the expression of that character and assuming that the alleles concerned are heterozygous in both parents.

from 262,648 seeds) (11).

The tubers of such monohaploids (Figure 2a) are of no practical value; the plants are very weak and flower only occasionally (Figure 2b). They show, however, meioses (Figure 2c) and even produce berries (Figure 2d), which are up until now without seeds. For practical purposes they gain interest only after chromosome doubling.

Potatoes from Botanical Seeds

Before discussing our breeding scheme, a side line opened up by such monohaploids should be mentioned. One of the strongest obstacles in vegetative propagation of potato is the necessity to store tubers in order to maintain seeds for the following season.

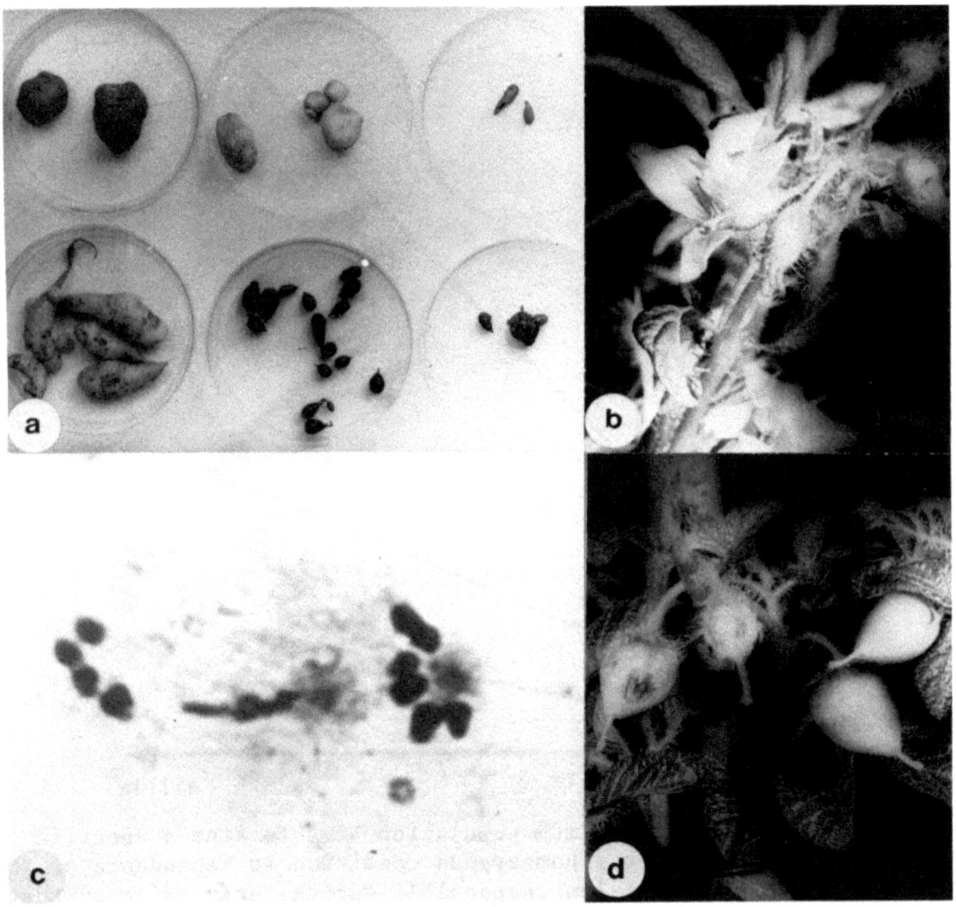

Figure 2. Monohaploid Solanum tuberosum plants with 2n = 1x= 12 chromosomes. (a) tubers; (b) flowers; (c) meiosis in a pollen mother cell; (d) berries.

In moderate climates this is no problem, in the tropics, however, it is the main reason why potatoes are costly. This problem might be overcome by the propagation via true seeds (TPS). Today several programs are started along this line (12). These programs work with highly heterozygous TPS and treat potato as an outbreeding species. Monohaploids, doubled two times, make the possibility attractive to work with homogeneous homozygotes and to use potato as an inbreeder. We could demonstrate that such homozygous tetraploids don't show strong inbreeding depression, although our experimental material is still limited. The other prerequisite for this breeding approach, i.e. self-fertility, could be improved as well. In 1980 we selected

3% self-fertile clones from random populations of dihaploids; this ratio could be increased to 8% in 1981 (Meyer and Wenzel, unpublished data).

Genome Doubling via Colchicine or Protoplast Fusion

Most monohaploids doubled spontaneously during the first tissue culture cycle. Doubling of the stable monohaploids via colchicine was not successful, neither worked the tissue culture doubling procedure (13). From 74 different monohaploid clones callus was induced and plated on auxin containing medium. From 5 clones plants could be regenerated, none of which became purely diploid (Meyer and Wenzel, unpublished data). This may indicate that clones staying monohaploid during the first *in vitro* culture process, possess such an efficient genome that chromosome doubling is of no selective advantage. Isolation and regeneration of protoplasts of monohaploids stopped at the stage of microcolonies, not enabling the doubling via monohaploid cell fusion. From dihaploids, however, protoplasts could be regenerated into functional plants. The tubers of such clones were rather homogeneous, in contrast for example to the tubers regenerated from microspores (14). Protoplasts of such clones could be fused and resulted in tetraploids.

Analytical Synthetic Breeding Scheme

The combined application of parthenogenesis, androgenesis, classical combination breeding, and protoplast fusion results in an analytical synthetic breeding scheme (14). For initial trials, we concentrated on virus and nematode resistance (15). These resistances follow monogenic (extreme resistance to PVX and PVY) as well as oligogenic modes of inheritance (resistance to the white potato cyst nematode *Globodera pallida*, or field resistances). So far we have been able to demonstrate that all steps needed for a tissue culture application in potato breeding function principally, but they are restricted to clones, which respond in tissue culture. As most of these clones do not contain valuable agronomic characters, we produced sexual hybrids amongst them and clones carrying the desired resistance. The F_1 hybrids were screened for the presence of the resistance and then used as anther donor plants. Parallel to the segregation of the resistance, we found segregation in relation to the positive response in anther culture. Several families consisted of clones which gave hardly any response; in other instances the resistant fraction of the F_1 population segregated into responding and non-responding clones (15). Accordingly, it is probable that the regeneration capacity or tissue culture ability is genetically determined. This view can be specified even further now: the switch towards morphogenesis via callus and organogenesis or via embryogenesis is also under genetic control. There exist clones delivering high frequencies of macroscopic structures all of which develop as embryoids, while others primarily form callus. By combin-

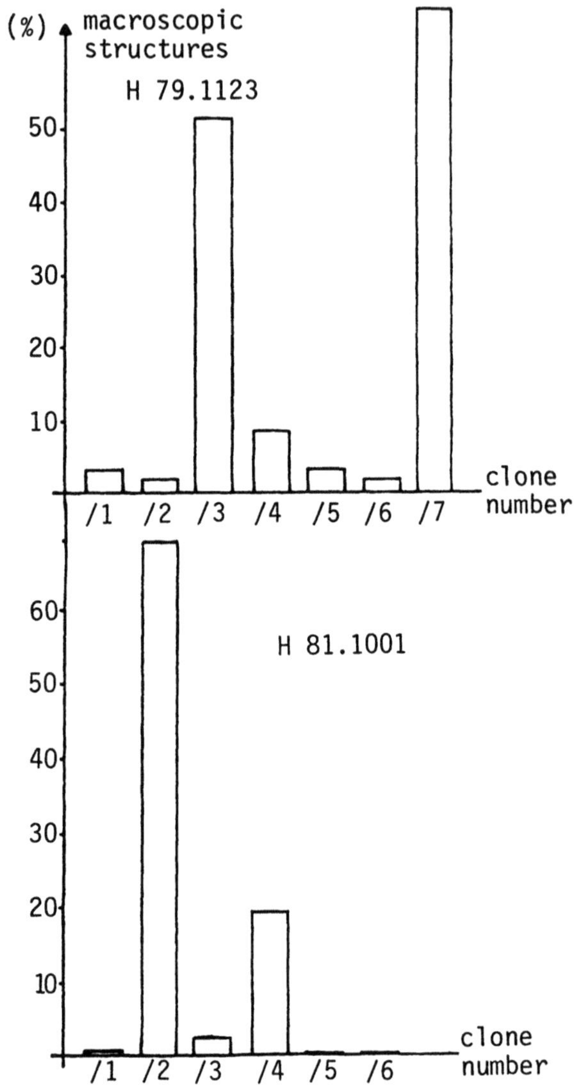

Figure 3. Segregation of macroscopic structure formation from F_1 hybrid families, one parent of which had passed already an anther culture cycle.

ing anther donor plants with a high embryogenic capacity, with only callus forming ones, the embryogenic pathway in androgenesis could be significantly improved.

According to the genetic bases of tissue culture ability, we do not have enough data to formulate a hypothesis. What can be stated is that plants carrying the genes for regeneration capacity in homo-

Table 2. Data for Clones Regenerated via Anther Culture (H = dihaploid; A = androgenetic doubled haploid clones).

Anther donor clone	Resistant to	number of anthers plated	number of macroscopic structures	number of functional clones total	1x	2x	4x
H 3-703	-	17.910	11.185	1.441	-	1.021	120
H 78.01/10	PLRV	2.452	1.483	323	63	260	10
H 78.01/27	PLRV	3.765	2.880	2.564	401	1.091	30
H 78.2020/26	PCN	5.200	199	60	-	60	-
H 78.2022/14	PCN	19.300	983	44	-	44	-
H 79.1038/18	PVY	831	1.009	48	12	31	4
H 79.1038/37	PVY	833	5.053	393	2	229	34
H 80.1088/8	PVX extr.	1.682	551	149	-	93	25
H 81.1001/4	PVY extr.	2.107	472	41	-	35	6
H 81.1004/53	PVX extr.	761	502	23	-	23	-
A 79.01/41		15	69	65	-	54	1
A 79.01/68		75	163	48	1	42	1

zygous conditions, e.g. androgenetic doubled monohaploids, lead to a more distinct segretation of this character in subsequent anther culture experiments (Figure 3). This view is further supported by the high frequency of androgenesis arising in anthers taken from A_1 plants (Table 2). In addition, during the anther culture cycle most deleterious genes are lost. The strong influence of genotype is also underlined by the rather different frequencies of functional monohaploid potatoes descending from different donor clones (the frequency ranges from 1 to 20%; Table 2). Furthermore, we found that during the transfer of androgenetic macroscopic structures from the induction to the regeneration medium, an average of 90% died. This survival rate can hardly be influenced by the medium (15), but a specific type will grow on a wide range of media. Therefore, we suggest that this behavior is caused by the genome. One factor is the presence of sublethal genes in the diploid condition which become lethal in the haploid. As long as the macroscopic structures are in close contact with each other and with the anther tissue, the plants are able to survive because of cross feeding; but after isolation and transfer to the second medium, the plantlets possessing lethal genes will die.

Evaluation of Androgenetic Offspring

In 1979, almost 2,000 androgenetic clones were grown in a field under high virus pressure. All plants with insufficient virus resistance should have been infected. According to tuber yield and shape, 10% of the plants were harvested and regrown in 1980. From 230 clones planted again, one third did not grow at all, and were probably totally infected by PLRV and PVY synergistically; one third revealed heavy PVY symptoms on the indicator plant. This can be interpreted as a segregation into resistant and non-resistant clones originating from a heterozygous field resistant anther donor plant. After anther culture of heterozygous G. pallida clones, 107 A_1 clones were obtained from which 15 were screened in 3 repetitions for the level of their resistance. All clones tested showed a good reduction of cyst numbers (15), ranging from a P_f/P_i value of 0.06, equivalent to the donor clone, to 0.90. This means that the quantitatively inherited resistance against G. pallida was not lost during the ploidy level reduction. Therefore, the resistance is now probably present in the homozygous condition. Consequently, there is no reason to assume that polygenically inherited characters will be lost by the reduction of the ploidy level during the random meiotic segregation.

Protoplast Fusion

The final step in the analytical synthetic breeding scheme is the fusion of heterozygous dihaploids to result in a highly heterozygous tetraploid clone in which the complete genomes of both parents are added. Ideally, several valuable characters have been considered. In relation to the increasing need for quantitative resistances this combination is of outstanding importance, as it is superior to any classical combination breeding technique; in the latter the chance to combine polygenic characters after the random distribution in meiosis is severely reduced.

Although the fusion process itself is no problem, the selection of secure fusion products turned out to be extremely difficult. In the successful fusion experiments published normally wide fusions with clear phenotypic or genotypic markers are performed. This enabled a precise and often early identification of the hybrids. Under applied aspects, the fusion partners will be more similar. When both parents carry different levels of field resistance for example, the differentiation of homokaryotic from heterokaryotic fusion hybrids is problematic. At first we used in our experiments the hybrid vigor selection system (16) and regenerated about 2,000 presumptive fusion products, all containing 48 chromosomes. According to the nodal band marker, which was incorporated in one parent, and according to the parental resistances, none of the plants was a somatic hybrid. On the other hand the pattern of total proteins showed significant deviations, but without allowing a clear classification (Stegemann and Wenzel, unpublished data). With the same technique the selection of wider somatic hybrids was possible between *S. tuberosum* (x) *S. stenotomum* (Schieder, unpublished data). To increase the success rate in *S. tuberosum* (x) *S. tuberosum* hybrid selection, Uhrig (17) developed a reversible bleaching technique. It could be demonstrated that the herbicide SAN 6706 bleaches dihaploid potato shoots without affecting the regeneration capacity of protoplasts isolated from herbicide treated material. If only one fusion partner was bleached, fusion products can efficiently be identified, as they consist of a predominantly cytoplasmic cell with a localized green chloroplast area. They can be fished and fixed, according to the procedure of Gleba and Hoffmann (18). From such experiments calluses exist in potato, but as they do not show morphogenesis yet, we have to await final confirmation of their hybrid nature.

As a more specific way for hybrid selection we work in the direction of fusing protoplasts from plants resistant to different exotoxins, e.g. the toxin of *Phytophthora infestans* or of *Fusarium solani*. Fusion products may be resistant to both exotoxins. Along this line it was demonstrated that resistant calluses can be isolated (19). Protoplasts, however, do not survive even on lower toxin levels. Therefore, we are trying to insert the selection step after cell wall or microcolony formation.

DISCUSSION

After the proposal of an analytical synthetic breeding scheme (14), it could be demonstrated now that central steps of this system work, i.e., parthenogenesis, androgenesis and protoplast fusion. The comparison of the androgenetic and parthenogenetic ploidy level reduction shows for the 4x - 2x reduction a clear advantage of parthenogenesis, while for the second reduction the androgenetic method is in our hands more successful. Of general importance may be the finding that besides monogenically inherited (qualitative) characters, like extreme resistances, also quantitative characters can be transferred from 4x to 1x and that the efficiency of these characters may even be increased. With increasing importance of polygenically inherited quantitative resistances the haploid breeding step should gain significance, as it allows a substantial reduction of the population size. For applied use of protoplasts in potato, only preliminary success has been achieved. The fusion of _S. tuberosum_ with the wild variety _S. stenotmum_ shows that this system principally functions. Consequently, the new strategy for breeding potatoes mainly demands an easier system for the secure selection of _S. tuberosum_ (x) _S. tuberosum_ hybrids.

According to Larkin and Scowcroft (20), tissue culture is not only a safe instrument for cloning but also a tool for the production of new variation. We found predominantly homogeneous populations after protoplast regeneration, but there existed some variation with regard to the total protein pattern. This slight variability, which can be explained by spontaneous mutations, must not be caused by the tissue culture itself, but probably by the increased sensibility of the genome in the haploid condition.

Taking into consideration the success obtained using tissue culture techniques in breeding procedures of other plant species, like barley and rye (21) on the one hand, and on the other side the increasing need for breeding superior varieties, we are convinced that the experimental field will be supplemented by the petri dish.

REFERENCES

1. J.G.Th. Hermsen and J. Verdenius, Selection from _Solanum tuberosum_ group phureja of genotypes combining high frequency haploid induction with homozygosity for empryo spot, _Euphytica_, 22:244 (1973).
2. S.K. Sopory, E. Jacobsen and G. Wenzel, Production of monohaploid embryoids and plantlets in cultured anthers of _Solanum tuberosum_, _Plant Sci. Lett._, 12:47 (1978).
3. T. Murashige and F. Skoog, A revised medium for rapid growth and bioassays with tobacco tissue culture, _Physiol. Plant._, 15:473 (1962).
4. K.N. Kao and M.R. Michayluk, Nutritional requirements for growth of _Vicia hajastana_ cells and protoplasts at a very low population density in liquid media, _Planta_, 126:105 (1975).
5. H. Binding, R. Nehls, O. Schieder, S.K. Sopory and G. Wenzel,

Regeneration of mesophyll protoplasts isolated form dihaploid clones of Solanum tuberosum, Physiol. Plant., 43:52 (1978).
6. W.A. Keller and G. Melchers, The effect of high pH and calcium on tobacco protoplast fusion, Z. Naturforsch, 28c:737 (1973).
7. K.N. Kao and M.R. Michayluk, A method for high-frequency intergeneric fusion of plant protoplasts, Planta, 115:355 (1974).
8. G. Wenzel, Protoplast techniques incorporated into applied breeding programs, in: "Advances in Protoplast Research," L. Ferenczy and G.L. Farkas, eds., Pergamon Press, London (1980).
9. S.S. Chase, Analytical Breeding in Solanum tuberosum L. A scheme utilizing parthenotes and other diploid stocks, Can. J. Genet., 5:359 (1963).
10. J.M. Dunwell and N. Sunderland, Anther culture of Solanum tuberosum L., Euphytica, 22:317 (1973).
11. G. Wenzel, C. Meyer, T. Przewozny, H. Uhrig and O. Schieder, Incorporation of microspore and protoplast techniques into potato breeding programs, in: "Plant Regeneration and Genetic Variability," E. Earl, ed., Praeger Publishers, New York (in press), (1982).
12. M.D. Upadhya, Potential for potato production from true seed under developing country condition, in: "Production of Potatoes from True Seed," CIP, Lima (1979).
13. E. Jacobsen, Doubling dihaploid potato clones via leaf tissue culture, Z. Pflanzenzuchtg, 80:80 (1978).
14. G. Wenzel, O. Schieder, T. Przewozny, S.K. Sopory and G. Melchers, Comparison of single cell culture derived Solanum tuberosum L. plants and a model for their application in breeding programs, Theor. Appl. Genet., 55:49 (1979).
15. G. Wenzel and H. Uhrig, Breeding for nematode and virus resistance in potato via anther culture, Theor. Appl. Genet., 59: 333 (1981).
16. O. Schieder and I.K. Vasil, Protoplast fusion and somatic hybridization, Intl. Rev. Cytol. Suppl., 11B:21 (1980).
17. H. Uhrig, Regeneration of protoplasts of dihaploid potato plants bleached by a herbicide (SAN 6706), Mol. Gen. Genet., 181:403 (1981).
18. Y.Y. Gleba and F. Hoffmann, Hybrid cell lines Arabidopsis thaliana + Brassica campestris: No evidence for specific chromosome elimination, Molec. Gen. Genet., 165:523 (1978).
19. M. Behnke, Selection of potato callus for resistance to culture filtrates of Phytophthora infestans and regeneration of resistant plants, Theor. Appl. Genet., 55:69 (1979).
20. P.J. Larkin and W.R. Scowcroft, Somaclonal variation a novel source of variability from cell cultures for plant improvement, Theor. Appl. Genet., 60:197 (1981).
21. B. Foroughi-Wehr, W. Friedt and G. Wenzel, Field experiments with anther derived barley and rye, this volume, (1982).

APPLICATION OF ANTHER CULTURE TECHNIQUE TO CROP IMPROVEMENT IN CHINA

Zeng Junzhi

Institute of Genetics
Academia Sinica
Beijing
CHINA

INTRODUCTION

Since Guha and Maheshwari reported the direct development of haploid embryoids from microspores of *Datura innoxia* by anther culture (1, 2), many researchers have been interested in the induction of androgenetic haploids. Researchers in China began these studies in 1970. Soon pollen-derived plants were obtained of rice and wheat. Subsequently, studies on anther culture were also carried out in many species such as maize, rye, Triticale, barley, tobacco, cotton, soybean, rapeseed, rubber tree, Chinese cabbage, pepper and so on. Induction of pollen-derived plants of many species were first successfully achieved in China. Table 1 lists these species.

Concurrently experiments on the application of anther culture technique to the improvement of crops, particularly cereals, were done in China. They included the following: 1) the study of various factors influencing the response of anthers in culture in order to increase the induction frequency of pollen-derived plants; 2) the possibility of applying homozygotic pollen plants to haploid breeding; and 3) breeding new varieties and strains through anther culture technique.

VARIOUS FACTORS INFLUENCING THE RESPONSE OF ANTHERS IN CULTURE

Selecting Anthers with Suitable developmental Stage of Pollen for Inocultaion

Although tetrads and mature pollen can develop into pollen

Table 1. Species in Which Pollen-Derived Plants were First Obtained in China.

Species	Year	Reference
Triticale	1971	34
Capsicum annuam L.	1971	35
Triticum aestivum	1971	8
Beta vulgaris L. 2n = 4x = 36	1973	36
Brassica pekinensis	1973	16
T. aestivum x Agropyronglaucum	1973	37
Populus nigra L.	1974	38
Zea mays L.	1975	9
Brassica chinesis L.	1977	39
Hevea brasiliensis	1977	40
Sorghum vulgare	1978	41
Rehmannia glutinoza	1978	42
Beta vulgare 2n = 2x = 18	1979	43
Citrus microcarpa	1979	44
Linum ustitalissimum	1979	45
Medicago denticulata	1979	46
Saccharum sinensis	1979	47
Colix lacryma	1980	48
Glycine max	1980	49
Fragacia orientalis	1981	50
Lycium barbarum	1981	51
Vitis vinifera	1981	52

plants (3 - 6), the best yield of pollen plants cannot be obtained at every developmental stage of pollen. For example, in wheat the pollen-derived calli could be induced in cultured anthers inoculated at stages from pollen mother cell to mature pollen. Usually when the induction frequencies of callus were the highest, the anthers with mid- or late-uninucleate microspores were inoculated (7). In fact many other species are also similar (Table 2).

Improving Culture Media

The culture medium is the main environmental factor which induces and controls the development of intact plantlets from pollen grains.

Raising sucrose concentration. Initially we noticed that the concentration of sucrose was an important factor in inducing pollen plants. When the concentration of sucrose was raised from 3% to 6%, pollen plants were first successfully obtained in wheat (8). Suc-

Table 2. Appropriate Pollen Phase for Inoculation in Several Cereal Crops.

Species	Development phase of pollen	Reference
O. Sativa subsp. Keng	Mid-late uninucleate phase	31, 53, 54
O. Sativa subsp. Shien	Uninucleate or late uninucleate phase	24, 55
Triticum aestivum	Mid-late uninucleate phase	8, 56
Hordeum vulgare	Uninucleate or late uninucleate phase	13, 57
Triticum aestivum X Agropyrom glaueum	Late uninucleate phase	37
Seceale cereale	Late uninucleate phase	58
Lolium	Uninucleate phase	13
Triticale	Uninucleate phase	59
Coix lacryma	Uninucleate phase	48

rose may not only regulate the osmotic pressure of the medium, but also may be the most effective carbohydrate source. Later researchers also took notice of sucrose concentration. Up to now it had been defined that 9% sucrose is appropriate for anther culture of wheat, and 12% for maize (9 - 12). In other species higher levels of sucrose are used; for example in rapeseed, Triticale and barley it was generally raised from 6% to 12% (13 - 16).

Optimal regulation of the ammonium nitrogen:nitrate nitrogen ratio. Clapham (17) and Chu et al. (14) found that high concentration of ammonium ions could inhibit formation of pollen calli in barley and rice. Based on this result, a new N_6 medium was developed by considerably reducing the concentration of ammonium ions and regulating the ratio of ammonium nitrogen to nitrate nitrogen to an optimal condition. It was proved that it is more efficient than other synthetic media for anther culture of rice and other cereals (12, 18 - 22).

Developing potato medium. Natural extracts such as potato, sweet potato, yam, tomato, sprouted wheat, and endosperm from rice and maize in grain filling stage were tested to replace the components in synthetic medium. Such natural extracts have two merits: high induction-frequency and simplicity in operation. Among the aforementioned natural extracts, potato aqueous extract attracted much attention. Potato aqueous extract was first successfully used in anther culture of tobacco (23). Subsequently, it was used in anther culture of rice to replace principal components of Miller's medium and satisfactory results were obtained (15, 24).

Both potato media showed approximately the same effects on the induction of pollen embryoids or pollen callus as their corresponding synthetic media. Since the contents of the major salts of potato tubers vary widely, sometimes the inducing effects of this medium were not stable enough. Recently potato media were further improved by chuang chiachum (8). The new potato medium, which is called potato-2 medium, contains 10% aqueous potato extract, 1/2 macroelements of W_5 medium, and Fe-salt and thiamine as in MS medium. The frequency of induced callus in this medium was much higher than that in other synthetic media (Table 3).

Improving Culture Condition

The physical factors of culture such as temperature, light, etc., influence the frequency of induced pollen callus.

Temperature condition in culture. When the temperature of cultures was elevated or changed the frequencies of callus of wheat and rapeseed would increase (25, 26). For example, we found that if wheat anthers were incubated at elevated temperature for a few days prior to transfer to normal temperature, the frequency of induced

Table 3. Effect of Different Media on Induction of Pollen-derived Plants of Wheat Variety "Orefen."

Media	No. of anthers inoculated	Calli		Plantlets	
		No.	%	No.	%
N_6	630	48	7.62	8	1.26
Potato-2 (Sichuan Potato)	630	149	23.65	27	4.28
Potato-2 (Ke-1 Potato)	630	162	25.71	32	5.07
W_5	630	46	7.30	10	1.58

callus might increase. Yields of green plants have obviously increased (Table 4) (27). In Table 4, when anthers were cultured at 33°C for 8 days then transferred to 24°C, not only the frequency of induced callus increased but also the frequency of differentiating green plantlet significantly increased. From this Table we may seen that the physiological condition of donor plant also has an influence on the induction of callus and differentiating plants. The frequencies of induced callus and differentiating plantlets from anthers of main spike are more than that from tillering spike.

Cold pretreatment. A cold pretreatment before inoculation may increase the frequency of induced callus. The frequency increases about two times when the wheat anthers were pretreated at 1°C-4°C for 48 hr. In rice and other cereal crops, the effect of cold pretreatment is also very effective, so it has been widely applied in many species (28). Table 5 shows the proper temperature and duration of cold pretreatment in several cereal crops.

At present, induction frequencies of green plantlets on the basis of anther number are more than 5% for *Oryza sativa* L. Subsp. Keng, 0.4% for *O. sativa* L. Subsp. Hsien and about 5% for *Triticum aestivum* L.

THE POSSIBILITY OF USING POLLEN-DERIVED PLANTS FOR CROP IMPROVEMENT

Stability of Homozygous Pollen Plants

In order to apply anther culture technique to breeding as soon as we began the anther culture work, anthers of F_1 hybrids of intervarietal crosses of wheat or rice were used for culture. Then genetic expression of the haploid pollen plants from hybrid pollen and

Table 4. Effect of Incubation Conditions and Physiological State of Donor Plants on Induction of Pollen-derived Plants of the Variety "Orefen."

Treatment	No. of anthers inoculated	Calli		Green Plantlets	
		No.	%	No.	%
Main Spikes 33°C for 8 days after inoculation	1186	715	60.3	194	16.4
24°C	1186	516	43.5	92	7.8
Total	2372	1231	51.9	286	12.1
Tillering Spikes 33°C for 8 days after inoculation	1092	451	41.3	156	14.3
24°C	1134	240	21.2	33	2.9
Total	2226	691	31.1	189	8.4
TOTAL	4598	1922	41.8	475	10.2

Table 5. Procedure of Cold Pretreatment of Flower Buds.

Species	Temperature	Duration	Reference
Secale cereale	6°C	3-15 days	21
Hordeum vulgare	3°C	48 hours	4
Oryza sativa	10°C	48 hours	54
	10°C	4-7 days	60
Triticum aestivum	3-5°C	48 hours	61
	4°C	48 hours	29
Triticale	3-5°C	72 hours	22

the progeny of their homozygous diploid plants (spontaneously or artificially doubled) were studied. From a series of experiments (8, 29, 30) we proved that homozygous diploid plants may be obtained by means of anther culture, and that their characters could be stabilized rapidly. This reduces the generations of breeding required and raises the selective efficiency in breeding, thus shortening the breeding cycle. Table 6 analyzes the three characters (plant height, ear length and 1,000 seed weight) of pollen-derived lines (H_3) and their parents in a wheat cross, and shows that the C.V. of three characters of pollen-derived lines (H_3) is less than or approximately equal to that of the parents. This suggests that the characteristics of pollen-derived plants are uniform, similar to stable varieties. According to the statistical data, about 90% diploid pollen plants (spontaneously or artificially doubled) are complete homozygotes. Their characters do not appear to segregate.

Diversity of Phenotypes of Pollen Plants (H_1) from F_1 Hybrids

Since the pollen grains from F_1 hybrids are heterozygous, different gene combinations of both parents of a cross are present in every grain. If pollen grains of F_1 hybrids will be induced into pollen plants by anther culture, the pollen plants (H_1) should show various phenotypes of both parents. This phenomenon has been proved in our experiments with rice and wheat. For example, from culture of F_1 anthers of wheat cross between "Xian nog 5675" (a variety with white glume color, top awn, red grain color, clavate spike shape and short stalk) and "Jili" (a variety with red glume color, awn, red grain color, fusiform spike shape and tall stalk) we have obtained several diploid pollen plants (15). According to the characters of spike, they were divided into seven forms (Table 7). The seven forms reflect the gene recombinations of both parents. In rice the

Table 6. Analysis of Characteristics of Pollen-derived Lines (H_3)* and their Parents in the Cross "Kotung 58 X 7497."

Characters	Kotung 58 (♀) n = 20	7497 (♂) n = 20	Kotung 58 X 7497 H_3 n = 34
Plant height	87.00 ± 3.69	95.15 ± 4.25	96.00 ± 3.00
C.V. %	4.24	4.47	4.06
Ear length	8.42 ± 0.40	8.63 ± 0.39	8.52 ± 0.37
C.V. %	4.75	4.52	4.34
1,000 seed weight	50.90 ± 3.15	45.02 ± 3.54	55.22 ± 2.69
C.V. %	6.19	7.86	4.87

Table 7. Diversity of Characters of H_1 Pollen Plants in "Xian nong 5675 X Jili."

Plant material	Glume color	Character of awnness	Grain color	Spike type	Plant height (cm)*
Xian nong 5675	white	top	red	clavate	41
Jili	red	awn	red	fusiform	115
H_1 form					
1	white	top	red	clavate	60-69
2	white	awn	red	fusiform	41-49
3	white	non-awn	red	clavate	19-24
4	white	awn	red	oblong	23-32
5	red	top	sterile	oblong	52
6	red	non-awn	sterile	clavate	
7	red	top	sterile	clavate	

* Plant height of every form is rather short; it may be associated with the weak growth of H_1 pollen plants.

result was similar to wheat (31); for example, in a cross "Nipponbare X Chienchunpang" the ear characters of H_1 pollen plants exhibited that some of them are like the paternal, short and rather compact, some like the maternal, long and rather lax, and other combined the characters of both parents: long and compact, or short and lax.

Easier Expression of Recessive Characters

The frequency of recessive characters within pollen plant population is much higher than that within the F_2 population in conventional breeding. Since haploid plant only contains a set of chromosomes, there is no such phenomenon that dominant characters mask recessive characters, so recessive characters emerge rather frequently. For example, in a wheat cross between "Sonora 62" with red grain color and "Hongtu" with white grain color, red grain color is dominant character. We compared the segregation ratio of grain color between F_2 and H_2 population in the cross and found that the emerging frequency of white seeds is much higher in H_2 population than that in F_2 population. Table 8 shows the results of two crosses. In the cross of "Sonora 62" X "Hongtu", there are 413 plants

Table 8. Comparison of the Segregation Ratio of Grain Color Between F_2 and H_2 Population.

Material		No. plants/ lines with red seed	No. plants/ lines with red seed	No. plants/ lines with white seed	red:white
Sonora 62	F_2	413 plants	313 plants	100 plants	3.1:1
X					
Hongtu	H_2	32 lines	14 lines	18 lines	0.8:1
Orefen	F_2	147 plants	116 plants	31 plants	3.7:1
X					
Xianoyan 759	H_2	55 lines	27 lines	28 lines	0.9:1

in F_2 generation. The plants with red grain are 313, the plants with white grain are 100, the segregation ratio is 3.1:1, but in H_2 plant lines the ratio of red to white grains is 0.8:1. The former tallies with the rule 3:1 ($X^2 = 0.1364$, $p > 0.05$), the latter tallies with ratio 1:1 ($X^2 = 0.5$, $p > 0.05$). In another cross "Orefen" X "Xianoyan 759", the result is approximately the same. The experiments thus fully proved that haploid breeding can raise the selective efficiency of recessive characters. These results provide reliable grounds for establishing the technique of haploid breeding.

APPLICATION OF ANTHER CULTURE TECHNIQUE TO CROP IMPROVEMENT

Development of New Varieties

In Figure 1 both procedures of haploid and conventional breeding are compared. It shows that haploid breeding may usually reduce 3-4 generations. For example, varieties of rice "Hua Yu No. 1 and No. 2" only took five years from making the cross to releasing the new varieties (32). In comparing the developing processes of Hua Yu 1 and 2 and their parents, it was interesting that their development used three kinds of breeding: anther culture, bulk method and pedigree method. Among the three methods the duration needed for the anther culture is the shortest, four generations.

In rice there have been 81 rice varieties and strains developed through anther culture. Hua Yu No.1 and No. 2 are two of them. The two varieties were selected by the Institute of Genetics of Chinese Academy in collaborating with Niu Jing, Liu Ruifu et al. of Tianjin Agricultural Research institute, which possess the characteristics of high yield (7500 kg/ha), resistance to bacterial blight and wide adaptability. They have been released in the suburbs of Beijing and Tiangjing. It is expected that the area of cultivation will reach

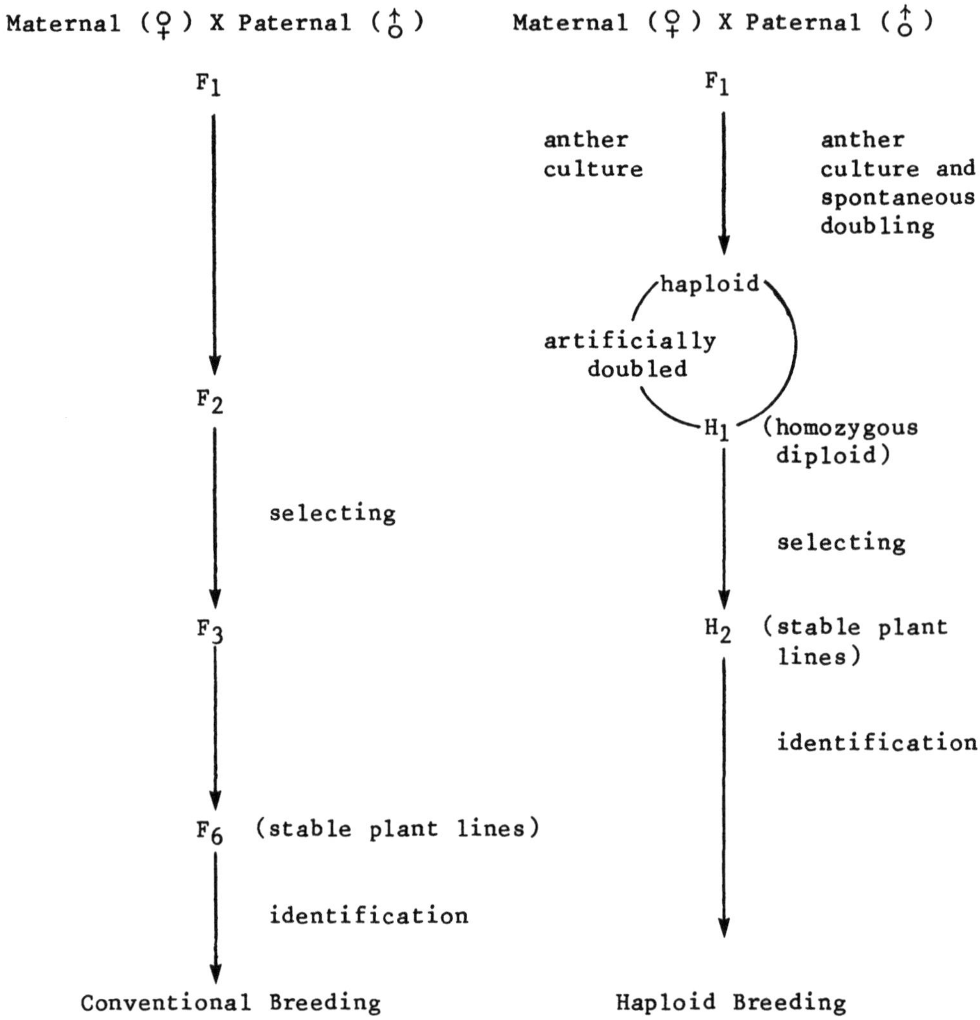

Figure 1. Comparison of procedures between haploid and conventional breeding.

several thousand hectares in Tiangjing suburb in the coming years.

New varieties of rice "Xin Xiu" and "Tang Huo No. 2" are another two of them. "Xin Xiu" was developed by Zhang Zhen-hua at Shanghai Academy of Agricultural Science. It has been released over 100,000 hectares in the eastern part of China. "Tang Huo No. 2" was developed by Wu Xiaoyu at Agricultural Research Institute of Tonglin county, An Hui province in China. It has been cultivated on about

6,000 hectares. Both varieties possess the characteristic of high yield.

In wheat, using hybrid anthers of winter type wheat as material, through anther culture a new strain of winter wheat, "Jingdan 2288," was selected by Hu Daofun et al. This new strain possesses the characteristics of large spike and many grains, vigorous tillering, and stripe rust and powdery mildew resistance, short stem and resistance to lodging. It is a promising new strain of winter wheat in the Beijing area in the coming years. It took only about five years to develop this new strain.

In tobacco, as early as in 1974, the variety "Danyu No. 1" was developed by Shandong Tobacco Research Institute in China. Then "Danyu No. 2 and No. 3" were subsequently developed. They combined the merits of both parents and possess characteristics of high yield and disease resistance. It took about three years from inoculating F_1 anthers to releasing the product (23).

The Development of Inbred Lines of Maize and Asexual Lines of Rubber Tree and Sugarcane

Since 1975, a lot of maize pollen-derived plants have been obtained in China, and 25 inbred lines have been selected. Among them a line named "Qun Hua" selected by Wu Jialing et al. is a notable one and has good combining ability (33). In 62 crosses about 90% (56 crosses) increased their yield. Among them the cross "411 X Qun Hua" was the best, which showed high yield and disease resistance. The selection process of this pure line was reduced by ten generations in comparison with conventional process.

In addition, the use of pollen plants from the rubber tree and sugarcane in improving varieties has been studied. Now the highest pollen plant of rubber tree has reached more than five meters and the asexual lines have been propagated. In sugarcane a pollen plant containing high sugar content and tall stem was selected. The asexual line of the pollen plant has been propagated for further identification (27).

Combining Application of Anther Culture Technique with Conventional Cross Method and Hybrid Rice in China

In hybrid rice, six strains have been selected from pollen plants derived from F_1 anther culture of hybrid rice "Shan You No.2". The phenotype and some economic characters of four lines among them are very like the donor plants. They are named "Shan Hua No. 7706, 78-1, 791, 792." Researchers attempted to use these strains to replace hybrid rice "Shan You No. 2", and to save the trouble of producing hybrid seed.

In Conventional Cross Method

Zhang Zhen-hua et al. at Shanghai Academy of Agricultural Science (62) combined the anther culture with the multiple crosses and selected variety "Huo Han Zao" and other new strains such as Strain 77001. In this way, the "selection blindness" of a multiple cross could be overcome. Therefore, anther culture not only increased the correctness of selection of multiple crosses, but also reduced the number of years significantly.

In summary we believe that further great successes will be achieved in the application of the anther culture technique to crop improvement when this technique is further developed.

ACKNOWLEDGEMENTS

We wish to thank Profs. Hu Han, Chen Yi, C.H. Chen and Z.H. Chang and Drs. C.C. Chu, J.L. Wu, X.Y. Wu and D.F. Hu for their data.

REFERENCES

1. S. Guha et al., Nature 204:297 (1964).
2. S. Guha et al., Nature 212:97-98 (1966).
3. M.M. AboEl-Nil, Can. J. Bot. 54:2107-2109 (1973).
4. A. Bouharmont, Caryologia 30:351-360 (1977).
5. T. Kameya et al., Jap. J. Breed. 20:82-87 (1970).
6. I. Nakata and M. Tanaka, Jap. J. Genet. 43:65-71 (1968).
7. H. Dinggang et al., in: "Annual Report of the Institute of Genetics," Academia Sinica, Peking (1980).
8. T.W. Ouyang et al., Sci. Sin. 16:79-95 (1973).
9. Anon., Acta Genet Sin. 12(2):138-142 (1975).
10. M.K. Ku et al., in: "Proc. Symp. on Plant Tissue Culture," Science Press, Peking (1978).
11. S.H. Miao et al., in: " Proc. Symp. on Plant Tissue Culture," Science Press, Peking (1978).
12. Anon., Acta Genet. Sin. 4:361 (1975).
13. D. Clapham, Z. Pflanzenzuchtg. 65:285 (1971).
14 C.C. Chu et al., Sci. Sin. 18:659-668 (1975).
15. Anon., in: "Collection of Works of Haploid Breeding," Science Press, Peking (1977).
16. L.P. Teng et al., in: "Proc. Symp. on Anther Culture," Science Press, Beijing (1978).
17. D. Clapham, Z. Pflanzenzuchtg. 69:142-155 (1973).
18. C.C. Chu et al., Acta Bot. Sin. 20:6-12 (1978).
19. A. Genovesi et al., Crop Sci. 19:662-664 (1979).
20. Anon., Acta Genet. Sin. 3(2):169-170 (1976).
21. Nitzsche et al., Haploid in plant breeding, Z. Pflanzenzuchtg.

Vol. 8 (1977).
22. C.S. Sun et al., Acta Bot. Sin. 20:6-12 (1980).
23. Anon., Acta Bot. Sin. 16(4):301-303 (1974).
24. Anon., Acta Genet. Sin. 1:81-89 (1975).
25. T.F. Ho et al, in: "Proc. Symp. on Anther Culture," H. Hu, ed., Science Press, Peking (1978).
26. S.K. Sopory et al., J. Exp. Bot. 27:58-68 (1976).
27. H. Han et al., (in press).
28. C.C. Chu et al., in: "XIII Intl. Bot. Congress," Sydney, Australia (1981).
29. Anon., Heredity and Breeding 4:24-25 (1977).
30. H. Han et al., Acta Genet. Sin. 3:322-330 (1979).
31. Anon, Sci. Sin. 2:209-221 (1974).
32. Anon., in: "Collection of Works of Haploid Breeding," Science Press, Peking (1977).
33. W. Jialing et al., Acta Phytophysiol. Sin. 6(2):221-224 (1980).
34. C.S. Sun et al., Acta Bot. Sin. 15:165-173 (1973).
35. Y.Y. Wang et al., Sci. Sin. 16:147-151 (1973).
36. Anon., in: "Proc. Symp. on Anther culture," Science Press, Beijing (1978).
37. C.C. Wang et al., Acta Genet Sin. 2:72-80 (1975b).
38. C.C. Wang et al., Acta Bot. Sin. 17:56-62 (1975a).
39. C.H. Chung, in: "Proc. Symp. on Anther Culture," H. Hu, ed., Science Press, Beijing (1978).
40. C.H. Chen et al., Acta Genet. Sin. 5:99-107 (1978).
41. W.B. Zhou, Acta Genet Sin. 5(4):337-338 (1979).
42. H.Z. Shandong, Plant Mag. 4:8 (1978).
43. M.W. Shao, Zhongguo Tiancai 1:27-30 (1979).
44. Z.G. Chen et al., Acta Genet. Sin. 7:189-191 (1980).
45. H.T. Sun, Kexue Tongbao 24(20):948-950 (1979).
46. X. Su, Plant Mag. 6:5 (1979).
47. Z.H. Chen et al., in: "Annual Report of the Institute of Genetics," Academia Sinica, Peking (1979).
48. C.C. Wang et al., Acta Bot. Sin. 22:316-322 (1980).
49. G.C. Yin et al., Kexue Tongbao 25(11):976 (1980).
50. G.R. Xue et al., Acta Hort. Sin. 8(1):9-12 (1981).
51. S.R. Gu, Acta Bot. Sin. 23(2):246-248 (1981).
52. C.J. Zhou et al., Acta Bot. Sin. 22:71-81 (1981).
53. S. Guha et al., Curr. Sci. 39:174 (1970).
54. C.C. Wang et al., Acta Bot. Sin. 1:43-54 (1974).
55. S. Guha and S. Mukherjee, J. Exp. Bot. 24:139-144 (1973).
56. P. Jingli et al., Acta Bot. Sin. 2:122-128 (1978).
57. C. Zhou et al, Act Bot. Sin. 2:122-128 (1980).
58. C.S. Sun et al., in: "Proc. Symp. on Anther Culture," Science Press, Beijing (1977).
59. C.S. Sun et al., Sci Sin. 2:168-173 (1974).
60. Y. Chen et al., in: "Annual Report of the Institute of Genetics," Academia Sinica (1979).
61. P. Jingli et al., Acta Bot. Sin. 17:2 (1975).

THE USE OF *IN VITRO* TECHNIQUES IN GROUNDNUT IMPROVEMENT

D.C. Sastri, J.P. Moss and M. Nalini

International Crops Research Institute for the Semi-Arid Tropics (ICRISAT)
Patancheru
P.O. 502 324
A.P.,
INDIA

INTRODUCTION

Genetic improvement of crops by utilization of characters from wild species, which often have desirable traits not present in the cultivated forms, has been the subject of serious attention. The interspecific transfer of genes has not always been simple and straightforward because of reproductive barriers and genetic incompatibility. This is true in the genus *Arachis* where a number of species have been identified as sources of resistance to fungal and viral pathogens as well as to insect pests (1). A few species have been crossed with *A. hypogaea* and their derivatives produced (1 - 4), but there are a number of species which are not crossable with groundnut (5). This limits the scope for crop improvement, and the situation necessitates the adoption of existing methods, or development of new ones for the production of hybrids following these desirable wide crosses.

The genus is predominantly diploid (2n = 20) with its species grouped into seven sections (5). The cultivated species is a tetraploid in the section ARACHIS and cannot easily be crossed with other tetraploid species except *A. monticola*, the only other tetraploid in section ARACHIS. The other tetraploid species belong to the section RHIZOMATOSAE, and the attractive feature of these rhizomatous species is that they possess multiple resistance (6). Current knowledge of the genus is insufficient to decide on any specific method for breaking incompatibility. In earlier attempts to cross four wild species including *A. glabrata*, the one named tetraploid species of the section RHIZOMATOSAE, with three cultivars of *A. hypogaea*, 25 pods were obtained which were 1/2 to 3/4 of the normal size with aborted seeds about 1/16" long (7), although no details of number of pollinations were given. Gregory (8) also reported pod production in these crosses, but no embryos or seeds were formed. Our results show the frequency of peg formation in such crosses was up to a maximum of 21.3%, none of which bore pods with well formed seeds (3). Johansen and Smith (9) observed 20 ovules of *A. hypogaea*, 2-7 days after pollinations with *A. glabrata*, and found that there were no pollen tubes in the ovaries, and fertilization had failed in all. Recently Murty et al. (10) have recorded similar observations and reported that the time taken for fertilization varied from 6 - 48 hours, and the maximum frequency of fertilization observed was 56%.

It should be emphasized here that *Arachis* is one of the few geocarpic taxa. After fertilization, a stalk (the gynophore) develops between the ovary and the node on which the sessile flowers are borne. This stalk elongates downwards until it has pushed the ovary into the soil. Seed development is almost absent during the aerial growth of teh gynophore and is resumed only after the peg is buried in the soil.

Therefore, it was decided that priority should be given to methods to produce plants from the fertilized ovules that were formed after wide crosses, but did not develop *in vivo*.

EMBRYO CULTURE

Culture of embryos in incompatible crosses in *Arachis* was suggested by Gregory in 1946 but was not taken up. Dissecting and culturing proembryos from about 1.5 mm long ovules is technically laborious and not always successful. Gynophore development and pod initiation and development from these incompatible crosses have been induced by hormone treatments (11, 12). Embryos in such cases grow to late globular or heart stages. Embryos at later stages were also observed in very rare instances. The younger embryos, when cultured, occasionally formed callus on MS with IAA or NAA (1 or 2 ppm) with Kinetin or BAP (0.5 or 1 ppm), but then became brown and necrotic. The embryos later than heart stages formed healthy callus and shoots. The rare well-developed embryos grew normally when cultured on White's as well as on MS medium. Culture conditions favoring the growth of embryos at globular stages are being investigated.

IN OVULO EMBRYO CULTURE

In ovulo culture of embryos has been useful in a number of plants where the dissection and culture of proembryos is difficult (13). Martin (14) cultured ovules from aerial gynophores of *A. hypo-*

gaea and obtained plants, using MS with Heller's microelements, vitamins, amino acids and casein hydrolysate as in Dulieu's medium, and 0.5 - 1.0 mg per litre of kinetin.

Sastri et al. (15) used a number of media but were unable to culture A. hypogaea ovules, even on Martin's (14) medium, and attempts at ovule cultures were discontinued. However, in vitro pollination of ovules had been achieved (15). Culture of whole ovaries, or gynophore tips, was, therefore, attempted.

OVARY AND GYNOPHORE TIP CULTURE

Ziv and Zamski (16) were the first to successfully culture the terminal segments of aerial gynophores and to obtain pods. Using their medium with various concentrations of kinetin and NAA and dark incubation, we failed to induce pod formation in peg tip cultures of A. monticola and A. hpyogaea. However, MS supplemented with 2 mg of NAA and 0.5 mg of BAP per litre has induced pod formation in dark incubated peg tip cultures of A. hypogaea and A. pusilla (a diploid wild species incompatible with all other species of Arachis). Some pegs from incompatible pollinations were cultured on this medium but without any success. Attempts with ovary cultures have resulted in callus formation and details have been reported earlier (15). Another observation worthy of mention is that on this medium pegs form callus which differentiates into plants. Rangaswamy et al. (17) reported root formation from callus in pericarp cultures.

THE FUTURE

The importance of groundnut as a crop in developing countries, and as a source of edible oil and protein for developing and developed countries, necessitates an objective realization of the problems reducing productivity and demands diversification of approaches to crop improvement, particularly when conventional methods of breeding have been unable to alleviate certain problems. The recent interest in, and application of, tissue culture methods in the improvement of some crops, notably potato, rapeseed and sugarcane, has stimulated groundnut scientists to turn their attentions to a wider range of approaches. Some of the initial problems associated with utilization of wild species have been overcome. Despite the ploidy differences between some wild species and the cultivated groundnut, Spielman et al. (18) produced fertile interspecific hybrids in large numbers. More recent reports of plant regeneration in tissue and organ cultures (19, 20) have now created hopes of regeneration from single cells and application of these techniques to some of the refractory problems in groundnut breeding.

The importance of haploids in plant breeding is now too well recognized to elaborate. The success of Mroginski and Fernandez (21, 22) in inducing plantlets in anther cultures from the cultivated and a few diploid species of Arachis has stimulated our efforts to produce haploids at ICRISAT. Plantlets have been obtained from anther cultures of A. pusilla and A. sp. P.I. No. 276233.

Production of plants form meristem cultures (19, 23) and their survival after freezing for 28 days (24) should be utilized for generation of disease-free populations, long term storage and international transfer of germplasm. Whilst our present knowledge is being utilized to realize these objectives, efforts to explore other possibilities will be initiated. Foremost amongst these must be spontaneous and induced variability in cultures and selection of cell lines with desirable traits. In a very recent report Bajaj et al. (25) have obtained genetically variable plants from the anther derived calli of A. hypogaea and a diploid wild species, A. villosa.

The recent developments in methodologies should, therefore, soon lead to realization of hybrids in large numbers from incompatible crosses as was achieved in compatible species (2, 18). In the meantime optimization of conditions for culturing a range of tissues should stimulate efforts to produce agriculturally desirable plants by other methods.

REFERENCES

1. J.P. Moss, Wild species in the improvement of groundnuts, in: "Advances in Legume Science," J. Summerfield and A.H. Bunting, eds., Royal Botanic Gardens, England (1980).
2. V.S. Raman, "Cytogenetics and Breeding in Arachis," Today and Tomorrow Printer and Publishers, New Delhi (1976).
3. A.K. Singh, D.C. Sastri and J.P. Moss, Utilization of wild Arachis species at ICRISAT, in: "Proceedings of the International Workshop on Groundnuts," R.W. Gibbons, ed., Patancheru, India (1980).
4. H.T. Stalker, Cytogenetic investigations in the genus Arachis, in: "Proceedings of the International Workshop on Groundnuts," R.W. Gibbons, ed., Patancheru, India (1980).
5. M.P. Gregory and W.C. Gregory, Exotic germplasm of Arachis L. Interspecific hybrids, J. Hered. 70:185-193 (1979).
6. R.W. Gibbons, ed., "Proceedings of the International Workshop on Groundnuts," ICRISAT, Patancheru, India (1980).
7. F.H. Hull and W.A. Carver, Peanut improvement, Florida Agric. Exp. Sta. Ann. Report 1938:39-40 (1938).
8. W.C. Gregory, Peanut breeding program underway. Research and farming, 69th Ann. Rep. N.C. Agric. Exp. Sta. 1946:42-44 (1946).
9. E.L. Johansen and B.W. Smith, Arachis hypogaea x A. diogoi. Embryo and seed failure, Amer. J. Bot. 43:250-258 (1956).

10. U.R. Murthy, N.G.P. Rao, P.B. Kirti and M. Bharathi, Cytogenetics and groundnut improvement, IARI Reg. Sta. Rep. Hyderabad (1980).
11. D.C. Sastri and J.P. Moss, Hormonal requirement for incompatible crosses in the genus Arachis (in preparation) (1982).
12. Anon., "ICRISAT Annual Report 1979-1980," ICRISAT, Pantacheru, India (1981).
13. V. Raghavan, "Experimental Embryogenesis in Vascular Plants," Academic Press, London (1976).
14. J.P. Martin, Culture in vitro d'ovules d'Arachide, Oleagineaux 25:155-166 (1970).
15. D.C. Sastri, M.S. Nalini and J.P. Moss, Arachis ovule and ovary culture in vitro, in: "Proceedings of Symposium on Plant Tissue Culture, Somatic Hybridization and Genetic Manipulation in Plant Cells," P.S. Rao et al., eds., BARC, Bombay, India (1980).
16. M. Ziv and E. Zamski, Geotropic responses and pod development in gynophore explants of peanut, Arachis hypogaea cultured in vitro, Ann. Bot. 39:579-583 (1975).
17. N.S. Rangaswamy, T.S. Rangan and P.S. Rao, Morphogenesis of peanut (Arachis hypogaea L.) pericarp in tissue cultures, Naturwiss. 22:1-2 (1965).
18. I.V. Spielman, A.P. Burge and J.P. Moss, Chromosome loss and meiotic behavior in interspecific hybrids in the genus Arachis L. and their implications in breeding for disease resistance, Z. Pflanzenzuchtg. 83:236-250 (1979).
19. D.C. Sastri, M.S. Nalini and J.P. Moss, Tissue culture and prospects of crop improvement in Arachis hypogaea and other oilseed crops, in: "Proceedings of Symposium on Tissue Culture of Economically Important Crops in Developing Countries," National University of Singapore (in press) (1981).
20. L.A. Mroginski, K.K. Kartha and J.P. Shyluk, Regeneration of peanut (A. hypogaea) plantlets by in vitro culture of immature leaves, Can. J. Bot. 59:826-830 (1981).
21. L.A. Mroginski and A. Fernandez, Cultivo in vitro de anteras de especies de Arachis (Leguminosae), Oleagineaux 34:243-248 (1979).
22. L.A. Mroginski and A. Fernandez, Obtencion de plantulas por cultivo in vitro de anteras de especies silvestres de Arachis (Leguminosae), Oleagineaux 35:89-92 (1980).
23. K.K. Kartha, K. Pahl, N.L. Leung and L.A. Mroginski, Plant regeneration from meristems of grain legumes: soybean, cowpea, peanut, chickpea and bean, Can. J. bot. 59:1671-1679 (1981).
24. Y.P.S Bajaj, Freeze preservation of meristems of A. hypogaea and Cicer arietinum, Indian J. Exp. Biol. 17:1405-1407 (1979).
25. Y.P.S Bajaj, A.K. Ram, K.S. Ram, K.S. Labana and H. Singh, Regeneration of genetically variable plants from anther deriv-

ed callus of *Arachis hypogaea* and *A. villosa*, *Plant Sci. Lett.* 23:35-39 (1981).

SUGARCANE VARIETAL IMPROVEMENT THROUGH TISSUE CULTURE

T.V. Sreenivasan and N.C. Jalaja

Sugarcane Breeding Institute
Coimbatore 641007
INDIA

The present day sugarcane varieties grown all over the world are complex species hybrids synthesized from about 25 basic genotypes belonging to the five species of *Saccharum*. Considering the vast assembly of sugarcane germplasm in the world collections from a wide range of geographic and environmental conditions, the sugarcane breeder at present is in a very sound position to breed varieties for the changing needs of the farmer and industry. High yielding varieties with built-in pest and disease resistance have been developed during the past 68 years through conventional breeding and selection programs, which requires 10 to 15 years to release a variety. Co 7704 is a recently released variety from the Sugarcane Breeding Institute, combining high yield and high sucrose with resistance to diseases like smut and one strain of red rot. The variety is very much liked by the farmers and the early maturity (10 months) and high sugar content is beneficial also. However, this variety suffers from a genetic defect - a characteristic pattern of leaf drying between 80 and 120 days after planting. The crop at this stage gives a dry appearance and, hence, the spread of the variety is restricted. Because of the genetic complexity of the variety, recovery of a similar variety will be impossible if attempts are made to rectify this defect through conventional breeding methods. Sugarcane cell and tissue culture work done so far has shown that the technique can be utilized profitably in such specific areas.

MATERIALS AND METHODS

Expanding leaves surrounding the growing point from field grown plants of the variety Co 7704 was the material used. The medium used was that of Murashige and Skoog (1). Solidification of the medium was done by 9 g/l agar. The basal medium was supplemented with 20 g/l sucrose, 3 mg/l 2,4-D, 100 mg/l mesoinositol and 10% by volume coconut milk for the induction of the callus. The pH was adjusted to 5.8. Plantlet development and growth were obtained in MS medium with the following additions: sucrose 20 g/l; glycine 2 mg/l; IBA 2 mg/l; kinetin 2 mg/l; mesoinositol 100 mg/l; nicotinic acid 0.5 mg/l; pyridoxine HCl 0.5 mg/l and thiamine HCl 0.1 mg/l. For rooting plantlets were transferred to White's basal medium on filter paper supports.

The leaves encircling the growing point were unwhorled, dissected out and used for inoculation. For the induction of the callus the flasks were incubated in dark at 26°C. The callus was subcultured three to four times. For differentiation of plants, callus bits were transferred to differentiating medium and kept in light for 8 hours. The plantlets which differentiated in clumps initially were separated into smaller groups and were transferred to fresh medium. When the shoots were a few inches in height, they were separated individually and kept for rooting. After sufficient root formation, they were planted in mildly sterilized soil in plastic containers. After the initial establishment, the subclones were planted in bigger pots before they were planted in the field. During the first year, each subclone was planted at 60 cm distance, five plants in a 3 m row. In subsequent years, each subclone was planted in a 3 m row by stem cuttings, at two locations. After the initial observations and screening at one row trial stage, replicated experiments were conducted by utilizing promising genotypes.

The standard cytological procedures were followed for the study of chromosome number and meiosis.

RESULTS

The callus initiated after 7 to 10 days of inoculation from cut ends. The growth of the callus was rapid and the callus was ready for subculture after one month. Differentiation of the shoots from the callus was noticed after 10 days of transfer. The process of differentiation was continuous and plantlets of different sizes were present in the same flask. Rooting was poor in the solid medium. Root initials were visible at the four to five leaved stage. Mortality was very high when rooted plants were directly planted from solid agar medium to soil. Profuse rooting occurred in White's liquid medium and the survival was 85% to 90% from liquid medium to soil.

A total of 200 plants were established initially, but 165 plants survived finally. They were produced in three batches, the first batch consisted of 15 subclones, the second batch 24 subclones, and the third batch 136 subclones. The first batch is in the advanced stage of testing, the second batch in the one row trial and the third batch in large pots awaiting transplantation in the field.

Variation in morphology in subclonal population was noticed for leaf length, leaf width, erectness of leaves, stalk diameter, number of stalks per clump, color of the root zone and the time of flowering (Figure 1). Variation in cane weight, brix % and sucrose % was also noticed between the clones. However, only 15.38% of the subclones markedly differed from the parent in their morphology. Data on number of stalks per row, stalk diameter, brix % and sucrose % and leaf drying from a replicated experiment involving eight subclones along with the parent are presented in Table 1.

Figure 1. Subclones extablished in the field from Co 7704 callus culture.

Table 1. Data on Co 7704 Subclones Trial.

Subclone number	No. of canes per row	Cane diameter	Brix %	Sucrose %	No. of leaves dried	Chromosome number 2n
1	41	3.35	19.36	16.08	27	113
2	35	3.32	19.78	15.98	19	112
3	27	3.50	19.46	16.60	nil	108
4	40	3.25	18.41	15.08	7	114
5	36	3.14	18.96	15.04	32	108
6	35	3.15	19.11	15.62	nil	112
7	42	3.06	15.98	14.94	nil	112
8	51	3.17	19.96	16.78	4	114
Co 7704 (control)	38	3.35	19.61	16.36	28	114

Mean of four replications. Data collected at nine month stage.

Out of the 39 field planted subclones screened for leaf drying at intervals of 15 days starting from the 80th day after planting, 3 subclones from the first batch and 5 from the second batch were free from leaf drying. Out of these 8 subclones, two resembled the parent in all the morphological characters. One subclone without leaf drying differed markedly in having reduced leaf length, leaf width, and stalk diameter, profuse tillering, and a fan shaped carriage.

Cytology of the Parent

The chromosome number of the variety Co 7704 was studied from root tip cells, shoot apices and at meiosis. In 90% of the root tip cells 2n = 114 chromosomes were observed. chromosome number ranged from 2n = 110 to 113 in 10% of the cells. In the shoot apices the frequency of aberrant cells were more. In 30% of the cells the chromosomes ranged from 100 to 113. Maximum variation was noticed in pollen mother cells at meiosis. Here the 2n number ranged from 101 to 121. The somatic chromosome number of 114 as determined in root tip cells was observed only in 28% of PMC's. The number of bivalents ranged from 48 to 59 and univalents from 1 to 11 at metaphase I.

Pollen mother cells with the normal chromosome number showed 54 bivalents and 6 univalents.

Cytology of the Subclones

Out of the 16 subclones studied cytologically, only 6 plants showed 2n = 114 chromosomes. The remaining 10 plants had chromosome numbers ranging from 2n = 105 to 2n = 123. Studies on meiosis of many of these subclones showed further cell to cell variation in pollen mother cells within the plant. In addition to this chromosomal numerical variation, higher chromosome associations indicating structural changes were also recorded in some of the subclones.

DISCUSSION

Occurrence of extensive chromosomal numerical variation which results in the production of genetically variable subclonal populations from callus and cell cultures have been a common feature in many plant genera (2) and has been well documented in sugarcane (3). Taking advantage of this phenomenon, it was possible to isolate resistant subclones from susceptible sugarcane varieties for two fungal diseases and one virus disease (4, 5). Fiji disease resistant subclones of variety Pindar 70-31 is being released for commercial cultivation in Fiji (6). As subclones without leaf drying isolated from Co 7704 are morphologically similar to the parent and in preliminary yield trials they are comparable to the parent clone, they may have the potential of becoming commercial clones of importance.

Sugarcane varieties are highly balanced genetic systems. Even in conventional breeding programs, the genetic variability released through sexual reproduction is the result of genetic recombination and whole chromosome segregation. Isolation of the similar variety by the addition of a desirable character or elimination of a defect by normal breeding techniques is almost impossible in sugarcane. By chromosome manipulation at the cellular level it may be possible to get sufficient genetic variability to rectify certain specific defects without completely disrupting the genetic balance of a highly adapted variety. The cytological information presented, though limited, indicates that the extent of numerical variation varies in different tissues of the same plant. It may be possible to increase the frequency and spectrum of chromosomally variable populations by careful selection of tissues for callus induction. It is known that in sugarcane, part of the variation in cultures is due to the chromosomal variation existing in the tissues utilized for inoculation (7, 8). Heinz et al. (7) have reported that greater variation in all respects was obtained in plantlets derived from H.50-7209 known to be a chromosomal mosaic. In Co 7704 also, it looks that part of the variation observed in the chromosome number of the subclones have derived from the parent itself.

ACKNOWLEDGEMENTS

The authors are grateful to Dr. K. Mohan Naidu, Sugarcane Breeding Institute, Coimbatore for his keen interest and encouragement. The excellent technical assistance provided by Mrs. S. Shantha is acknowledged. They are also thankful to Shri. R. Umeshwar Lala for assistance in field work.

REFERENCES

1. T. Murashige and F. Skoog, A revised medium for rapid growth and bioassays with tobacco tissue cultures, Physiol. Plant. 15:473-497 (1962).
2. R.M. Skirvin, Natural and induced variation in tissue culture, Euphytica 27:241-266 (1978).
3. D.J. Heinz, M. Krishnamurthy, L.G. Nickell and A. Maretzki, Cell tissue and organ culture in sugarcane improvement, in: "Applied and Fundamental Aspects of Plant Cell, Tissue and Organ Culture," Springer-Verlag, Berlin (1977).
4. D.J. Heinz, Sugarcane improvement through induced mutations using vegetative propagules and cell culture techniques, in: "FAO/IAEA Panel Proc.," PL 501/5:53-59 (1973).
5. M. Krishnamurthy and J. Tlaskal, Fiji disease resistant *Saccharum officinarum* var. Pindar subclones form tissue culture, in: "Proc. Intl. Soc. Sugarcane Technologists," XV:130-137 (1974).
6. M. Krishnamurthy, personal communication (1981).
7. Heinz et al., (1969).
8. T.V. Sreenivasan and N.C. Jalaja, Production of subclones from *Saccharum* - *Zea* callus culture, Plant Sci. Lett. (in press) (1981).

PRODUCTION OF ESSENTIAL OIL FROM CALLUS CULTURES OF *OCIMUM BASILICUM* L.

P.V. Purohit and P. Khanna

Laboratory of Pl. Physiology and Biochemistry
Department of Botany, University of Rajasthan
Jaipur,302015, INDIA

INTRODUCTION

Biological role of monoterpene glycosides and its role in production and accumulation of allylphenols and phenyl propanoid compounds have been discussed by Skoop and Horster (1) for the first time in lamiaceae species *Thymus vulgaris* L. Preliminary phytochemical investigations of essential oil from undifferentiated tissue culture of *Matricaria chamolia* L..had shown no similarity, neither to the oil of *in vitro* differentiated leaves, nor to the oil of the flower of intact plant (2). Reinhard and co-workers (3) have investigated production of essential oil and coumarins by tissue cultures of *Ruta graveolens* L.. They found that when cultures were well illuminated, the essential oil is produced and deposited in excretory cavities while in dark cultures, essential oil is produced too, however, in ordinary cells. Nagel and Reinhard (4) have reported that root tissue cultures of *Ruta graveolens* L. however maintained their chemical differentiation in all conditions. Whether cultivated in light or dark they produced an oil that resembles that of the roots of the whole plant. But in stem tissue cultures of *Ruta graveolens* L. the synthesis and composition of volatile oil depends on light intensity and its wave length and further more on the kind of oil secretion in oil cells or secretory passages.

We have investigated twenty-four months old callus tissues of *Ocimum basilicum* Linn., grown as static and suspension cultures for its production of essential oil. The effect of continuous illumination on static cultures has also been studied on production and accumulation of allylphenols and phenylpropanoids, constituents, of essential oil.

MATERIALS AND METHODS

Unorganized callus cultures of O. basilicum L. was raised from young internodal segments on revised (5) Murashige and Skoog's (6) RT medium supplemented with 1 ppm of 2,4-D and 1% agar. Tissue grown as static cultures when transferred to RT liquid medium containing 0.1 ppm of 2,4-D; (60ml of media in 150ml flask on rotary shaker at 100 rpm) resulted a uniform cell suspension culture. The static and suspension cultures were maintained by frequent subculturings every 4 to 6 weeks in fresh RT medium. The tissues were kept in dark and maintained at 26 $\pm$ 1°C with relative humidity of 55%. One lot of static cultures was transferred to continuous light (about 1600 lux).

The tissues were harvested at 2,4,6 and 8 weeks in case of static and 2,4,6 weeks in case of suspension cultures and their growth indices (GI) were calculated (final dry weight of tissue - initial dry weight of tissue/initial dry weight of tissue) separately.

The fresh tissue samples were macerated with small quantity of distilled water and slurry was hydrodistilled with modified clevenger apparatus and oil samples were collected. The stale media was also checked. Distillation was continued for 6 hours in each case.

The samples of oil as well as volatile constituents of reference samples were chromatographed on silicage 1 G plates (20x20cm and 0.3mm thick) using n.Lexane - ether (4:1) as well as Benzene-Ethylacetate (95:5). The chromatograms were then inspected under UV light and located with chromogenic spray (5% W/V Vanillin in sulfuric acid) before and after heating at 105° for 5 min. Two dimensional TLC was also performed using Benzene-Ethylacetate (95:5) and Petroleum ether (60°-80°) - Ethylacetate (85:15) respectively.

The oil samples were also subjected to Gas chromatography using Hewlett Packard model: 5712 A, having stationary phase column SE-30, with support of chromosorb W-60.80, temp. 170°; TCD: 250°; carrier gas: hydrogen at flow rate of 60ml/min.; lenght of the column: 2m with ¼" O.D.; attenuation 10x10^{-2}; chart speed 0.25"/min; sample size: 0.4 1. The quantitative estimations were done by using enrichment technique. This technique was comprised of adding to the tested oil a known amount of compound assumed to be present so that the concentration thereof was increased. The quantitative determination was based on the peak area measurements.

RESULTS AND DISCUSSION

The static cultures maintained in dark was fragile in nature and yellowish in color. The maximum growth index was 7.5 (Table 1.a)

Table 1. a. Production of essential oil from static cultures *Ocimum basilicum* L. Kept in dark.

Age of tissues in weeks	GI*	Oil yield % on fresh weight basis	% of linalool	% of Methyl chevicol
2	2.5	0.027	12	35
4	4.0	0.035	18	54
6	7.5	0.022	15	42
8	7.0	0.018	09	24

Table 1. b. Production of essential oils from static cultures of *Ocimum basilicum* L. kept under continuous illumination.

Age of tissues in weeks	GI*	Oil yield % on fresh weight basis	% of linalool	% of Methyl chevicol
2	2.3	0.022	23	41
4	3.8	0.046	42	58
6	7.2	0.033	30	38
8	6.8	0.016	18	22

Table 1. c. Production of essential oils from suspension cultures of *Ocimum basilicum* L.

Age of tissues in weeks	GI*	Oil yield % on fresh weight basis	% of linalool	% of Methyl chevicol
2	1.6	0.014	15	23
4	3.2	0.026	24	34
6	2.4	0.018	11	09

*(Final dry weight of tissue - initial dry weight of tissue/ initial dry weight of tissue).

in 6 week old tissue. The maximum oil yield was 0.035% on fresh weight basis having 18% linalool and 54% methyl chevicol in 4 week old tissue.

The static cultures maintained in light was compact in nature and creamish white in color. The maximum growth index was 7.2 (Table 1. b) in 6 week old tissue. The maximum oil yield was 0.046% on fresh weight basis having 42% linalool and 58% methylch-evicol.

The suspension cultures maintained in dark was light brown in color. The maximum growth index was 3.2 (Table 1. c) in 4 week old tissue which also showed maximum oil yield of 0.026% on fresh weight basis having 24% linalool and 34% methylchevicol.

The media was also hydrodistilled where only traces of oil was found.

While comparing the data of light grown and dark grown cultures, it was observed that there is not much change in the growth indices in both the cultures. However, maximum oil yield was obtained in 4 week old tissue in static as well as suspension cultures. This indicates that the rate of bio-synthesis and production of oil is quite fast during a 4 week period. Subsequently there is a decline in the yield of oil in all the cases, which indicates its degradation or possible interconversion into some other secondary products.

The cultures grown under continuous illumination has shown quite notable increase in % of constituents as well as total yield of oil which clearly emphasize the role of light on the production and accumulation of oil components.

REFERENCES

1. H. Skoop, H. Horster, Planta medica, 29:208 (1976).
2. J. Reichling, J. Becker, Planta medica, 30:258 (1976).
3. E. Reinhard, G. Corduan, O.H. Volk, Planta medica, 16:8 (1968).
4. M. Nagel, E. Reinhard, Planta medica, 27:263 (1975).
5. B. Kaul, E.J. Staba, Lloydia, 31:171 (1968).
6. T. Murashige, F. Skoog, Physiol Pl., 15:473 (1962).

EXPRESSION OF 5-METHYLTRYPTOPHAN RESISTANCE IN REGENERATED *DATURA INNOXIA* PLANTS

J.P. Ranch and J.M. Widholm

Department of Agronomy
University of Illinois
Urbana, IL 61801
U.S.A.

Suspension cultures of *Datura innoxia* initiated from haploid plants were mutagenized with nitrosoguanidine and other compounds for 24 h, were rinsed into fresh medium and allowed to recover for four to seven days and then were plated in 46 M 5-methyltryptophan. After incubation for three to five months, growing colonies were placed into liquid medium and were tested for resistance to 5-methyltryptophan (Figure 1). In several experiments the untreated cells produced resistant colonies at the rate of 1.9×10^{-7} while nitrosoguanidine treatment increased this to 3.3×10^{-6}.

As shown in Figure 1, the two selected cell lines MT-1 and MT-2 are clearly more resistant to growth inhibition by 5-methyltryptophan than the wild type line. Most 5-methyltryptophan-resistant cells selected previously contained an altered anthranilate synthase which showed less sensitivity to feedback inhibition by both 5-methyltryptophan and the natural feedback inhibitor tryptophan (1). This was also true with the 5-methyltryptophan-resistant *D. innoxia* lines selected here (Figure 2).

Previously when 5-methyltryptophan-resistant tobacco cells were regenerated into plants, the resistance was passed back into cultures reinitiated from leaves, but the altered enzyme trait was not expressed in the leaves (3). Likewise shoots would not form roots in medium containing 5-methyltryptophan. The *D. innoxia* shoots regenerated from several resistant lines did, however, form roots in medium containing 5-methyltryptophan at concentrations which inhibit root formation from wild type shoots.

Subsequently, when anthranilate synthase activity was measured

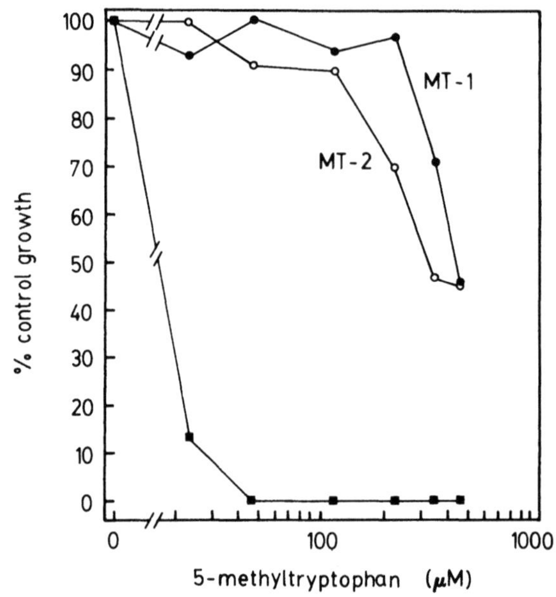

Figure 1. Growth of D. innoxia suspension cultures in the presence of 5-methyltryptophan. 0.2 g fresh weight cells were incubated in 100 ml medium for 12 days before collecting on filters and weighing. Wild type (■) and resistant (o, ●) lines.

in extracts from the leaves of regenerated plants, the feedback inhibition characteristics were found to be similar to those of the cultures from which the shoots were regenerated (Figure 3).

The analysis of free amino acids from both cultured cells and leaves of plants regenerated from these cells, showed that free tryptophan was elevated in the 5-methyltryptophan-resistant ones (Table 1). This evidently results from the alteration in the feedback

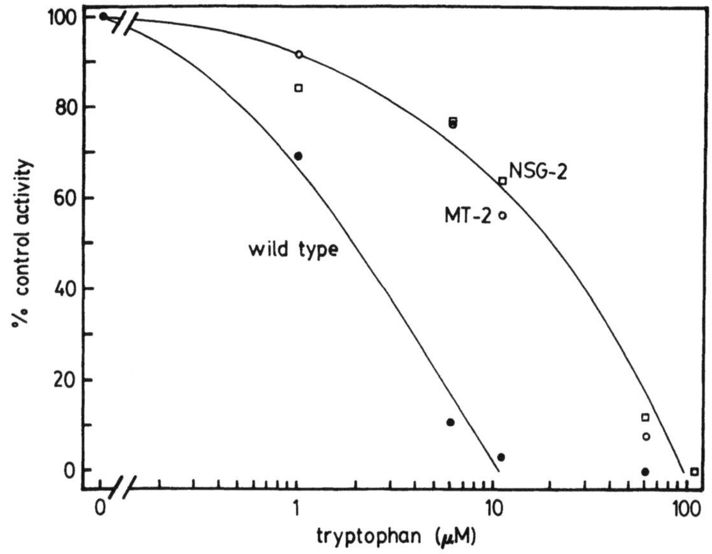

Figure 2. Inhibition of anthranilate synthase activity from suspension cultures by tryptophan. Activity was measured in partially purified preparations (2). Wild type (o) and resistant (■, □) lines.

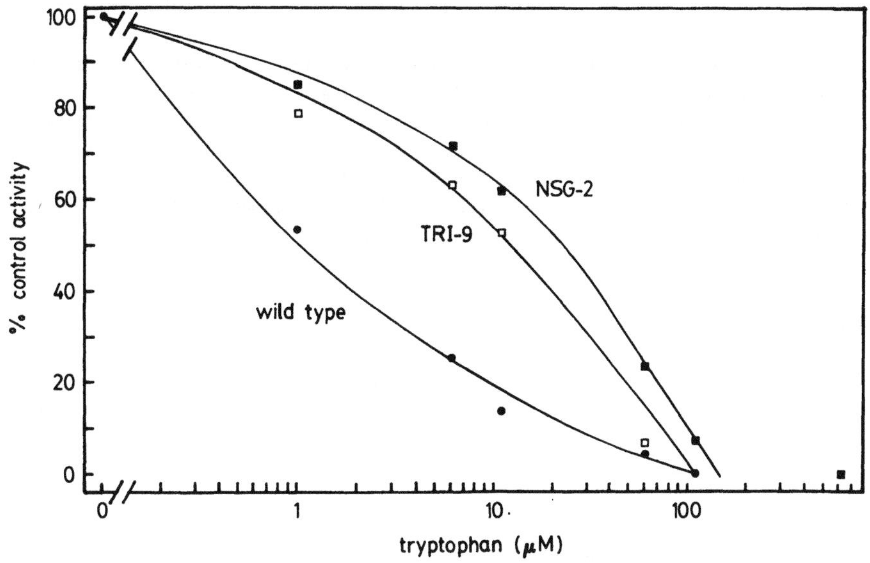

Figure 3. Inhibition of anthranilate synthase activity from leaves of regenerated plants by tryptophan. Activity was measured in partially purified preparations (2). Wild type (o) and resistant (□, ■) lines.

Table 1. Free Amino Acid Levels in Suspension Cultures and Regenerated Plants (nmoles/g fresh weight).

Amino acid	Suspension Cultures		Plant Leaves	
	Wild type	NSG-2	Wild type	NSG-2
Tryptophan	23	437	224	802
Phenylalanine	326	437	260	172
Tyrosine	180	681	173	57
Glutamine	1,744	5,440	2,574	1,786
Serine	544	303	2,291	1,470
Isoleucine	44	49	379	163
Histidine	23	55	172	71
Total all free amino acids	5,185	11,546	17,114	12,487

control properties of an anthranilate synthase as has been noted before with 5-methyltryptophan-resistant cell lines (1).

In the experiments described here, however, plants have been regenerated which express 5-methyltryptophan-resistance due to the presence of an altered anthranilate synthase which causes an accumulation of free tryptophan. Several lines were selected and all have given plants which express the trait. Some of the plants have the diploid chromosome numbers and are producing seed so a determination of the inheritance of the trait should be forthcoming.

ACKNOWLEDGEMENTS

This work was carried out with funds from the National Science Foundation grant PCM 80-10927 and USDA/SEA grant Agr 59010410801390.

REFERENCES

1. J.M. Widholm, The selection of agriculturally desirable traits with cultured plant cells, in: "Propagation of Higher Plants Through Tissue Culture - A Bridge Between Research and Application," K.W. Hughes, R. Henke and M. Constantin, eds., Technical Information Center, U.S. Department of Energy, Washing-

ton, D.C. (1979).

2. T. Kameya, M.E. Horn and J.M. Widholm, Hybrid shoot formation from fused *Daucus carota* and *D. capillifolius* protoplasts, *Z. Pflanzenphysiol.* 104:459-466 (1981).

3. J. Widholm, Differential expression of amino acid biosynthetic control isoenzymes in plants and cultured cells, *in*: "Plant Cell Cultures: Results and Perspectives," F. Sala, B. Parisi, R. Cella and O. Ciferri, eds., Elsevier/North-Holland Biomedical Press, Amsterdam (1980).

IN VITRO ISOLATION OF BIOCHEMICAL MUTANTS IN HAPLOID CELL CULTURES OF *NICOTIANA TABACUM*

N. Gupta and S. Bhaskaran

Nuclear Research Laboratory
Indian Agricultural Research Institute
New Delhi - 110012 INDIA

INTRODUCTION

Biochemical mutants in higher plants have several attractions. 1. They provide genetic markers for the selection of heterokaryons in studies with somatic hybridization and transformants in studies with DNA uptake (2). 2. They provide useful material for cell biological studies. 3. Some of the mutants like the pathotoxin resistant, amino-acid-analogue resistant (3) and herbicide resistant (4) could be of direct applied value.

The higher plant cells present some difficulties in the selection of mutants particularly the auxotrophs (5, 6) and thus what seems to be most desirable to develop suitable rechniques for isolation of mutants in a workable system. The present studies were undertaken to make an attempt to isolate auxotrophs and mutants resistant to PFP in haploid cell cultures of *Nicotiana tabacum*.

MATERIALS AND METHODS

Micro-colonies obtained by culturing protoplasts of androgenically obtained haploid plants of *Nicotiana tabacum* on liquid Ohyama and Nitsch's (7) medium were used as the source of the cells for mutant selection. Mutagen treatment to the shake cultures was offered in the doses of 0.025%, 0.05% and 0.1% with EMS and 1.0 kR, 1.5 kR and 2.5 kR with x-rays. The cells were resuspended in liquid Ohyama and Nitsch's medium after washing them in the medium. The suspension was distributed in a number of petri dishes containing liquid Ohyama and Nitsch's medium.

After six weeks colonies from those petri dishes which showed approximately 50% lethality were separated. The mutagen doses which showed about 50% lethality were 0.05% of EMS and 1.5 kR of x-rays. The 0.5% EMS treated colonies were used for isolation of auxotrophs for vitamin 'B' components, i.e., thiamine Hcl, pyridaxine Hcl, biotin, folic acid and nicotinic acid, while those treated with 1.5 kR x-rays used for isolation of mutants resistant to the amino acid analogue paraflurophenyl alanine. The procedure of selection of the two types of variants was as follows.

Selection of Auxotrophs

The EMS (0.05% treated colonies, growing on the complete Ohyama and Nitsch's liquid medium, (hereafter referred to complete medium or the CM) were individually transferred to test-tubes containing Ohyama and Nitsch's agar medium without vitamins, (hereafter referred to as the minimal medium or MM). After six weeks, the colonies which had ceased to grow were separated and individually transferred to tubes containing solid CM. When these colonies grew into callus masses each callus mass was sub-divided into a number of small parts and separately cultured on the following media:

(I) CM-thiamine Hcl referred to as $thia^-$ (II) CM-biotin, bio^- (III) CM-pyridoxine Hcl, pyr^- (IV) CM-nicotinic acid, NA^- and (V) CM-follic acid, FA.

After 8 weeks of incubation colonies which did not show growth on these media were picked up and further tested for the requirement by plating callus pieces from them on MM and MM + the suspected requirement and comparing their growth.

The process of screening was repeated for 10 sub-cultures and only those variants which consistently exhibited growth inhibition on the deficient media while growing normally on the supplemented media were selected. The calli differentiated into plantlets, during the course of sub-cultures. Cytological examination of these plantlets was carried out by the leaf squash technique of Burns (8).

Selection of Variants Resistant To Parafluorophenyl Alanine PFP

Colonies treated with 1.5 kR x-rays were plated on Ohyama and Nitsch's medium with agar containing increasing concentrations of PFP ranging from 0 to 15 µg/ml. The colonies which showed continued growth on these doses were further carried to media containing 20 µg/ml to 50 µg/ml of PFP. Controls were also grown on media without PFP. A number of sub-cultures of those calli which showed consistent resistance were made on these media. Plantlets got differentiated during the course of these sub-cultures. Chromosome numbers of the variants showing resistant to these concentrations of PFP were studied using Burn's (8) leaf squash technique. Free phenylalanine content in the shoot tissue of these plantlets was determined on the automatic amino acid analyzer.

RESULTS

Auxotrophic Variants

Out of a total of 212 surviving colonies 14 showed complete arrest of growth on the MM (Fig. 1). These colonies resumed growth on sub-culture to the CM. Since MM lacked the five vitamin components, namely thiamine HCl, Pyridoxine HCl, biotin, folic acid and nicotinic acid the non-growing colonies evidently carried auxotrophy for at least one of these vitamins. Callus pieces from each of the slow growing colonies were further sub-cultured on MM and MM + the suspected single requirement and growth compared with the controls. Those calli which did not show normal growth on the supplemented media perhaps had multiple auxotrophy and were carried forward. Nine calli, possibly showing single auxotrophies were thus recovered. Of these 3 were thiamine requirers, 2 biotin requirers, one folic acid requirer, 2 nictonic acid requirer and one pyridoxine requirer. During the course of sub-cultures the calli got differentiated. Thus plantlets with various auxotrophies were obtained. Growth of the variants was very much stunted on the MM but on the supplemented medium growth was as good as in the case of control (Fig. 2).

Ploidy status of the auxotrophs

Of the 9 auxotrophic variants studied cytologically 6 were found to be haploids and 3 diploids. The three thia$^-$ one bio$^-$ and NA$^-$ ond one pyr$^-$ and one NA$^-$ were diploids.

PFP Resistant Variants

Growth of colonies on medium containing 15 μg/ml PFP was very much inhibited. A few colonies continued to grow though at a slower pace. Calli from such colonies were transferred to media containing 20, 30, 40 and 50 μg/ml to PFP. These concentrations of PFP drastically inhibited growth of these colonies. However, a few of them continued to grow, though slowly. These calli were further sub-cultured on media containing the same concentration of PFP. The resistant calli were tested through a number of sub-cultures on PFP free and then on media containing PFP. During the course of sub-cultures the calli differentiated into plantlets (Fig. 3). The variants were found to show resistance consistently.

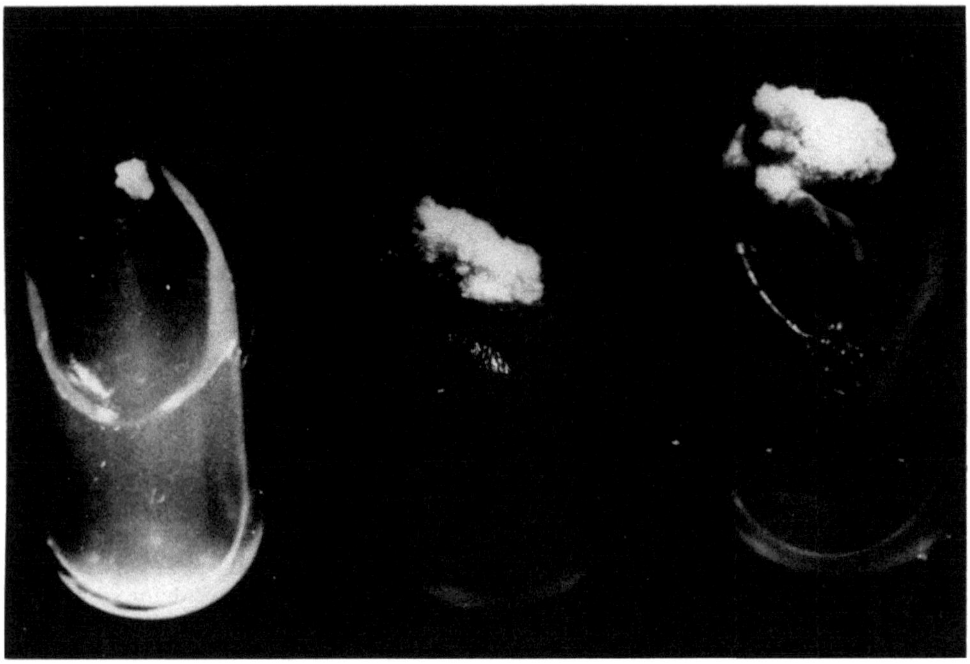

Figure 1. Growth of the auxotrophic callus of haploid *Nicotiana tabacum* on minimal medium and complete medium (a) callus on minimal medium (b) callus on complete medium (c) control.

Ploidy Status of the PFP Resistant Variants

Cytological examination of the variants and the normal plantlets showed that 9 out of 10 of the controls were haploids and one was diploid. The number of plantlets studied in each case was ten. Plantlets showing resistance to 20 and 30 μg/ml of PFP were predominently diploids, i.e., six out of ten in the case of former and seven in the case of latter, were diploids and rest haploids. Plants resistant to 40 μg/ml PFP were predominently triploids (Fig. 4), and those resistant to 50 μg/ml tetraploids (Fig. 5). Mitotic index was extremely poor in the latter case.

Free Phenylalanine Assay in PFP Resistant Variants

Assay of free phenylalanine in the shoot tissues of the normal haploids, diploids and the variants showed an increase in the amount

Figure 2. Auxotrophic variant plantlet of haploid N. tabacum growing on minimal medium and the supplemented medium. (a) control (b) pyridoxine auxotroph growing on the supplemented medium (i.e., MM + pyridoxine) (c) Pyridoxine auxotroph on minimal medium.

of free phenylalanine in the variants resistant to 30, 40 and 50 μg/ml. (Table 1). The amount was highest in the variant resistant to 40 μg/ml of PFP.

DISCUSSION

Auxotrophic Mutants

Reports on isolation of auxotrophic mutations in higher plants are rare, primarily because of the non-availability of suitable techniques. The BUdR suicidal technique followed by Carlson (9) was subsequently found unworkable by Aviv and Galun (10) and Savage

Table 1. Free phenylalanine content in the PFP resistant variants (shoot tissues).

Material	Free phenylalanine μg/g fresh wt.
Normal haploid	Traces
Normal haploid	Traces
PFP 20 μg/ml ($1.10X10^{-4}$M)	Traces
PFP 30 μg/ml ($1.65x10^{-4}$M)	10.9
PFP 40 μg/ml ($2.20x10^{-4}$M)	55.2
PFP 50 μg/ml ($2.75x10^{-4}$M)	31.1

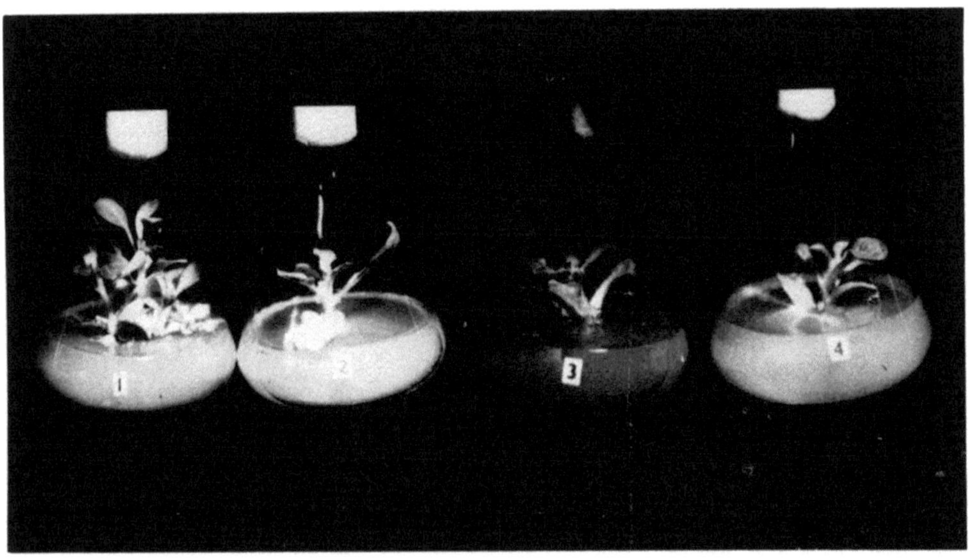

Figure 3. PFP resistant variants (1) Resistant to 20 μg/ml ($1.10 x10^{-4}$M) PFP (2) Resistant to 30 μg/ml ($1.65x10^{-4}$M) PFP (3) Resistant to 40 μg/ml ($2.20x10^{-4}$M) PFP (4) Resistant to 50 μg/ml ($2.75x10^{-4}$M) PFP

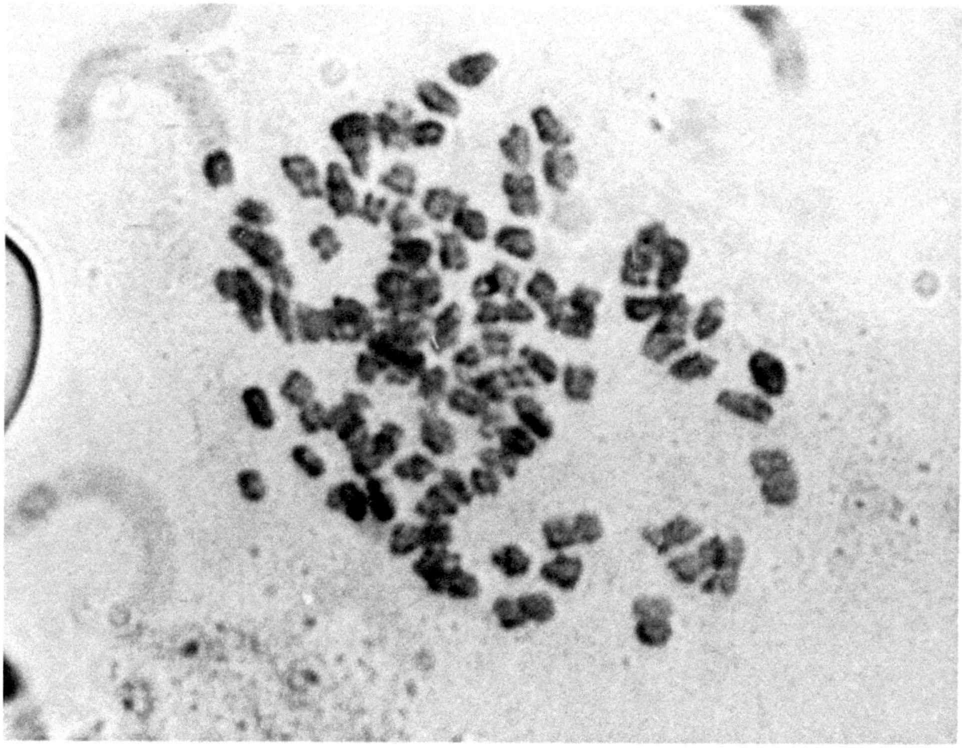

Figure 4. Chromosomes (3n = 72) of a mutant resistant to 40 μg/ml (2.75×10^{-4}M) PFP

et al., (11). The genome of a higher plant cell is complex and thus prevents easy detection of the auxotrophic mutation. The wild type gene product continues to be produced in the mutant perhaps by other copies of the gene present elsewhere in the genome resulting in the leakiness of the mutants (3, 9). Thus one way of isolating auxotrophic mutants could be by comparing growth of the suspected mutants with normal unmutagenized calli on the minimal medium and subsequently on supplemented medium.

The fact that a number of auxotroph mutants could be isolated using such criteria in the present study indicates that this technique may be useful in isolating auxotrophic mutants in cell-cultures of higher plants. Savage et al., (11), successfully isolated calcium pantothanate auxotrophic cell line in haploid cell cultures of Datura innoxia using a similar technique which they have called as the non-selective screening procedure.

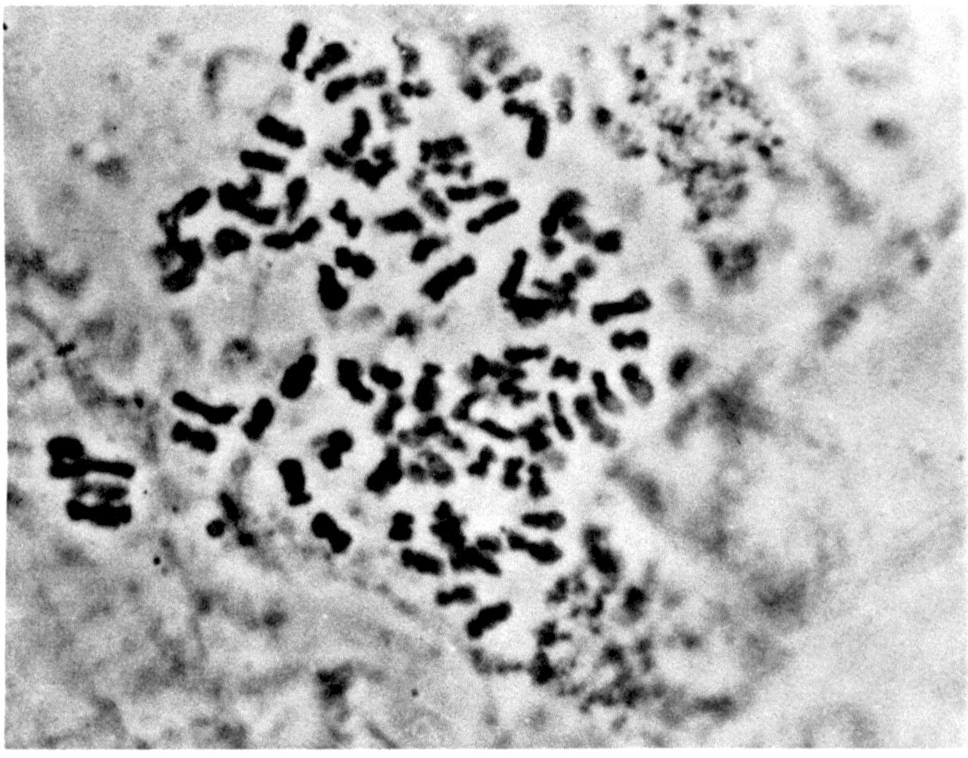

Figure 5. Chromosomes (4n = 96) of a mutant resistant to 50 µg/ml ($2.75x10^{-4}$M) PFP.

Ploidy status of the auxotrophs

Instability of chromosome number of cells in culture is well known (3, 6, 9, 12, 13, 14, 15). Occurrence of haploids in high frequency, in the present study, indicates that change in ploidy started quite late. This could be due to frequent sub-cultures necessitated for testing the auxotrophy.

PFP Resistant Variants:

PFP has been found to be inhibitory to growth at doses of 10-15 µg/ml. In the present study variants resistant to more than three times the inhibitory doses were isolated though they grew at a slower rate.

On assaying, free amino acid accumulation of large amounts

of phenylalanine was observed. Widholm (16) and Palmer and Widholm (17) found phenylalanine accumulation of phenolics, the breakdown product of the PFP resistant variant cells of *Nicotiana*. This was accompanied by induction of the enzyme phenylalanine ammonia lyase (PAL) in the cells. Thus the mutants in their case though overproduced the amino acid phenylalanine but did not accumulate it. Mutants reported in the present study overproduced and accumulated the amino acid. This accumulation of phenylalanine perhaps blocks uptake of the analogue, thus imparting resistance in the mutants.

Ploidy Status of the PFP Resistant Variants

PFP is routinely used in the fungus *Aspergillus* for selecting haploid mycelia in diploidized ones. In higher plants also there are reports of preferrential selection of haploid cells on PFP (15, 18, 19, 20). However, in the present study predominance of increased ploidy with increased resistance to PFP was encountered in the variants. Reasons for this are not known and needs to be investigated further.

ACKNOWLEDGEMENTS

The authors wish to express their thanks to Dr. B.R. Murthy, Project Director, N.R.L., for his interest and Dr. S.R. Chatterjee for help in amino acid analysis.

REFERENCES

1. S. Bhaskaran, S.K. Mukherjee, R. Majid, N. Gupta and B. Vatsya, *In vitro* genetic manipulation of higher plants, *Curr. Sci.* 50(15):653-661 (1981).
2. E.C. Cocking, M.R. Davey, D. Pental and J.B. Power, Aspects of plant genetic manipulation, *Nature* 293:265-269 (1981).
3. P.S. Carlson, The use of protoplasts for genetic research, *Proc. Natl. Acad. Sci.* USA 70:598-602 (1973).
4. R.S. Chaleff and M.F. Parsons, Direct selection *in vitro* for herbicide-resistant mutants of *Nicotiana tabacum*, *Proc. Natl. Acad. Sci.* USA 75:5104-5107 (1978).
5. J.M. Widholm, Selection and characterization of biochemical mutants of cultured plant cells, *in*: "Tissue Culture and Plant Science," H.E. Street, ed., Academic Press, New York (1974).
6. J.M. Widholm, Problems in selecting mutant cells, *IAPTC Newsletter* 23:2-5 (1978).
7. K. Ohyama and J.P. Nitsch, Flowering haploid plants obtained from protoplasts of tobacco leaves, *Plant Cell Physiol.* 10:917-921 (1972).
8. J.A. Burns, A technique for making preparations of mitotic

chromosomes from Nicotiana, Tobacco Sci. 8:22-23 (1964).
9. P.S. Carlson, Induction and isolation of mutants in somatic cell cultures of Nicotiana tabacum, Science 168:487-489 (1970).
10. D. Aviv and E. Galun, An attempt at isolation of nutritional mutants from cultured tobacco protoplasts, Plant Sci. Lett. 8:299 (1977).
11. A.D. Savage, J. King and O.L. Gamborg, Recovery of a pantothenate auxotroph from a cell suspension culture of Datura innoxia Mill., Plant Sci. Lett. 16:367-376 (1979).
12. R.D. Iyer and S.K. Raina, The early ontogeny of embryoids and callus from pollen and subsequent embryogenesis in anther cultures of Datura metel and rice, Planta 104:146-156 (1972).
13. H.H. Smith, Model genetic systems for studying mutation, differentiation and somatic cell hybridization, in: "Polyploidy and Induced Mutations in Plant Breeding," IAWA, Vienna (1974).
14. H.E. Street, Plant cell culture: present and projected applications for studies in genetics, in: "Genetic Manipulation of Plant Material," L. Ledoux, ed., Plenum Press, New York (1975).
15. P. Mathews and I.K. Vasil, The dynamics of cell proliferation in haploid and diploid tissues of Nicotiana tabacum, Z. Pflanzenphysiol. 77:222-236 (1975).
16. J.M. Widholm, Selection and characterization of biochemical mutants of cultured plant cells, in: "Tissue Culture and Plant Science," H.E. Street, ed., Academic Press, New York (1974).
17. J.E. Palmer and J.M. Widholm, Characterization of carrot and tobacco cell cultures resistant to p-fluorophenylalanine, Plant Physiol. 56:233-238 (1975).
18. N. Gupta and P.S. Carlson, Preferrential growth of haploid cells in vitro, Nature (New Biol.) 230:86 (1972).
19. L. George and S. Narayanaswamy, Anther culture of Atropa belladonna L. Effect of certain chemicals, amino acids and x-irradiation on yield and ploidy of pollen plantlets, in: "Abstracts Symp. Plant Tissue culture, Genetic Manipulation and Somatic Hybridization of Plant Cells," Bhabha Atomic Research Centre, Trombay, Bombay (1980).
20. V. Padmanabhan, W.R. Sharf and E.E. Paddock, Tomato anther culture - effect of PFP and salt concentration on callus formation, in: "Abstracts XII Bot. Cong.," 2:306 (1975).

SELECTION OF ANTHER DERIVED RESISTANT CELL LINES OF *SOLANUM KHASIANUM* TO CULTURE FILTRATE OF *FUSARIUM* SPECIES

E. Ramnath, S.K. Ghose*, S. Sen* and S.K. Sen

Programme in Genetical Research
Department of Botany
Bose Institute
Calcutta 700009
INDIA

* Department of Genetics and Plant Breeding
B.C.K.V.V.
Kalyani
West Bengal
INDIA

INTRODUCTION

Anther culture and its utilization has become a rapidly expanding field of research in recent years. In China, new varieties of tobacco, rice and wheat have been bred by means of this technique (1). Successes no doubt have attracted attention for the use of haploids in mutation studies for positive selection of resistant lines to herbicides, fungicides and pathogen toxins (2). It is known that the biochemical interaction between tissue cultures and pathogen toxins is usually identical to the whole plant response (3, 4). Mutant cell lines in haploids have been obtained in *Nicotiana tabacum* and *Petunia hybrida* for antibiotic resistance (5, 6). Albeit methods for producing haploids in Solanaceous plants have been established in a large number of species and interspecific hybrids, development of mutant lines against fungal pathogen has been reported in only *Solanum tuberosum* (7). Resistant maize plants against its parasite, *Helminthosporium maydis* (8) have been evolved.

S. khasianum plants, a rich source for the steroidal alkaloid solasodine (9) are highly susceptible to the wilt disease caused by the fungus *Fusarium solani*. The present report deals with the standardization of various cultural conditions for obtaining resistant cell lines of S. khasianum through mutation and subsequent screening against the relevant fungal toxin.

MATERIALS AND METHODS

Culture Conditions

Anthers from susceptible plants at uninucleate microspore stage were cultured on NN (10), N6 (11), BL (12), SH (13) and a modified Blaydes' medium (MBL) by substituting the organic constituents with that of SH medium and supplementing with casein hydrolysate: 0.5 g/l; 2,4-D: 5 mg/l; sucrose: 40 g/l; and NAA and BAP each at 1×10^{-6} mol concentration. The final pH was adjusted to 5.8. The cultures were maintained at 2500 lux with 16 h photoperiod at 26°C ± 1°C.

Mutagenic Treatment

The dissected anthers were immersed in the liquid medium containing varying doses of NG (N-methyl-N'-nitro-N-nitroso-guanidine), 100, 200 and 300 µg/ml and treated for 12, 24 and 36 h. These flasks were shaken gently on a gyratory shaker at 40-50 rpm. Subsequently, the anthers were washed with the culture medium for 4-5 times and placed in petri dishes (12-16 anthers/plate) containing solid MBL medium.

Preparation of Fungal Culture Filtrate

Vigorously growing fungal pathotype (Fusarium solani) was transferred from PDA (Potato Dextrose Agar) medium to the Richard's nutrient medium (14) and incubated at 28°C ± 2°C for 10 days. The fungal culture was then filtered to separate the mycelial mat. This culture filtrate was used for the screening of resistant/tolerant cell lines.

Screening of Resistant Lines

Originally 10 cell lines (600 callus pieces/line) obtained after treatment with the mutagen were transferred to the petri dishes containing fungal culture filtrate with final concentration of 0.025-1.5%. In the preliminary selection cycles, the callus lines were exposed to the medium containing sub-lethal to moderately lethal concentrations (0.025-0.1%) of the culture filtrate. In each subsequent selection cycle (one subculture transfer), callus lines were gradually subjected to higher toxin concentrations (0.2, 0.3, 0.5, 1.0, 1.5%). The calli retained for each new selection cycle were selected visually, judged by their ability to grow at the stepwise higher toxin level. Subculturing was done at 20-25 days interval. The callus masses were multiplied for 3 subcultures before proceeding for the actual selection procedure.

RESULTS

Among the various media tested, MBL was the most suitable in which as high as 37.8 percent anthers responded favorably for callusing (Table 1). This was followed by BL and NN medium, whereas SH medium showed the poorest response (7.2%). After about 20-25 days, the swollen anthers showed emergence of white friable callus. Light was an indispensable factor for the induction of callus. It was found that cultures kept in dark after inoculation failed to induce callusing in all the media tested.

The treatment of anthers with the mutagen was observed to affect the potentiality of callus induction. Results show that out of the three mutagenic concentrations used, maximum response of 25.6% was obtained with 100 μg/ml followed by 200 (12.62%) and 300 g/ml (6.85%) at 12 h treatment. Enhancement of treatment period always reuslted in decreased survivability (Table 2).

During the screening of resistant lines, we observed a varying degree of resistance in the cell lines (Figure 1). By the initiation of the third cycle, three cell lines were eliminated (Table 3). The ability of these cell lines to grow in the preliminary selection cycle was inhibited at toxin concentrations of 0.025-0.1%. By visual estimation it was observed that growth was inversely related to toxin concentration. Out of the original 10 cell lines, we obtained two cell lines (TR1 and TR7) resistant up to 1.5% concentration. Yield of about 5% callus masses from each line was recovered after the end of the 7th selection cycle, which in turn showed stability in their expression of resistance.

Table 1. Evaluation of Different Culture Media for *Solanum khasianum* anthers.

Media	No. of anthers cultured	No. of anthers callusing	Percentage
NN	1,240	255	20.6
N6	2,150	234	10.8
BL	1,880	485	25.7
MBL	2,225	842	37.8
SH	1,912	138	7.2

Table 2. Response Effect of Anthers after Mutagenic Treatment.

Mutagen Conc. µg/ml	Mutagen Treatment time (h)	No. of anthers cultured	No. of anthers responded	Percentage
100	12	160	41	25.6
	24	350	86	24.5
	36	280	53	18.9
200	12	293	37	12.62
	24	326	35	10.73
	36	185	16	8.64
300	12	203	14	6.85
	24	225	10	4.44
	36	307	10	3.25

Table 3. Survivability of Cell Lines During Selection Cycles for Toxin Resistance.

Selection cycle	Toxin conc. (%)	Cell lines	Survivability number
3	0.1	TR 1	62
3	0.1	TR 2	28
3	0.2	TR 4	16
3	0.2	TR 5	31
3	0.2	TR 7	53
3	0.2	TR 10	49
4	0.2	TR 1	48
4	0.3	TR 7	42
4	0.3	TR 10	14
5	0.5	TR 1	32
5	0.5	TR 7	33
5	0.5	TR 10	11
6	1.0	TR 1	28
6	1.0	TR 7	30
7	1.5	TR 1	26
7	1.5	TR 7	28
8	2.0	-	-

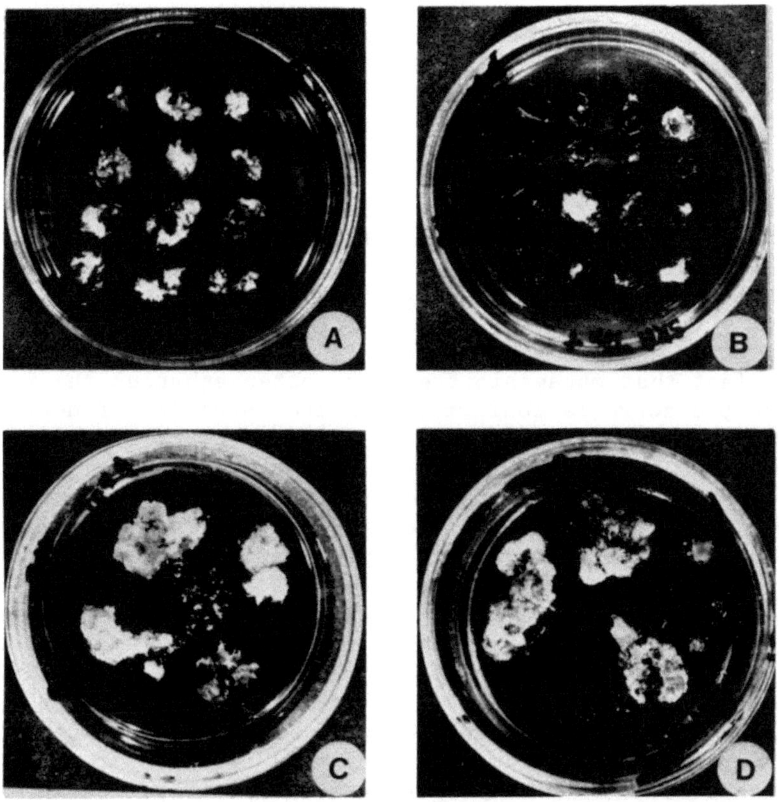

Figure 1. A: S. khasianum anther calli after 3 weeks of culture.
B: Selection of resistant callis in the presence of fungal culture filtrate.
C and D: Response of calli during various selection cycles.
C - calli after 4th selection cycle;
D - calli after 7th selection cycle.

DISCUSSION

Evolution of disease resistant lines obtained through anther culture and subsequent selection against the congenial pathogen is known to be a reliable technique (15, 16). On similar lines, we were able to define the various cultural conditions required for obtaining a fairly large number of anther derived cell lines. Our results showed that a combination of BL and SH medium with additional hormones, enhances callus induction at a relatively higher rate than their individual use. The indispensability of light for callus induction is contrary to the earlier reports on other solanaceous plants.

Although the selection of resistant calli to the toxin can be done even without the mutation of microspores (4), we preferred to induce increased genetic variability of the microspore cells prior to callus induction. The mutagen, NG, is known to be comparatively less lethal and has been widely used for bacterial as well as plant systems (17). Our results regarding the efficiency of this mutagen indicate that concentration of 200 μg/ml is the optimum dose for recovering supposedly mutant microspores. This is confirmed by the fact that less than this dose resulted in higher survivability of calli lines in the presence of fungal culture filtrate. Prolonged treatment always affected the callus inducing capacity of the anthers.

The fact that mutagenic treatment often enhances the result of selection procedure is confirmed with the retention of nearly 5% callus masses of two lines, whereas the control callus did not survive at all at this level of concentration. At the initial phase of selection, a modest selection pressure was applied since the effect of severe toxin inhibition on the survivability of the callus was not known. Earlier observations revealed that preliminary selection at high toxin levels might hamper the capacity of resistant cell lines to further growth. However, there may exist a possibility that the selected calli could still be a mixture of resistant and susceptible cells unless regeneration of the plants is obtained. But the expectancy of changed gene activity of cells after selection in a callus (18) can be ruled out here, because the resistant calli could maintain their survivability up to 7 passages at lethal doses. Gengenbach et al. (8) could regenerate resistant plants in maize after 5 selection cycles. Further experiments are underway for regenerating these calli lines into complete plants in which the specific nature of the resistance could be studied.

ACKNOWLEDGEMENTS

Grants from the Department of Science and Technology, Government of India to S.K. Sen and from the C.S.I.R., New Delhi to E. Ramnath in the form of a fellowship are acknowledged herewith.

REFERENCES

1. H. Han, Advances in anther culture investigations in China, in: "Proc. Symp. Plant Tissue Culture," Science Press, Peking (1978).
2. E. Thomas, P.J. King and I. Potrykus, Improvement of crop plants via single cells *in vitro* -- An assessment, *Z. Pflanzenzuchtg*. 82:14-16 (1979).
3. D.S. Ingram, Applications in plant pathology, in: "Plant Tissue and Cell Culture," H.E. Street, ed., Blackwell Science

Publ., Oxford (1977).
4. E.D. Earle, Phytotoxin studies with plant cells and protoplasts, in: "Frontiers of Plant Tissue Culture 1978," T.A. Thorpe, ed., The International Association for Plant Tissue Culture, Alberta, Canada (1978).
5. P.S. Carlson, Methionine sulphoximine resistant mutants of tobacco, Science 180:1366-1368 (1973)/
6. P. Maliga, A. Sz-Breznovits and L. Marton, Streptomycin resistant plants from callus cultures of haploid tobacco, Nature (New Biol.) 144:29-30 (1973).
7. M. Behnke, Selection of potato callus for resistance to culture filtrates of Phytophthora infestans and regeneration of resistant plants, Theor. Appl. Genet. 55:69-71 (1979).
8. B.G. Gengenbach, C.E. Green and C.M. Donovan, Inheritance of selected pathotoxin resistance in maize plants regenerated from cell cultures, Proc. Natl. Acad. Sci. USA 74:5113-5117 (1978).
9. J.D. Mann, Production of solasodine for the pharmaceutical industry, in: "Advances in Agronomy," N.C. Brady, ed. (1978).
10. J.P. Nitsch and C. Nitsch, Haploid plants from pollen grains, Science 163:85-87 (1969).
11. C.C. Chu, C.C. Wang, C.S. Sun, C. Hsu, K.C. Yin and C.Y. Chu, Establishment of an efficient medium for anther culture of rice through comparative experiments on the nitrogen sources, Sci. Sinica 18:659-668 (1975).
12. D.F. Blaydes, Interaction of kinetin and various inhibitors in the growth of soybean tissue, Physiol. Plant. 19:748-753 (1966).
13. R.U. Schenk and A.C. Hildebrandt, Medium and techniques for induction and growth of monocotyledonous and dicotyledonous plant cell cultures, Can. J. Bot. 50:199-204 (1972).
14. R. Heitefuss, M.A. Stahmann and J.C. Walker, Production of pectolytic enzymes and fusaric acid by Fusarium oxysporum F. Conglutinass in relation to cabbage yellows, Phytopath. 50:367-370 (1960).
15. D.J. Heinz, Sugarcane improvement through induced mutations using vegetative propagules and cell culture techniques, in: "FAO/IAEA Panel Proc. Induced Mutations in Vegetatively Propagated Plants," PL 501/5:53-59 (1973).
16. M. Krishnamurthi and J. Tlaskal, Fiji disease resistant Saccharum officinarum var. Pindar subclones from tissue culture, in: "Proc. Intl. Society Sugarcane Technologists," XV:130-137 (1974).
17. J.P. Nitsch, Haploid plants from pollen, Z. Pflanzenzuchtg. 67:3-18 (1972).
18. P. Maliga, G. Lazar, Z. Svab and F. Nagy, Transient cycloheximide resistance in a tobacco line, Mol. Gen. Genet. 149:267-272 (1976).

ESTABLISHMENT OF CALLUS AND CELL SUSPENSIONS AND ISOLATION OF MUTANT CELL LINES IN MUNG BEAN (VINGA RADIATA VAR. AUREUS)

V.K. Chowdhury, P.K. Sareen, D.R. Sharma, J.B. Chowdhury and V.K. Gupta

Tissue Culture Laboratory
Department of Genetics
Haryana Agricultural University
Hissar 125004
INDIA

INTRODUCTION

Mung bean is an important pulse crop of India and forms the chief source of protein in vegetarian diet. Improvement in quality of pulses through mutations using tissue culture techniques is being given much attention these days. The present communication deals with establishment of callus and cell suspension cultures in mung bean.

MATERIALS AND METHODS

Seeds of mung bean were surface sterilized by immersing in 0.2% (w/v) mercuric chloride with 4.5 drops of detergent for 10 min. followed by 5-6 changes of sterile distilled water and placed on agar medium for germination. Explants from 2-3 cm long roots were cultured on various media (Table 1). The cultures were incubated in the culture room at 27°C $\pm$ 1°C in dark as well as under light. Cultures thus obtained were subcultured on day 25 on a fresh medium. To study the effect of gamma irradiation on growth, cultures were exposed to 1, 2 and 3 kR of gamma rays (source ^{60}Co). Growth rate of the cultures was determined on fresh weight basis by recording weight on day 12 and day 25 after culturing.

Suspension cultures were initiated by transferring several pieces of actively growing, friable callus to different media (Table 2) in 250 ml Erlenmeyer flasks. The flasks were continuously shaken

Table 1. Composition of the Media Tested for Callus Initiation and Subsequent Growth.

Sr. No.	Original medium	Modified medium	Composition and reference
1	MS		Murashige and Skoog (7)
2		MS_2	MS + 2,4-D (2.0 mg/l)
3		MS_5	MS + 2,4-D (5.0 mg/l)
4		MS_2-C	MS_2 + Casein hydrolysate (1.0 g/l)
5	B_5		Gamborg et al. (8)
6		B_5-C	B_5 + Casein hydrolysate (1.0 g/l)
7		B_5-CN	B_5-C + NAA (5.0 mg/l)
8	Wl		Liau and Boll (9)
9	PC-L2		Phillips and Collins (1)
10		PC-L2(V)	PC-L2 with NAA (0.5 mg/l), 2,4-D (0.5 mg/l) and BAP (1.0 mg/l); Picloram omitted.

Table 2. Composition of Media Tested for Growth and Maintenance of Cell Suspension Cultures.

Sr. No.	Original medium	Modified medium	Composition and reference
1	L_1		Linsmaier and Skoog (10)
2		L_3	L_1 + Casein hydrolysate (1.0 g/l)
3	B_5		Gamborg et al. (8)
4	PC-L2		Phillips and Collins (1)
5		PC-L2(V)	PC-L2 with NAA (0.5 mg/l), 2,4-D (0.5 mg/l) and BAP (1.0 mg/l); Picloram omitted.

on a slow motion rotary shaker in light (5000 lux) at 27°C ± 1°C. The growth rate was measured in terms of fresh weight and packed volume. For determination of packed volume, a known amount of cell suspension was poured into centrifuge tubes and centrifuged at 4000 rpm for 5 min. Packed cells were then transferred to 100 ml of fresh medium and again packed volume recorded on day 8.

RESULTS AND DISCUSSION

Root explants of Vinga radiata, when cultured on different media, showed callus initiation on MS_2, B_5, PC-L2 and PC-L2(V) media only. However, for subsequent growth PC-L2(V), a modification of PC-L2 medium (1), was found to be the most suitable. PC-L2(V) medium had 2,4-D (0.5 mg/l), BAP (6-benzylaminopurine) (1.0 mg/l) and NAA (naphthalene acetic acid) (0.5 mg/l) instead of Picloram (0.06 mg/l). The growth of cultures continued for 25 days, after which they turned dark brown without further growth. Thus, in order to maintain fresh cultures, subculturing was needed on every 25th day.

Effect of 2,4-D, BAP and KNO_3 on callus cultures was also recorded (Table 3). Cultures on medium devoid of 2,4-D showed approximately 50% reduction in fresh weight compared to the standard medium. Similarly higher amounts of BAP and KNO_3 in the medium resulted in enhanced growth. In addition, the cultures showed greening under higher concentrations (3000 mg/l) of KNO_3.

Table 3. Effect of Photoperiod and Media on Callus Growth in Terms of Fresh Weight/g Initial Tissue Weight of Calli.

		Photoperiod	
Medium*	Dark	12 h	24 h
V_1	13.94	21.12	19.06
V_2	19.44	17.97	22.68
V_3	23.74	15.49	27.91
V_5	8.28	7.19	10.07

* V_1 - PC-L2(V)
V_2 - PC-L2(V) + KNO_3 (3000 mg/l)
V_3 - PC-L2(V) + BAP (2.0 mg/l)
V_5 - PC-L2(V) - 2,4-D

Differences in growth rate were also recorded under different light conditions. The cultures exposed to 12 h light showed maximum growth in terms of fresh weight (Table 3), only on PC-L2(V) medium.

As is evident from Table 4, 1 kR of gamma irradiation stimulated growth under all light conditions. Though higher doses (2 and 3 kR) also stimulated growth, the effect was not so pronounced as with 1 kR. Bajaj et al. (2) observed a significant stimulation in bean cultures at 0.5 kR and a gradual decrease at higher doses (1 to 10 kR). There are also some other reports on stimulation of growth and development by low doses of gamma rays (3, 4, 5). This stimulation can lead to an increase in plant height, root length and seed germination exhibiting an increase in cell number and/or volume (6).

Initiation of cell suspension cultures was observed only in PC-L2(V) medium on day 3 of incubation. The cultures turned brown and further growth stopped when incubated in the medium for more than 7 days. It is contemplated that the browning could be due to the oxidation of polyphenols, but this possibility was discounted by separating the medium and the cells. It was observed that the medium was devoid of any pigmentation, whereas the cells had turned brown. Data on packed volume of cells showed that there was a 6.5-fold increase in 197 h. Borwning was thus the consequence of the depletion of nutrients from the medium. Frequent subculturing (between 4-6 d) prevented browning of the cells.

From these suspension cultures, mutant cell lines have been isolated which are resistant to amino acid analogues, viz., p-fluorophenylalanine, DL-5 methyltryptophan, DL-5 fluorotryptophan and DL-methionine-DL-sulfoximine.

Table 4. Effect of Gamma-Rays and Photoperiod on Callus Growth in Terms of Fresh Weight/g Initial Tissue Weight of Calli Cultured on PC-L2(V) Medium.

Dose of Gamma Rays (kR)	Dark	Photoperiod 12 h	Photoperiod 24 h
0	13.94	21.13	19.06
1	19.61	29.26	23.77
2	20.89	23.03	14.50
3	17.64	23.47	17.37

REFERENCES

1. G.C. Phillips and G.B. Collins, In vitro tissue culture of selected legumes and plant regeneration from callus cultures of Red clover, Crop Sci. 19:59 (1979).
2. Y.P.S. Bajaj, A.W. Saettler and H.W. Adams, Gamma irradiation on seeds, seedlings and callus tissue cultures of Phaseolus vulgaris L., Radiation Bot. 10:119 (1970).
3. N. Degani and D. Pickholz, Direct and indirect effect of gamma irradiation on the differentiation of tobacco tissue culture, Radiation Bot. 13:381 (1973).
4. H.K. Shama Rao and S. Narayanaswamy, Effect of gamma irradiation on cell proliferation and regeneration in explanted tissue of pigeon pea (Cajanus cajan L. Mill), Radiation Bot. 15:301 (1975).
5. P. Spiegel-Roy and J. Kochba, Stimulation of differentiation in orange (Citrus sinensis) ovular callus in relation to irradiation of the media, Radiation Bot. 13:97-103 (1973).
6. A.P. Casarett, "Radiation Biology," Prentice Hall Inc., New York (1968).
7. T. Murashige and F. Skoog, A revised medium for rapid growth and bioassays with tobacco tissue cultures, Physiol Plant. 15:473-497 (1962).
8. O.L. Gamborg, R.A. Miller and K. Ojima, Nutrient requirements of suspension cultures of soybean root cells, Exp. Cell Res. 50:151 (1968).
9. D.-F. Liau and W.G. Boll, Callus and cell suspension cultures of bush bean (Phaseolus vulgaris), Can. J. Bot. 48:119 (1969).
10. E.M. Linsmaier and F. Skoog, Organic growth factor requirements of tobacco tissue cultures, Physiol. Plant. 18:100 (1965).

AMINO ACID ANALOG RESISTANCE IN THE SPECIES OF SOYBEAN AND TOBACCO

P.K. Das* and J.M. Widholm

Department of Agronomy
University of Illinois
Urbana, IL
USA

* Present address:

Department of Genetics and Plant Breeding
B.C.K.V.V.
Kalyani
West Bengal
INDIA

INTRODUCTION

Interest for isolating plant cell culture mutants has been increasing. Many cell lines with resistance to growth inhibition by amino acid analogs are now reported. Often they produce higher than normal amounts of the corresponding natural amino acids, and, hence, growth in specific amino acid analog is used as a selection tool (1). Several mechanisms can give rise to analog resistance, but most recovered variants appear to possess altered amino acid levels caused by changes in regulatory enzymes resulting in relaxed feed back control (1, 2). We attempted to select several amino acid analog resistant cell lines from plant suspenion cultures and a few of them derived from soybean and tobacco cell suspension are characterized. We present in this communication a brief report of our study.

MATERIALS AND METHODS

The cell lines used in the experiment were DX (*Datura innoxia* M.), TXD (*Nicotiana tabacum* L), SB (*Glycine max* L.), CCh (*Daucus carota* L.) and ALF (*Medicago sativa* L.). They were maintained as batch culture in liquid basal Murashige and Skoog (MS) meidum (3) with 0.4 mg/l 2,4-dichlorophenoxy acetic acid. Five amino acid analogs were used. They were A2C (Azetidine-2-carboxylic acid) a proline analog, AEC [S-(2-Aminoethyl)-cysteine] a lysine analog, 5MT (DL-5-methyl tryptophan) a tryptophan analog, and Eth (S-ethylhomocystein) a methionine analog. The concentrations used for A2C and 5MT were 10, 100 and 500 μM, for AEC and Eth were 20, 200 and 1000 μM. Amino acid analogs of different concentrations were dissolved in culture media, pH 5.8 and autoclaved. Growth studies were done by inoculating 0.5 g fresh weight cells into 50 ml of analog - culture media in 125 ml Erlenmeyer flasks. The cultures were incubated on reciprocating shakers (105-110 rpm) at 27°C ± 1°C for 7-12 days, collected by vacuum filtration on Miracloth and weighed. Following determination of complete growth inhibitory concentration of amino acid analogs, 1.0 g fresh weight of cells were inoculated in 100 ml culture medium containing inhibitory concentration of amino acid analogs in 250 ml Erlenmeyer flasks for screening resistant lines. The flasks were incubated in reciprocating shakers for up to 10 weeks since the growth of one resistant cell should be detectable within that time, assuming a doubling of 2-3 days. The resistant lines were further reinoculated in different concentrations of analogs and their growth was compared with corresponding normal cell lines by measuring fresh weight (g) after a certain periof of time. The resistant lines were also grown away from analogs for 25 cell generations and again reinoculated in the media containing analogs to examine whether resistant cell lines continued to maintain resistance. Free amino acids of resistant cell lines and of corresponding normal cell lines were extracted from exponentially growing cells which had been cultured away from inhibitory analogs for at least 10 cell generations, using methanol:chloroform:water (12:5:3) as described by Singh et al. (4). The aqueous extract was taken to dry mass under vacuum at 45°C. The residue was analyzed by physiological fluid methodology with Beckman 119Cl amino acid analyzer.

RESULTS AND DISCUSSION

Of the amino acid analogs the growth inhibitory effect of AEC was most striking. Cell lines of ALF (Alfalfa), SB (Soybean), CCh (Carrot) and DX (Datura) needed higher concentration of AEC for 50% growth inhibition than TXD (Tobacco) and the corresponding lysine level in these species in general was high. For ALF, the 50% growth inhibition concentration of 5 MT was low and so was the corresponding amino acid tryptophan level. The correlation study, however, indicated no relationship between amino acid analog concentration for 50% growth inhibition and the corresponding free amino acid level (Table 1).

Following screening in liquid medium, no cell lines of DX

Table 1. Relationship of Amino Acid Analog Concentration (μM) for 50% Growth Inhibition with Corresponding Free Amino Acid Level (nmoles/g fresh weight).

	Analog				Amino acid			
	(1) A2C	(2) AEC	(3) 5MT	(4) Eth	(5) Proline	(6) Lysine	(7) Tryptophan	(8) Methionine
DX	56.0	112.1	39.0	4.5	4.50	139.84	32.58	14.64
TXD	21.0	5.6	3.4	9.0	7.24	59.60	3.06	3.64
SB	27.0	316.2	32.0	6.8	25.08	73.65	0	0
CCh	15.5	200.0	9.2	17.0	24.87	57.37	164.01	184.01
ALF	38.0	500.0	0.5	24.3	824.16	196.25	3.20	89.08

Correlation: 1 & 5: 0.210; 2 & 6: 0.634; 3 & 7: 0.130; 4 & 8: 0.690.

(Datura), CCh (Carrot) and ALF (Alfalfa) were recovered resistant to any of the four amino acid analogs used. Three from SB (Soybean) and five from TXD (Tobacco) resistant to A2C (100 μM) and Eth (200 μM), respectively, were recovered. The frequency of appearance was 5.5×10^{-8} for TXD and 4.0×10^{-8} for SB. The concentration of A2C required to reduce growth by 50% was about 20 times higher for SB resistant cells than normal cells (Figure 1, top). The concentration of Eth required to reduce growth by 50% was about 50 times higher for TXD resistant cells than for normal cells (Figure 1, bottom).

The resistant trait was stable, since lines remained resistant even after 25 mass doublings in a medium lacking A2C for SB and Eth for TXD respectively and one line in each case is presented graphically (Figure 1, broken line).

Total free amino acids in SB resistant lines showed about a two- to three-fold increase over the control, and a few amino acids such as asparagine, histidine, arginine, etc., were increased significantly. However, the proline level, the corresponding amino acid of A2C, did not show any appreciable increase (Table 2). In this case analog resistance may be due to decreased uptake of the analog, selective incorporation of the natural amino acid into protein or compartmentalization of the analog within the cell (1, 5). Total free amino acids in TXD resistant lines showed about three-fold increase over the control. The level of methionine, the corresponding amino acid of Eth, was increased significantly in TXD resistant lines. Two other amino acids in aspartic biosynthetic pathway, namely threonine and isoleucine, showed significant increase in TXD

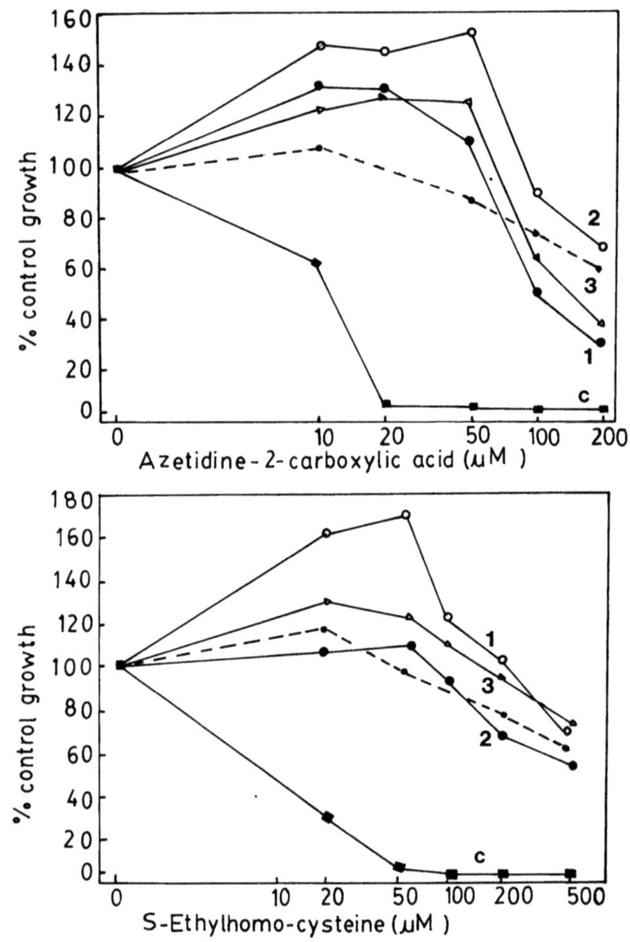

Figure 1. Growth of resistant lines (1, 2 and 3) and normal cells (C) of SB (top) and TXD (bottom) in different concentrations of A2C and Eth where resistant cells grown continuously with the analogs. Growth of resistant cells grown away from the analog for 25 generations (broken line).

Table 2. Amino Acid Analysis of Azetidine-2-carboxylic Acid (A2C) Resistant Lines of *Glycine max* (nmoles amino acid/g fresh weight).

Amino acid	Cell line			
	SBR1	SBR2	SBR3	Control
Asparagine	453.0	123.3	307.6	65.8
Proline	63.1	31.5	51.8	25.1
Tyrosine	329.3	118.3	273.7	92.7
Lysine	361.7	142.3	233.8	73.7
Histidine	1660.2	1122.5	448.8	191.7
Arginine	583.5	392.9	256.9	66.5
TOTAL AMINO ACIDS	28575.0	14638.3	24958.9	8966.7

Table 3. Amino Acid Analysis of Ethionine (Eth) Resistant Lines of *Nicotiana tabacum* (nmoles amino acid/g fresh weight).

Amino acid	Cell line			
	TXDR1	TXDR2	TXDR3	Control
Aspartic acid	132.8	396.7	304.4	24.2
Lysine	215.0	165.8	217.1	59.6
Methionine	203.1	44.9	146.3	3.6
Threonine	12230.0	5995.0	7783.8	369.5
Isoleucine	177.6	114.4	238.2	32.0
Glycine	1213.2	1020.0	3162.0	237.6
Tyrosine	122.1	264.4	851.4	35.5
TOTAL AMINO ACID	24814.0	20972.3	24986.3	7983.0

resistant lines and at the same time a manyfold increase on aspartic level was recorded (Table 3). The significance of high build up of aspartic acid level itself was not properly understood in terms of resistance. The increased levels of specific amino acids found in the resistant cell lines indicated that the feedback controls in the biosynthetic pathway in question might be altered, as was found previously for tryptophan and phenylalanine analog resistance and oversynthesis (6, 7, 8). Enzyme studies have not been carried out, but possible altered controls in the aspartate pathway which synthesizes three of the amino acids which were oversynthesized, methionine, threonine and isoleucine, should be explored. In this pathway aspartokinase and homoserine dehydrogenase are known to be two regulatory enzymes (9, 10). Other controls could also operate, but more study is needed. It would be of great interest to examine further how these elevated levels of free amino acids would influence the composition of cell protein.

ACKNOWLEDGEMENTS

We thank Miss Linda Wagner for her technical assistance. This work was supported by a Fulbright Indo-American Fellowship awarded to P.K. Das.

REFERENCES

1. J.M. Widholm, Selection and characterization of amino acid analog resistant plant cell cultures, Crop Sci. 17:597-600 (1977).
2. J.M. Widholm, Selection and characterization of cultured carrot and tobacco cells resistant to lysine, methionine and proline analogs, Can. J. Bot. 54:1523-1529 (1976).
3. T. Murashige and F. Skoog, A revised medium for rapid growth and bioassays with tobacco tissue cultures, Physiol. Plant. 15:473-497 (1962).
4. T.N. Singh, L.G. Paley and D. Aspinall, Stress metabolism. I. Nitrogen metabolism and growth in the barley plant during water stress, Aust. J. Biol. Sci. 26:45-56 (1973).
5. J.M. Widholm, Selection and characterization of biochemical mutants of cultured plant cells, in: "Tissue Culture and Plant Science, 1974," H.E. Street, ed., Academic Press Inc., New York (1974).
6. J.M. Widholm, Cultured *Nicotiana tabacum* cells with an altered anthranilate synthetase which is less sensitive to feedback inhibition, Biochim. Biophys. Acta 261:52-58 (1972a).
7. J M. Widholm, Anthranilate synthetase from 5-methyl tryptophan-susceptible and -resistant cultures of *Daucus carota* cells, Biochim. Biophys. Acta 279:48-57 (1972b).

8. J.E. Palmer and J.M. Widholm, Characterization of carrot and tobacco cell cultures resistant to p-fluorophenylalanine, Plant Physiol. 56:233-238 (1975).
9. J.K. Bryan, Studies on the catalytic and regulatory properties of homoserine dehydrogenase of *Zea mays* roots, Biochim. Biophys. Acta 171:205-216 (1969).
10. K.F. Wong and D.T. Dennis, Aspartokinase from wheat germ. Isolation, characterization and regulation, Plant Physiol. 51:322-326 (1973).

REGENERATION OF WHEAT, RYE AND TRITICALE PLANTS THROUGH TISSUE CULTURE

S. Eapen and P.S. Rao

Plant Morphogenesis and Tissue Culture Section
Bio-Organic Division
Bhabha Atomic Research Centre
Trombay
Bombay 400085
INDIA

INTRODUCTION

Recent developments in plant cell cultures have fostered the hope that this technique could be of invaluable help in crop improvement (1, 2). However, for effective utilization of this technique, methods have to be perfected for an efficient and rapid rate of plant regeneration from cultured cells.

The present report deals with the development of tissue culture techniques for bread, durum and emmer wheat, rye and triticale, and selection of amino acid analog resistant lines from mutagenized embryos of breadwheat.

MATERIALS AND METHODS

Breadwheat (*Triticum aestivum* L. Thell) cvs. NI-4, Hira, Moti, UP-310, Kalyan Sona and Chinese Spring, dudum wheat (*T. turgidum* L. Thell ssp. durum) cvs. Dwarf durum and Bijaga Yellow, emmer wheat (*T. turgidum* L. Thell ssp. dicoccum) cv. 2-9-18, rye (*Secale cereale* L.) and triticale (*T. aestivosecale* L. Mackey) were used as the source. Mature and immature embryos were used for callus initiation in breadwheat, rye and triticale while for durum and dicoccum only mature embryos were used. The basal medium comprised macro- and micro-elements of Murashige and Skoog (3), vitamin composition of Lin and Staba (4), 5.0 mg.l of 2,4,5-Cl_3 phenoxypropionic acid and 2.0% sucrose. For regeneration, calli were transferred to media devoid of or supplemented with different auxins and cytokinins.

For selection of AEC and ethionine resistant plants, embryos of Kalyan Sona were subjected to gamma radiation (1-3 kR at 3.3 kR/min) or 1.0% EMS (1 and 2 h) and cultured on media supplemented with 300 μM of AEC or 150 μM of ethionine.

RESULTS

When embryos of bread, durum and emmer wheat, triticale and rye were cultured on basal media supplemented with 5.0 mg/l of 2,4,5-Cl_3POP, vigorous callus growth was observed within 3-4 days and at the end of 20 days a good proliferating callus was obtained. At the end of 4-6 weeks, differentiation of shoot buds were observed in the same medium in all the species studied. For further development of shoot buds, transfer to media devoid of growth regulators was found essential in bread and durum wheat. Zeatin (0.5-5.0 mg/l) in combination with IAA (0.1-0.5 mg/l) in the basal medium enhanced the number of shoot buds produced in durum wheat and some of the cultivars of breadwheat. In emmer wheat, basal medium supplemented with coconut mild (15%) either alone or in combination with NAA (0.2 mg/l) was ellential for further development. In cultures of rye, zeatin (0.5-5.0 mg/l) alone in the basal medium promoted further development of shoot buds, whereas in triticale a combination of zeatin (5.0 mg/l) and IAA (0.1 or 0.5 mg/l) was essential for successful shoot bud regeneration.

6% of cultures of triticale, 10% of breadwheat cultivars of N1-4 and Hira and 50% of emmer wheat cultures gave rise to chlorophyll variants including albinos and striatas. The number of shoot buds per culture could be increased by the use of agitated liquid media. The regenerated plants obtained in vitro were transferred to field and reared to maturity.

Ethionine at 150 μM and AEC at 300 μM were found inhibitory for shoot length and germination percentage of Kalyan Sona embryos cultured under sterile conditions. The inhibition caused by ethionine was reversed by methionine and AEC by lysine. Embryos were subjected to gamma radiation or EMS and screened for resistance to AEC and ethionine. Two plants resistant to AEC and eight plants resistant to ethionine were selected from 4000 embryos screened. These plants were grown to the seed setting stage. The embryos collected from seeds of the resistant plants, when tested for growth on media supplemented with AEC or ethionine, were found capable of growth just like the parental line showing that the character was heritable.

DISCUSSION AND CONCLUSIONS

The present investigation using different wheat cultivars, rye and triticale has demonstrated that calli derived from mature and immature embryos are capable of regenerating whole plants. 2,4,5-Cl_3POP can be used as an effective growth regulator for induction and maintenance of callus cultures of wheat, rye and triticale. The usefulness of agitated liquid medium for increasing plantlet yield has been reported previously (5). Chlorophyll variants reported in this investigation has been obsrved in grass cultures (6).

The AEC and ethionine resistant traits selected through sterile embryo culture remained heritable in breadwheat.

REFERENCES

1. P.S. Carlson and J.C. Polacco, Plant cell cultures: genetic aspects of crop improvement, Science 188:622-625 (1975).
2. E. Thomas, P.J. King and I. Potrykus, Improvement of crop plants via single cells in vitro - an assessment, Z. Pflanzenzuchtg. 82:1-30 (1979).
3. T. Murashige and F. Skoog, A revised medium for rapid growth and bioassay with tobacco tissue cultures, Physiol. Plant. 15:473-497 (1962).
4. M. Lin and E.J. Staba, Peppermint and spearmint tissue cultures. I. Callus formation and submerged culture, Lloydia 24:139-145 (1961).
5. C.H. Chen, N.C. Stenberg and J.G. Ross, Clonal propagation of big bluestem by tissue culture, Crop Sci. 17:847-850 (1977).
6. P.F. Lo, C.H. Chen and J.G. Ross, Vegetative propagation of forage grasses through tissue culture, Crop Sci. 20:363-367 (1980).

MORPHOGENESIS IN CULTURES OF TAGETES PATULA L

S.L. Kothari and N. Chandra

Botany Department
University of Rajasthan
Jaipur 302004
INDIA

INTRODUCTION

Cell tissue culture techniques are being used for improvement of economically important crops (1) and in clonal propagation of ornamental plants as well (2). In the present communication successful induction of shoot buds and roots, from different explants of *Tagetes patula* L., has been reported.

MATERIALS AND METHODS

Seeds of *T. patula* were collected from the Botanic Garden, University of Rajasthan, Jaipur, and germinated aseptically in culture tubes on Murashige and Skoog (3) medium. Cotyledon and hypocotyl explants were excised from 8-10 day old seedlings while stem and leaf explants were taken from young plants growing in the garden. 5-10 mm long explants were sterilized with 0.1% mercuric chloride, thoroughly washed with sterile distilled water, and then incubated on MS medium supplemented with auxins and cytokinins. The pH of the medium was adjusted to 5.8 before autoclaving. All the cultures were grown under continuous light from fluorescent tubes and incandescent bulbs at 26°C ± 2°C.

OBSERVATIONS AND DISCUSSION

When cotyledons, hypocotyl and leaf explants were grown on Murashige and Skoog (3) medium containing indole butyric acid (IBA, 5.0 mg/l) or naphthalene acetic acid (NAA, 0.5-2.0 mg/l) root forma-

Figure 1. Negatively geotropic green roots from hypocotyl on MS + IBA (1.0) + kinetin (1.0 mg/l).

tions were induced. However, when the same explants were grown on MS medium supplemented with IBA (1.0-4.0 mg/l) and kinetin (1.0-4.0 mg/l) negatively geotropic green roots were formed (Figure 1). These roots were green, short and stout and showed profuse branching resulting in anastomosis (Figure 1). A rhizome like structure developed after 30-35 days of incubation and this structure also developed green negatively geotropic roots. When subcultured on the same medium, they continued to regenerate their types up to third passage, however, in the subsequent cultures their regenerating capacity declined sharply. Microscopic examination of these roots revealed different types of root hair, viz., normal single celled, single celled with knobbed apices, 2-3 celled, and 2-3 celled branched hair with or without knobbed apices.

Many shoot buds were differentiated from stem internode, leaf, hypocotyl and cotyledons on MS medium enriched with indole acetic acid (IAA, 5.0 mg/l) and benzyl adenine (BA, 5.0 mg/l) after 10-15 days of incubation (Figures 2 and 3). Relatively more shoot buds were possible to be induced from cotyledon explants.

In vitro regeneration of negatively geotropic roots is quite interesting as it has not been reported elsewhere in culture. *In vitro* induction of negatively geotropic roots may be used to investigate biochemical and physical factors responsible for tropic movements in roots. The regenerating capacity of these roots up to third passage and a sharp decline thereafter is indicative of presence of certain substance(s) in the explants which passes in subse-

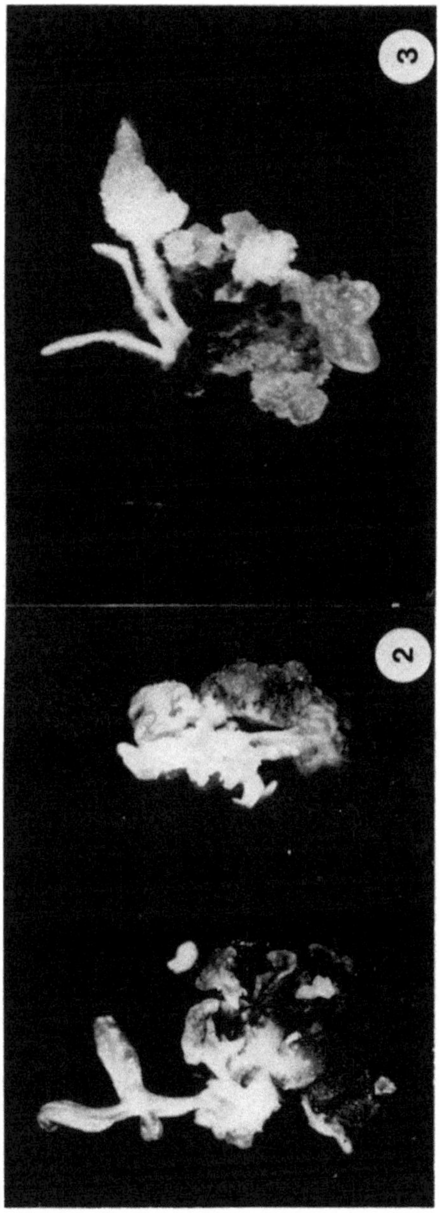

Figure 2. Shoot bud regeneration from cotyledon on MS + IAA (5.0 mg/l) + BA (5.0 mg/l).

Figure 3. Shoot and root regeneration from hypocotyl on MS + IAA (5.0 mg/l) + BA (5.0 mg/l).

quent cultures and which may be responsible for negative geotropism.

REFERENCES

1. G. Wenzel, Recent advances in applied tissue culture of potato rape and rye, *in*: "Plant Cell Culture Results and Perspectives," F. Sala, ed., Elsevier/North-Holland Biomedical Press, Amsterdam (1979).
2. T. Murashige, Clonal crops through tissue cultures, *in*: "Plant Tissue Culture and Its Biotechnological Applications," W. Barz, E. Reinhard and M.H. Zenk, eds., Springer-Verlag, Berlin (1977).
3. T. Murashige and F. Skoog, A revised medium for rapid growth and bioassay with tobacco tissue cultures, *Physiol. Plant.* 15:473-497 (1962).

CALLUS INITIATION AND REGENERATION IN TISSUE CULTURE OF PAPAYA

R.M. Pandey and M.S. Rajeevan

Division of Horticulture and Fruit Technology
IARI
New Delhi 110012
INDIA

INTRODUCTION

Widespread cultivation of papaya is limited because of several factors: their dioecious nature, susceptibility to a large number of viral diseases and lack of commercial methods for clonal multiplication. There have been few attempts in the past to develop a tissue culture propagation system for papaya. Except for the procedure of Litz and Conover (1), others are not ideal as they went through callusing for plantlet formation and are restricted to seedlings (2 - 5). We report here our preliminary results to develop an ideal system for clonal multiplication of two papaya varieties.

MATERIALS AND METHODS

Seeds of the two varieties, Ranchi Selection 1-45D and Coorg Honey Dew, were supplied by the Horticultural Research Station, Pusa, Bihar and the Indian Institute of Horticultural Research, Bangalore, respectively. The various explants for callus induction and regeneration studies, viz. stem segments, petiole from the top 4-5 leaves and shoot tip, were taken from outdoor grown seedlings of 6 weeks old. Sterilization of the explants was done with 0.1% $HgCl_2$ for 5 minutes and was followed by washing in sterile water 3-5 times. The stem and petiole were made into segments of 5 mm. The shoot tips were also of 5 mm length and consisted of the top young unexpanded leaves. Prior to planting the explants in medium, they were continuously agitated in sterile water in a horizontal shaker at 160 rpm for 2-3 hr. This was done to minimize the flow of latex into the medium.

The following culture media were tried: MS (6), B_5 (7) and AS (5). AS medium contains inorganic components of MS medium except NaH_2PO_4 replaces KH_2PO_4 at 300 mg/l. This medium was supplemented with the following organic components: nicotinic acid, 5.0 mg/l; pyridoxine HCl, 1.0 mg/l; thiamine HCl, 1.0 mg/l; inositol, 100 mg/l; adenine sulfate 100 mg/l; malt extract 0.5 g/l; casein hydrolysate 1.0-1.5 g/l; gibberellic acid 1.0 mg/l; naphthalene acetic acid (NAA), 2.0 mg/l; and kinetin, 0.5 mg/l.

The pH of the media was adjusted to 5.7 prior to sterilization done at 15 psi for 15 min. All the media were solidified with 8.0 g/l Difco Bacto Agar. Cultures were maintained at 26°C with 16 hr light (3500 lux) and 8 hr dark. We tried 8 concentrations of growth regulators ranging from 0.5-100 μM of two auxins, namely NAA and 2,4-D, and two cytokinins, viz. kinetin and 6-benzylaminopurine (BAP) in all possible combinations with the first two basic media.

RESULTS AND DISCUSSION

Callus Initiation

The medium recommended by Arora and Singh (5) was not found satisfactory for both the varieties for initiating callus when stem segments were taken from seedlings grown in the field. It is likely that the optimum concentrations of various organic components, which they found individually, had a suppressing effect on callus initiation when put together. We feel that the final medium recommended by them has not been tested for the effectiveness on callus initiation and this defect adds substance to our conclusion.

Callusing in both the varieties was obtained from stem segments of seedlings grown in the summer months in B_5 medium supplied with NAA and kinetin. The effective concentration of NAA ranged from 10-25 μM and in the case of kinetin from 2.5-10 μM. The combination of NAA 25 μM and kinetin 10 μM gave the highest percentage (80%) of cultures with callus. Callusing was noticed even with combination of low concentrations of NAA (5-10 μM) and a high concentration of kinetin (50-100 μM) but the percentage of success was only 40%. Such a pattern of response could not be detected by Arora and Singh (5) as they noted the effect of only a limited number of combinations of auxins and cytokinins. The response of stem segments to such a wide range of NAA and kinetin rendered difficult to fix their optimum requirement for callusing. Moreover, when the stem segments were taken from seedlings grown in winter months, none of them did produce callus in the above range.

Callusing of petiole was observed in MS medium with NAA 0.5 μM and BAP 2 μM. This is in agreement with the result obtained by De Bruigne et al. (2) for the same explants of unspecified age. The growth of callus from petiole segments in this medium was, however, not satisfactory and needs further improvement. Our efforts to induce callusing from midrib of the young leaves of both seedlings and mature plants have not yet given satisfactory results. B_5 medium with NAA 25 μM and kinetin 2.5 μM resulted in a very low amount of callus growth.

Callus originated mostly from all over the surface of the various explants. They looked friable and white with light green tinge. Callus could be subcultured successfully at biweekly intervals in medium with NAA and kinetin, each at 10 μM. It is evident from these results and those reported by De Bruigne et al. (2) on callus initiation and its subsequent growth that papaya requires a higher level of cytokinin than what is recommended by Arora and Singh (5).

Regeneration

Root formation. During our study on callus production from stem segments we observed the tendency of such explants to give rise to roots from the nodal region in B_5 medium with NAA 10-15 μM and kinetin 10 μM. This was observed in 10% of the cultures of Coorg Honey Dew, and there were about 10 roots/explant after one month of culture period. The roots which measured about 1 cm in length were devoid of hairs and abnormal in shape. When they came in contact with the medium, they started callusing, and after a period of 20 days all the roots turned to a mass of callus. The regenerated roots when excised and placed in the same medium produced a large quantity of callus rapidly. In some of the cultures with shoot tips rapid growth of callus from regenerated roots and regeneration of tips could be noticed when the level of NAA was reduced by half. Callus obtained from root cultures were identical in morphological appearance as that of callus from stem segments and could be subcultured at biweekly intervals in the same medium.

Shoot tip regeneration. Shoot tip regeneration occured in B_5 medium with NAA 10 μM and kinetin 2.5 μM after 2 weeks with profuse callusing from the base. The growth of the shoot, however, continued for another 3 weeks only. Prolifically growing shoot cultures were obtained when shoot tips grown in MS medium with NAA 5 μM and kinetin 50 μM (establishment medium) were transferred after a period of 3 months to the same basic medium with NAA 0.5 μM and BAP 2.5 μM (proliferation medium). There were a number of well developed leaves, typical of papaya with normal petiole and internodal length after a period of 6 weeks. Litz and Conover (1) could achieve similar results with shoot tips with NAA 10 μM in the establishment medium and BAP 2 μM in the proliferation medium. We found that the establishment period of 3 months could be considerably reduced to 2 weeks when the kinetin level was doubled in the establishment medium. The proliferation of the explant occured by stimulation of axillary buds, whereby a small plantlet was formed in the axil of each culture. The axillary growth resulted in an extremely bushy appearance of the cultures. Subculturing of the axillary bud growth in the proliferation medium and subsequent rooting offers the possibility of propagating papaya in large numbers with identical characters.

ACKNOWLEDGEMENTS

We express our sincere thanks to Dr. S. Bhaskaran and his colleagues, Mr. Mansha Ram and Mr. S. Radhakrishnan for their assistance.

REFERENCES

1. R.E. Litz and R.A. Conover, _In vitro_ propagation of papaya, _Hort. Sci._ 13(3):241 (1978).
2. E. De Bruigne, E. Dehanghe and R. Van Rijek, Action of hormones and embryoid formation in callus cultures of _Carica papaya_, _in_: "Symp. Fytofarm.," Fytiat 26(1):637 (1974).
3. A.A. Mehdi and L. Hogan, tissue culture of _Carica papaya_, _Hort. Sci._ 11:311 (1976).
4. S.Y. Yie and S.I. Liaw, Plantlet regeneration from shoot tips and callus of papaya, _In Vitro_ 13:564 (1977).
5. I.K. Arora and R.N. Singh, Callus initiation in the propagation of papaya _in vitro_, _J. Hort. Sci._ 53:151 (1978).
6. T. Murashige and F. Skoog, A revised medium for rapid growth and bioassays with tobacco tissue culture, _Physiol. Plant._ 15:473 (1962).
7. O.L. Gamborg, R.A. Miller and K. Ojima, Nutrient requirements of suspension cultures of soybean root cells, _Exp. Cell Res._ 50:151 (1968).

DIFFERENTIATION OF SHOOT BUDS IN HYPOCOTYL EXPLANTS AND CALLUS CULTURES OF SOME LEGUMES

S. Bhargava, S. Upadhyaya, K. Garg and N. Chandra

Department of Botany
University of Rajasthan
Jaipur
INDIA

INTRODUCTION

The ability of cultured cells to produce fertile plants under controlled conditions is of special significance in relation to agriculture and forestry. However, differentiation in callus and cell cultures of legumes has proved difficult. Limited success has been achieved in regenerating shoot buds in callus and organ explants of a few species like *Medicago sativa* (1, 2), *Phaseolus vulgaris* (3), *Dalbergia sissoo* (4), *Cajanus cajan* (5), *Arachis hypogea* (6), and *Lathyrus sativus* (7). In the present communication, our findings on hypocotyl and callus cultures of two tree legumes, *Albizia lebbek* and *Dalbergia sissoo*, and one pulse crop legume, *Vigna aconitifolius*, are presented.

MATERIALS AND METHODS

Seeds of the two varieties of *Vigna aconitifolius* (vars. RDM 120 and IPCMO 926), *Dalbergia sissoo* and *Albizia lebbek* were surface sterilized with 0.1% mercuric chloride for five minutes, thoroughly washed with sterilized water and aseptically transferred to culture tubes containing water agar. Murashige and Skoog (MS) medium (pH 5.9) was used for all the cultures. All the cultures were incubated at 26°C ± 2°C under continuous light.

RESULTS AND DISCUSSION

Albizia lebbek

Callus was initiated when 2-5 mm long hypocotyl segments from 10 day old seedlings were cultured on MS + kinetin (0.2 mg/l) + IAA (indole acetic acid)(0.5 mg/l). Direct differentiation of shoots was observed from hypocotyl explants on a medium with kinetin (0.2-5.0 mg/l) and on combinations of kinetin (5.0 mg/l) and IAA (0.2-1.0 mg/l). Rhizogenesis occurred on MS medium with low kinetin (0.2 mg/l) and IAA (1.0 mg/l).

In callus cultures a large number of shoot buds differentiated on a medium with benzyl amino purine (BAP, 0.1-5.0 mg/l) within four to five weeks of incubation. The callus first became compact and nodulated and then differentiated into large number of shoot buds from each nodule. On subculturing, the shoot buds elongated and developed into full fledged shoots with typical pinnate leaves. The highest number of shoot buds were developed on medium with BAP (5.0 mg/l). The callus of hypocotyl origin was subcultured every 21 days and has been maintained for a year without any loss of morphogenic potential.

Dalbergia sissoo

Callus formation occurred in hypocotyl explants on MS medium with kinetin (0.2 mg/l) and 2,4-dichlorophenoxy acetic acid (2,4-D, 3.0 mg/l).

No direct differentiation of shoots was observed from hypocotyl explants. However, Mukhopadhyay and Mohan Ram (4) have reported direct regeneration of shoot buds from root explants. Regeneration of shoot buds was observed from callus on a medium with BAP (0.05-5.0 mg/l) in combination with IAA (0.05-3.0 mg/l). The callus turned green, compact and nodulated as in *A. lebbek*. Highest number of shoot buds were observed in the cultures supplemented with BAP (1.0-3.0 mg/l) and IAA (1.0 mg/l). The frequency of shoot buds was much less than in *A. lebbek*.

Vigna aconitifolius

Callus was initiated on hypocotyl and root explants of varieties RDM 120 and IPCMO 926. Soft friable and light yellow callus was isolated on MS medium with kinetin (0.2 m/gl) and 2,4-D (0.2 mg/l). Three week old callus turned brown and fine roots were initiated from this callus. The callus when subcultured on MS medium with various combinations of kinetin (1.0-5.0 mg/l) and IAA (1.0 mg/l) differentiated into tiny shoot buds. The maximum number of shoot buds was formed on a medium with 5.0 mg/l kinetin and 1.0 mg/l IAA. Shoot buds were, however, retarded in their growth even after

20-25 days of incubation. When subcultured on the same medium, the buds elongated into shoots with well formed leaves.

Hypocotyl and cotyledon explants of the variety (IPCM) 926 from 3-4 day old seedlings, when cultured on MS medium with BAP (0.1-0.5 mg/l), initiated callus which later regenerated shoot buds in 10-15 days and well formed shoots after 40-45 days of incubation. Only about 5-10% shoots showed formation of roots when cultured on medium with IAA (1.0-3.0 mg/l) or naphthalene acetic acid (NAA, 3.0 mg/l).

ACKNOWLEDGEMENTS

Thanks are due to U.G.C., New Delhi, for financial assistance.

REFERENCES

1. J.W. Saunders and E.T. Bingham, Production of alfalfa plants from callus tissue, Crop Sci. 12:804-808 (1972).
2. J.W. Saunders and E.T. Bingham, Growth regulator effects on bud initiation in callus cultures of *Medicago sativa*, Crop Sci. 62:850 (1975).
3. O.J. Crocomo, W.R. Sharp and J.E. Peters, Plantlet morphogenesis and the control of callus growth and root induction of *Phaseolus vulgaris* with the addition of a bean seed extract, Z. Pflanzenphysiol. 78:456-460 (1976).
4. A. Mukhopadhyay and H.Y. Mohan Ram, Regeneration of plantlets from excised roots of *Dalbergia sissoo*, Indian J. Exp. Biol. 19:1113-1115 (1981).
5. U. Mehta and H.Y. Mohan Ram, Regeneration of plantlets from the cotyledons of *Cajanus cajan*, Indian J. Exp. Biol. 18:800-802 (1980).
6. Y.P.S. Bajaj, P. Kumar, K.S. Labane and M.M. Singh, Regeneration of plants from seedling explants and callus cultures of *Arachis hypogaea*, Indian J. Exp. Biol. 19:1026-1029 (1981).
7. A. Mukhopadhyay, H.Y. Mohan Ram and S.S. Bhojwani, Tissue cultures of *Lathyrus sativus*, in: "Symp. Plant Tissue culture, Genetic Manipulation and Somatic Hybridization of Plant Cells," P.S. Rao, M.R. Heble and M.S. Chadha, eds., Bhabha Atomic Research Centre, Bombay (1980).

TISSUE CULTURE IN DIPLOID AND AUTOTETRAPLOID STRAINS OF METHA (*TRIGONELLA FOENUM-GRAECUM* L.)

D.S. Multani

Department of Genetics
Punjab Agricultural University
Ludhiana 141004
INDIA

INTRODUCTION

Trigonella foecum-graecum L. (fenugreek) a member of the Leguminosae is used as vegetable, forage and for preservative purposes. The seeds form an important source of diosgenin and yamogenin to be used in the steroid industry. In addition to the intact plant, cultured cells also yield secondary products. Callus formation and extraction of diosgenin or trigolline from static cultures has been demonstrated in *Trigonella* (1 - 6). Plant regeneration from leaf callus cultures have been reported in two species of *Trigonella* (7). However, no information is available on the callus growth and plant regeneration in different genotypes with various ploid levels. The present investigation was conducted to determine the response of five diploid and autotetraploid strains of *Trigonella foenum-graecum* L. for callus growth and plant regeneration from excised cotyledons, root and stem segments.

MATERIALS AND METHODS

The experimental material consisted of diploid and autotetraploid of five strains: T_8, Nagauri, IC74, IC7496 and IC5686. The seeds of these strains were sterilized twice in 0.25% mercuric chloride for five minutes each, rinsed three times with sterilized distilled water. These were grown aseptically in culture tubes containing solid MS medium (8). Seven days after seedlings were excised into three parts, viz., cotyledons, stem and root portions. Approximately 5 mm explants from each part were placed on modified MS medium containing 0.7% Bacto-agar, 3.0% sucrose further supplemented with 2,4-D (1, 2 and 5 mg/l), NAA (1.0 mg/l) and BA (1.0 mg/l). The pH of different media was adjusted to 5.6 prior to autoclaving. The media were sterilized at 15 lb/inch^2 at 121°C for 25 min. The cultures were maintained at 26°C $\pm$ 1°C and relative humidity 70% under light (500 lux). Calli were subcultured after every 3-4 weeks.

RESULTS AND DISCUSSION

The seedling tissue (cotyledons, root and stem portions) cultured on MS + 2,4-D (1, 2 and 5 mg/l) and NAA (1.0 mg/l) began to proliferate within a week, and a mass of callus was formed in 3-4 weeks. Callus growth was rapid and was produced in large quantities from cotyledon segments as compared to stem and root portions. Callus derived from cotyledons was friable and whitish. Further 2,4-D was found to be superior in callus induction than NAA. Fresh callus weight of five diploid and their autotetraploids grown on MS + 2,4-D is presented in Table 1. The autotetraploid strains had higher callus growth than their respective diploids at 2 ppm of 2,4-D. However, Nagauri autotetraploid strains showed significant increase (at $p = 0.01$ level) in callus growth at both 1 and 2 ppm concentrations of 2,4-D. The growth was reduced in all the five strains on lower (1 ppm) and higher (5 ppm) concentrations of 2,4-D, suggesting that 2.0 mg/l of 2,4-D is optimal for induction and growth of callus in this species. The calli initiated on 2,4-D was subcultured and maintained in medium containing 0.5 mg/l of 2,4-D.

Calli turned green when transferred to medium supplemented with auxin (NAA; 1.0 mg/l) and cytokinin (BA: 1.0 mg/l). Shoots were obtained from callus cultures derived from cotyledonary segments or calli formed on NAA and transferred to medium containing BA. Further subculturing these plantlets on medium with different concentrations of NAA and BA formed callus from lower portion. These calli were nodular, friable and dark green in color. When further subculturing was done on media with both auxin, cytokinin along with coconut milk (10% v/v) or casein hydrolysate (500 mg/l), these calli proliferate more rapidly without producing root or shoot. Sen and Gupta (7) observed that roots, shoots and development of isolated leaf occured in the calli of two species of <u>Trigonella</u> on MS medium containing NAA and CM (15% v/v) but omission of CM with subsequent addition of CH increased the number of differentiated organs per callus.

The effect of various phytohormones, when used singly or in combination with CM or CH, on callus growth is shown in Table 2.

Table 1. Mean Fresh Weight (gm) of Callus Cultures of Diploid and Autotetraploid Strains.

	Experiment I		Experiment II	
Material	Mean callus weight (gm) MS + 2,4-D (1.0 mg/1)	t-value	Mean callus weight (gm) MS + 2,4-D (2.0 mg/1)	t-value
T_8 (2n)	2.404 ± 0.38	1.3465	2.830 ± 0.33	0.53
T_8 (4n)	1.562 ± 0.95		3.02 ± 0.35	
Nagauri (2n)	1.188 ± 0.13	4.0476**	2.71 ± 0.05	6.24
Nagauri (4n)	2.191 ± 0.09		4.02 ± 0.17	
IC74 (2n)	1.631 ± 0.02	1.524	3.10 ± 0.21	1.55
IC74 (4n)	2.432 ± 0.33		3.41 ± 0.32	
IC7496 (2n)	2.380 ± 0.20	0.2712	1.13 ± 0.82	0.17
IC7496 (4n)	2.257 ± 0.43		1.22 ± 0.18	
IC5686 (2n)	1.336 ± 0.13	0.1472	0.84 ± 0.19	1.04
IC5686 (4n)	1.314 ± 0.05		1.16 ± 0.0	

** Significant at 1% level.

Table 2. Relative Callus Growth in Different Strains of *Trigonella foenum-graecum* L.

Phytohormone concentration (mg/1)			Cultivar									
			T_8		Nagauri		IC74		IC7496		IC5686	
2,4-D	NAA	BA	2n	4n	2n	4n	2n	4n	2n	4n	2n	4n
5	-	-	++	++	+	++	+	++	+	++	++	+
-	1	-	++	+++	++	+++	++	++	++	+++	+++	+++
1	1	1	++	+	+++	+++	#	+++	++	++	++	+
-	0.25	0.5 + CM	+	+	#	+	++	+++	+++	++	+++	+++
-	0.2	0.5 + CH (500 mg/1)	++	+	+++	+++	+++	++	++	++	++	++
-	0.5	- + kinetin (1.0 mg/1)	+++	++	#	+++	+++	+++	#	#	+++	#

- robust growth; +++ - normal growth; ++ - moderate growth; + - poor growth.

ACKNOWLEDGEMENTS

The author is thankful to C.S.I.R., New Delhi for Junior Research Fellowship and to Drs. D.S. Brar and J.L. Minocha for providing research facilities.

REFERENCES

1. P. Khanna and S.C. Jain, Effect of nicotinic acid on growth and production of trigonelline by Trigonella foenum-graecum tissue cultures, Indian J. Exp. Biol. 10:248-249 (1972).
2. P. Khanna and S.C. Jain, Diosgenin, gitogenin and tigogenin from Trigonella foenum-graecum tissue cultures, Lloydia 36:96-98 (1973).
3. P. Khanna, S.C. Jain and R. Bansal, Effect of cholesterol on growth and production of diosgenin, gitogenin, tigogenin and sterol in suspension cultures, Indian J. Exp. Biol. 13:211-213 (1975).
4. S.C. Jain, H. Rosenberg and S.J. Stohs, Steroidal constituents of Trigonella occulta tissue cultures, Planta Medica 31:109-111 (1977).
5. R. Hardman and R.G. Stevens, The influence of NAA and 2,4-D on the steroidal fractions of Trigonella foenum-graecum static cultures, Planta Medica 34:414-419 (1978).
6. S.S. Radwan and C.K. Kokate, Production of higher levels of trigonelline by cell cultures of Trigonella foenum-graecum than by the differentiated plant, Planta 147:340-344 (1980).
7. B. Sen and S. Gupta, Differentiation of callus cultures of leaf of two species of Trigonella, Physiol. Plant. 45:425-428 (1979).
8. T. Murashige and F. Skoog, A revised medium for rapid growth and bioassay with tobacco tissue cultures, Physiol. Plant. 15:473-497 (1962).

MORPHOGENESIS IN ORGAN, TISSUE AND CELL CULTURES OF SOME SPECIES OF BRASSICA

S. Singh, S. Banu, L.K. Pareek and N. Chandra

Department of Botany
University of Rajasthan
Jaipur
INDIA

INTRODUCTION

There has been a widespread interest in cell, tissue and organ culture techniques for clonal propagation (1), haploid production and breeding (2), and mutagenesis (3). In spite of the large amount of work being done in this field there are many limitations of the method. One of the most important limitations has been a lack of detailed knowledge regarding the precise conditions of culture and control of differentiation in cell and organ cultures of important crop plants. Different species of *Brassica* have attracted worldwide attention because of their economic importance as sources of oil and vegetables (4 - 10). The present report pertains to the control of differentiation in tissue cultures of two oil yielding species *B. campestris* var. DYS-1 (Rape), *B. juncea* var. Prakash (mustard) and two vegetable crop plants *B. oleracea* var. botrytis (cauliflower) and *B. oleracea* var. capitata (cabbage).

MATERIALS AND METHODS

Seeds of rape and mustard (*Brassica campestris* var. BYS-1 and *B. juncea* var. Prakash) were procured from I.A.R.I., New Delhi, and those of cabbage (*B. oleracea* var. capitata) and cauliflower (*B. oleracea* var. botrytis) from the Agriculture Research Station, Durgapura, Jaipur. Plants of all the species were grown in experimental plots. Cotyledon and hypocotyl explants were taken from aseptically grown seedlings raised in test tubes. Explants taken from field grown plants were surface sterilized with 0.5% mercuric chloride and then washed several times with sterilized distilled water. Murashige and Skoog (MS) meidum was used in the present study. All cultures were incubated in a growth room maintained at 26°C $\pm$ 2°C and 24 hour illumination with fluorescent tubes and incandescent lamps. Suspension cultures were maintained in Erlenmeyer flasks on a horizontal shaker at 100 rpm.

RESULTS AND DISCUSSION

Organ Cultures

Regeneration of shoot buds occurred in cotyledons of rape, mustard and cauliflower on MS medium with kinetin or benzylaminopurine (BAP, 1.0-5.0 mg/l). In rape the response was poor and only one or two shoots developed per cotyledon while in mustard a number of shoots developed from each cotyledon in all the replicates. George and Rao (7) have also reported regeneration of shoots on mustard cotyledons (var. rai 5). Bajaj and Nietsch (11) reported shoot bud regeneration in cotyledons of cabbage on coconut milk supplemented medium.

In hypocotyl explants regeneration of shoots was induced in rape only in presence ofkinetin (1.0-5.0 mg/l). Rhizogenesis occurred in all the species in response to low concentrations of BAP or kinetin (0.05-0.5 mg/l) or indole butyric acid (IBA, 1.0-5.0 mg/l). George and Rao (7) reported only callusing and root formation from hypocotyl segements of mustard (rai 5). Bajaj and Nietsch (11) also reported shoot bud regeneration in hypocotyl segments of cabbage.

Ovaries (pre- and post-fertilization) of rape and mustard were cultured. In rape only rhizogenesis or callus formation was observed while in mustard shoot buds differentiated on MS medium supplemented with K or BAP (0.5-5.0 mg/l). Addition of indole acetic acid (IAA, 0.5 mg/l) with BAP (3.0 mg/l) resulted in enlargement of shoot buds.

In anther explants of rape and mustard callus formation was observed in a medium with low cytokinins. On transferring this callus to a medium containing coconut milk (20% v/v) or higher levels of kinetin or BAP (1.0-5.0 mg/l) regeneration of shoot buds and embryo-like structures was observed. Shoot buds also differentiated from the connective region on higher cytokinin concentrations. The origin of the callus or plantlet could not be traced to the microspores. In all probability the callus originated from the cells of the wall or connective region. There are several reports on anther culture of different species of *Brassica* but the reported frequency of embryoid or plantlet formation (of microspore origin) is very low (2, 4, 5, 6, 12).

Callus and Suspension Cultures

Callus was isolated from hypocotyl segments of rape on MS medium supplemented with 0.5 mg/l of kinetin and 0.1 mg/l of IAA. In mustard and cauliflower it was obtained from cotyledons and mid-vein of leaf, respectively on a medium containing kinetin (0.5 mg/l) and IAA (1.0 mg/l). Suspension cultures were raised from callus cultures of rape and cauliflower in liquid media of the same composition. In all three plants regeneration of shoot buds occurred in callus cultures in the presence of kinetin or BAP (3.0-5.0 mg/l) alone or in combination with lower levels of IAA or IBA (0.5-1.0 mg/l). There were differences, however, in the regeneration potential, number and vigor of shoot buds. In rape callus a large number of shoot buds developed from hard and compact callus nodules. In mustard a few shoots developed from a callus mass. In cauliflower a large number of vigorous shoots developed from the callus mass and secondary shoots also differentiated, resulting in multiple shoot formation. In cabbage, however, only a few shoots differentiated. There are earlier reports of differentiation in callus cultures of cabbage and cauliflower (8, 11, 13).

In suspension culture of rape prolifice shoot bud formation was observed in the presence of kinetin (0.5-1.0 mg/l) and IAA (1.0 mg/l). Somatic embryogenesis occurred when IAA was omitted from the medium. Rhizogenesis occurred in the presence of IAA or IBA (1.0-5.0 mg/l). In suspension cultures of cauliflower somatic embryogenesis was induced on lower levels of kinetin or BAP (0.5-1.0 mg/l). Formation of shoot buds was observed on higher concentrations of BAP or kinetin (3.0-5.0 mg/l). Prolific rhizogenesis occurred in both rape and cauliflower in the presence of IAA or IBA with the maximal response on a medium with IBA 95.0 mg/l).

Complete plantlets were obtained in all the three species, when the regenerated shoots were transferred to a medium supplemented with auxins.

ACKNOWLEDGEMENT

Thanks are due to C.S.I.R., New Delhi, for financial assistance.

REFERENCES

1. T. Murashige, Clonal propagation through tissue culture, in: "Plant Tissue Culture and Its Biotechnological Application," W. Barz, E. Reinhard and M.H. Zenk, eds., Springer-Verlag, Berlin (1977).

2. G. Wenzel, Recent advances in applied tissue culture of potato rape and rye, in: "Plant Cell Culture Results and Perspectives," F. Sala, B. Parisi, R. Cella and O. Ciferri, eds., Elsevier-North Holland, New York (1979).
3. P. Maliga, Isolation, characterization and utilization of mutant cell lines in higher plants, in: "Perspectives in Plant Cell and Tissue Culture," I.K. Vasil, ed., International Review of Cytology (Suppl. 11A), London (1980).
4. T. Kameya and K. Hinata, Induction of haploid plants from pollen grains of Brassica, Jap. J. Breed. 20:82-87 (1970).
5. W.A. Keller, T. Rajhathy and J. Lacapra, In vitro production of plants from pollen in Brassica campestris, Can. J. Genet. Cytol. 17:655-666 (1975).
6. E. Thomas and G. Wenzel, Embryogenesis from microspores of Brassica napus, Z. Pflanzenzuchtg. 74:77-81 (1975).
7. L. George and P.S. Rao, In vitro regeneration of mustard plants (Brassica juncea var. Rai 5) on cotyledon explants from non-irradiated, irradiated and mutagen treated seed, Ann. Bot. 46:107-112 (1980).
8. L.K. Pareek and N. Chandra, Somatic embryogenesis in leaf callus from cauliflower (Brassica oleracea var. botrytis), Plant Sci. Lett. 11:311-316 (1978).
9. L.K. Pareek and N. Chandra, Induced regeneration in vegetative and flowering internodal segments and ovary explants of Brassica juncea coss., Indian J. Exp. Biol. 19:874-875 (1981).
10. S. Singh, S. Banu, L.K. Pareek and N. Chandra, Induced organ differentiation in hypocotyl segments, callus and suspension cultures of rape (Brassica campestris L.), Indian J. Exp. Biol. 19:658-660 (1981).
11. Y.P.S. Bajaj and P. Nietsch, In vitro propagation of red cabbage (Brassica oleracea L. var. capitata), J. Exp. Bot. 26:883-890 (1975).
12. W.A. Keller and K.C. Armstrong, Embryogenesis and plant regeneration in Brassica napus anther culture, Can. J. Bot. 55:1383-1388 (1977).
13. A.F. Mascarenhas, R.R. Hendre, A.L. Nadgir, D. Durga, M. Brave and V. Jagannathan, Differentiation in tissue culture of cabbage, Indian J. Exp. Biol. 16:122-125 (1978).
14. G. Wenzel, F. Hoffman and E. Thomas, Anther culture as a breeding tool in rape, Z. Pflanzenzuchtg. 78:149-155 (1977).

STUDIES ON GROWTH AND DIFFERENTIATION IN CULTURED CELLS OF RICE (*ORYZA SATIVA*)

S. Dawra, D.R. Sharma, J.B. Chowdhury and R.K. Jain

Department of Genetics
Haryana Agricultural University
Hissar - 125004 INDIA

INTRODUCTION

Callus induction and regeneration in rice (*Oryza sativa*) has been demonstrated by many workers (1). This capacity, however, declines with age of the cultures (4). Keeping in view the stimulatory effect of low doses of ionizing radiation on regeneration of plants from callus cultures (5), (6), we used gamma irradiation on old calli of rice which exhibited low potential to organize shoots. This communication deals with the effects of gamma irradiation on growth and differentiation of an old cell line of rice.

MATERIALS AND METHODS

About one cm long root explants of aseptically grown seedlings of a few cultivars of *Oryza sativa*, Jaya, IR-8 IR-36, Patnai, Basmati 370, PR 106, Getu and Ratna were cultured on MS (7), medium containing 2 mg/l of 2,4-D (designated as MS_2). Cultures were incubated in the medium of same composition after every four weeks. Growth rate of calli was determined by an increase in fresh and dry weights after a definite period of incubation.

The cultures were transferred to MS medium lacking growth hormones as well as to the medium supplemented with 2 mg/l of IAA and 6 mg/l of kinetin (designated as MSK). These were kept at $27 \pm 1^{o}C$ under constant illumination.

The long term cultures of vars. IR-8 and Patnai were exposed to gamma rays from ^{60}Co source at 0.5, 1.0 and 2.0 kR doses and were transferred to MS as well as to MSK media. The unexposed calli served as control.

Suspension cultures of var. IR-8 were prepared by agitating actively growing callus in liquid MS_2 medium on a low speed gyratory shaker. The cell clumps of suspension culture were transferred to solid MS_2 medium and growth rate was determined in irradiated as well as unirradiated calli. The viability of cells was ascertained with 0.05 Erythorosin B.

RESULTS AND DISCUSSION

Establishment and maintenance of callus

Callus initiation in Jaya and IR-8 started within a week whereas other varieties took 2-4 weeks for initiating the callus on MS_2 medium. However, in all cases calli showed slow growth. Callus initiation was from secondary root primorida. These calli later developed into soft nodular cell masses. Though cultures were maintained in the nodular form. occasional formation of roots could not be prevented. Rooting was more frequent in light than in dark. Growth rate on fresh weight was determined after 10,20 and 30 days while growth rate in terms of dry weight was taken after 20 days (Table 1).

Suspension culture

On agitating callus in liquid medium single cells and cell clumps were formed in suspension. About 50 of single cells were viable as revealed by exclusion of stain by these cells. As is evident from Table 2, the cell clumps on transferring to solid MS_2 medium showed increase in fresh weight after 30 days. But on irradiating the calli to 1.0 kR and 2.0 kR doses of gamma rays these showed decrease in fresh weight. On regenerating media, these calli showed limited growth and turned black.

Regeneration of whole plants

The calli readily became green on MS as well as MSK media and showed profuse rooting. In some cases, roots developed first and rooted calli then gave rise to shoots. However, some calli then gave rise to shoots. However, some calli produced rootless shoots. Greening of cultures took place earlier on MS medium by 2-3 days, whereas MSK medium proved better for further growth of plantlets. The long term cultures showed loss in regenerative potentiality. About 8 month old cultures were exposed to 0.5, 1.0 and 2.0 kR doses of gamma rays. At 0.5 kR dose calli showed perceptibly no difference in shoot formation. Most of the cultures turned black and exuded brown pigment in the medium. At

Table 1. Growth Rate in Rice Cultures on MS_2 Medium (on fresh and dry weight basis).

No. of passages	Var.	Initial weight (gms)	Final weight (gms)			Growth factor (Final/Initial weight)	
			10 days	20 days	30 days	20 days	30 days
1.	IR-8	.400 (.059)	.920	1.516 (.229)	2.274	3.79 (3.88)	5.68
	Pat	.317 (.049)	.734	1.642 (.201)	2.640	5.17 (4.10)	8.32
2.	IR-8	.513 (.036)	.990	2.798 (.206)	4.974	5.45 (5.72)	9.69
	Pat	.750 (.048)	.658	3.315 (.175)	5.155	4.42 (3.64)	6.87
3.	IR-8	.690 (.036	1.70	2.991 (.110)	4.870	4.33 (3.05)	7.05
	Pat	.650 (.042	1,180	2.30 (.179)	4.040	3.5 (4.97)	6.21

Figures in parenthesis are representing growth rate in terms of dry weight.

Table 2. Growth Rate of Callus Derived from Suspension of Var. IR-8 on MS_2 Medium on Fresh Weight Basis.

Dose of gamma rays	Initial weight (gms)	Final weight			Growth factor (Final/Initial weight)	
		10 days	20 days	30 days	20 days	30 days
0.kR	.210	1.040	1.857	3.225	8.84	15.35
1.0 kR	.295	.742	1.677	2.572	5.68	8.72
2.0 kR	.495	1.734	3.260	4.250	6.58	8.58

1.0 kR dose there was significant stimulation in organogenesis. An average of 8 plantlets regenerated after 40 days from a single callus of about 100 mg weight as compared to 0.3-0.4 plantlets in control. 2.0 kR and higher doses were found to be detrimental to regeneration.

It is concluded from this study that gamma rays can be used to initiate regeneration in recalciterant cells, at least in rice, maintained for a long period of time. The method requires further testing for its general application to other cell lines with low regeneration capacity.

REFERENCES

1. R.R. Henke, M.A. Mansur and M.J. Constantin, Organogenesis and plantlet formation from organ and seedling derived calli of rice, Physiol. Plant, 44:11-14 (1978).
2. P. Bhattacharya, S.K. Sen, Potentiality of leaf sheath cells for regeneration of rice plants, Theort. Appl. Genet., 58:87-90 (1980).
3. Y.P.S. Bajaj, S.S. Sainy and M. Gidani, Production of triploid plants from immature and mature endosperm of rice, Theort. Appl. Genet., 58:17-18 (1980).
4. T. Nishi, Y. Yamada, Y. Takahashi, Organ redifferentiation and plant restoration in rice callus, Nature, 219:508-509 (1968).
5. H.K. Shama Rao, B. Narayanaswamy, Effect of gamma irradiation in cell proliferation and regeneration in explanted tissues of pigeon pea, Cajnus Cajan (L.), Mills P, Radiat. Bot., 15:301-305 (1975).
6. R.K. Jain, N. Maherchandani D.R.Sharma and V.K. Chowdhury, Effect of gamma radiations on haploid cultured cells of Datura innoxia, in: "Proc. National Symposium "Plant Tissue Culture, Genetic Manipulation and Somatic Hybridization of Plant Cells", BARC, Bombay, India, (1980).
7. T. Murashige, F. Skoog, A revised medium for rapid growth and bioassays with tobacco tissue cultures, Physiol Plant, 15:473-497, (1962).

EFFECT OF PHLOROGLUCINOL ON SHOOT GROWTH AND INITIATION OF ROOTS IN CAROB TREE CULTURES GROWN *IN VITRO*

V. Thomas and A.R. Mehta

Tissue Culture Laboratory
Department of Botany
The M.S. University of Baroda
Baroda 390002
INDIA

INTRODUCTION

The number of reports on *in vitro* regeneration of tree species has risen rapidly in recent years (1, 2, 3). However, there have been few reports concerning multiplication of leguminous trees by tissue culture methods and hardly any success is achieved in regenerations from their callus tissues (4).

Ceratonia siliqua L., commonly known as carob tree (family Caesalpineae) holds great potential as a plantation crop in semi-arid conditions, in view of its minimum growth requirements and maximum economic uses (5, 6). The objective of this work has been to popularize carob and to explore applications of *in vitro* methods for its rapid propagation for commercial cultivation. Moreover, the plants being dioecious in nature, tissue culture methods would be handy for the selective multiplication of female trees.

It is likely that every plant tissue at some stage in its development produces and accumulates phenolic substances under both natural and experimental conditions. The phenolic compounds are now known to participate in plant growth regulation (7, 8). It has been demonstrated that phloridzin, the 2'-glucoside of phloretic acid and its degradation products, phloroglucinol (PG) and phloretic acid can significantly enhance the growth of both shoots and roots of apple rootstocks cultured *in vitro* (9, 10, 11). The specificity of these compounds in growth promotion has, however, remained controversial, since, although phloridzin is unique to species of *Malus* (12), it promoted the growth an regenerative potential of cultured shoots of other genera such as Prunus (13) and Rubus (14). Further support to the apparent non-specific growth promoting ability of these compounds has also been presented (15).

Since it is widely believed that both auxin and auxin synergists play fundamental roles in the formation of adventitious root primordia (14, 16, 17), the objective of the present investigation was also to examine the roles of auxin and phloroglucinol on shoot growth and in the induction of roots in carob tree cultures grown in vitro and comparison is made with the responses of M-26 rootstock cultures of apple.

MATERIALS AND METHODS

The primary plant material was either terminal shoot tips or nodal explants bearing one lateral bud collected in spring from four year old trees of carob (Ceratonia siliqua L.). The explants were washed with mild detergent solution followed by 70% ethanol for 3-5 min. Finally, surface sterilization was done with 0.1% (w/v) $HgCl_2$ solution for 2 min and rinsed in sterile distilled water at least 4 times. Explants measuring 1-1.5 cm in length were excised from the shoot tips or nodal segments and placed on a medium solidified with 0.5% agar in culture vessels. The number of explants placed depended on whether they were inoculated in Erlenmeyer flasks or test tubes. Initially, the explants were placed horizontally on the medium as it was observed that this arrangement had reduced the extent of browning of the medium due to leaching from the cut ends of the explants. This problem was further tackled by repeated transfer of explants to freshly made medium during the early period of incubation.

The culture medium used contained MS salts, 3% sucrose and a modified vitamin content. The hormonal additives which included 6-BA (6-benzyl adenine); K (kinetin): NAA (naphthalene acetic acid); IBA (indole butyric acid) and GA_3 (gibberellic acid) in the presence or absence of phloroglucinol (PG), were added as required in different experiments. All media pH was adjusted to 5.2 before sterilization by autoclaving.

RESULTS AND DISCUSSION

Bud Burst

Our preliminary experiments on carob tree had indicated that the sprouting of buds was directly controlled by auxin-cytokinin interaction. A combination of 2.0 mg/l 6-BA and 1.0 mg/l NAA oculd effect the highest frequency of bud burst (Table 1) within 4 to 5 weeks. It was essential to free the explants from oxidized phenol-

Table 1. Frequency of Response in the Induction of Bud Burst under the Influence of K*, 6-BA, NAA and IBA in Carob Explants Grown *In Vitro*.

No. additives to BM + 3% sucrose (mg/l)	Response	% Bud Burst	Remarks
1. Control (BM)	-	-	No response
2. 1.0 NAA + -	-	-	Callus formation
3. 1.0 NAA + 0.5 6-BA	-	-	Callus formation
4. 1.0 NAA + 1.0 6-BA	+	20	Produce callusing along with bud burst
5. 1.0 NAA + 2.0 6-BA	+	90	"
6. 1.0 NAA + 4.0 6-BA	+	30	Slight callusing; hard and brittle growth of buds
7. 1.0 IBA + -	-	-	Callus formation
8. 1.0 IBA + 0.6 6-BA	-	-	Friable callusing
9. 1.0 IBA + 1.0 6-BA	+	20	Slight callusing with bud burst
10. 1.0 IBA + 2.0 6-BA	+	30	"
11. 1.0 IBA + 4.0 6-BA	+	20	Buds turn hard and brittle

* K at the same concentrations could not substitute 6-BA in the induction of bud bursts.

ics leaching out of them. Further, kinetin could not be substituted for 6-BA in this woody tree. High concentration of 6-BA was inhibitory for the growth, as the tissues turned hard and brittle, while very low concentrations of 6-BA in the presence of 1.0 mg/l IBA or NAA gave rise to callus from the cut ends. Further growth of the sprouted buds was discouraged on the same medium as the conditions favored more of callus growth than organized development. As terminal apices gave rise to multiple shoots, they were preferred as primary explants over lateral buds which gave rise to solitary shoots.

Shoot Growth and Multiplication

The tissue mass containing sprouted buds was manually removed of their callus and transferred to another medium for further growth and multiplication. The MS medium supplemented with 2.0 mg/l 6-BA, 0.1 mg/l IBA and 2.0 mg/l GA_3 in presence of 162 mg/l PG was found optimal for carob. In the case of M-26 apple rootstocks, however, the levels of 6-BA and GA_3 were reduced to 1.0 mg/l each. Replacement of NAA by IBA (0.1 mg/l) reduced callus growth in the subsequent recultures. When PG was omitted from the medium, the shoots were only a few millimeters long. Incorporation of 162 mg/l PG, however, enhanced general growth and leaf expansion of shoots. Presence of PG also induced regeneration of many (12-15) shoots per culture (Figure 1A). Subsequent transfer of the regenerated shoot to a hormone-free medium with and without PG revealed marked differences

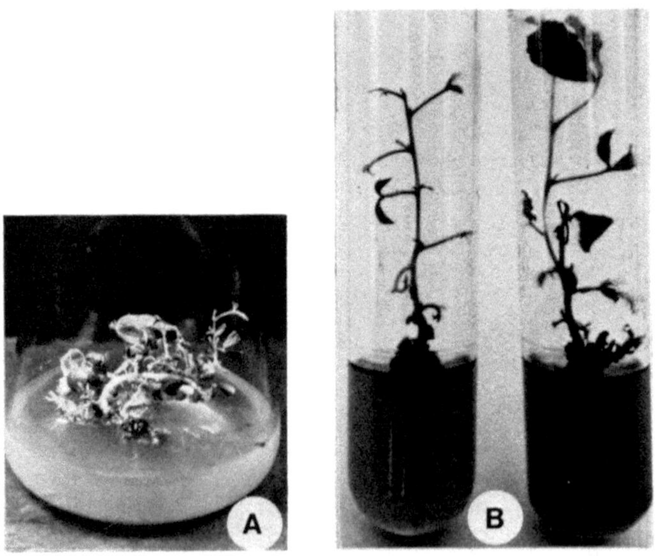

Figure 1. A: Shoot multiplication with PG in the medium.
B: Single shoot without and with PG.

in response by both carob (Figure 1B) and M-26 apple rootstock cultures. Shoot growth, especially leaf size, was significantly enhanced in the presence of PG. Of the three levels of PG (81, 162 and 324 mg/l) tested, 162 mg/l was found to be optimal. Long term exposure to higher doses of PG was inhibitory in both the cultures.

Root Initiation and Development

Induction of roots in the shoots harvested from multiple shoot culture was done in MS medium containing 1.0 mg/l IBA and 5 x 162 mg/l PG. After exposing the individual shoots of carob and apple for 4 and 6 days respectively to the above medium, they were transferred to a hormone-free medium containing only 3% sucrose and 162 mg/l PG. Earlier, NAA (1.0 mg/l) was tested for root induction, however, it produced callus before root primordia developed. Replacing NAA with IBA (1.0 mg/l) also did not induce rooting in any of the cultures, even after 4 weeks of incubation. Synergistic effect of PG was evident when in the presence of 1.0 mg/l IBA, three levels of PG (162, 630 and 1620 mg/l) were tested. Continuous incubation in higher levels of PG was, however, lethal for the shoot growth, negating the induction of roots. On the other hand, when shoots exposed to the rooting medium containing 1.0 mg/l IBA and 630 mg/l PG for 4 days and then transferred to a hormone-free medium containing only 162 mg/l PG, 30% cultures formed roots (Figure 1C). The same treatment could induce well formed root system in 80% of M-26 apple cultures when they were exposed for 6 days and transferred to a hormone-free medium. Further, high levels of IBA and PG could induce rhizogenesis, but their continued presence limited subsequent root development by promoting callus formation at the cut ends of both carob and apple cultures.

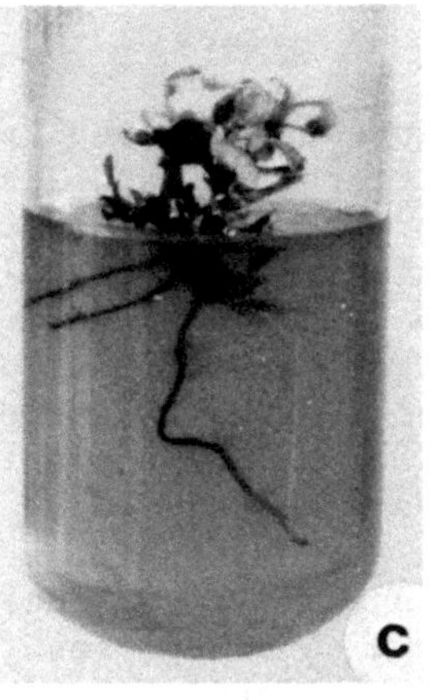

Figure 1. C: Root formation and development.

The present work showed that shoot apices or nodal explants of carob tree are capable of regenerating into complete plants when cultured in medium with appropriate growth regulators. Our experiments, like some other, emphasize the important role played by 6-BA in breaking the dormancy of buds and shoot multiplication of woody plants *in vitro* (1, 18, 19, 20).

Phloroglucinol had considerable promotary influence on *in vitro* growth as in other fruit trees like apple rootstocks (9, 10, 13, 14, 15).

The action of phenolics as auxin synergists in the promotion of rooting cuttings has been recognized for some time (7, 16, 17, 21). *In vitro* studies also demonstrated synergistic effect of IBA and PG in the control of rooting in apple rootstock M-9 (11), *Rubus* (14) and raspberry (15). Carob tree cultures studied here, responded similarly with synergistic enhancement of shoot growth and root production by IBA and PG. Use of other phenolic compounds as auxin synergists is reported in other species like jojoba (*Simmondsia chinensis*) (22) and tobacco (23). Though synergistic effect of auxin-phenol complexes are evident in quite a few species, unless their mechanism is elucidated properly we cannot exclude the possibility that phenolics act on plant growth by interference with processes not related to auxin, but also with actions of other hormones.

ACKNOWLEDGEMENTS

We thank Dr. D.J. James, East Malling Research Station, Kent, U.K. for supplying us M-26 apple rootstock cultures. One of us (VT) acknowledges the UGC, New Delhi, for financial assistance.

REFERENCES

1. L.L. Winton, Morphogenesis in clonal propagation of woody plants, *in*: "Frontiers of Plant Tissue Culture," T.A. Thorpe, ed., Calgary, Canada (1978).
2. J.M. Bonga, Applications of tissue culture in forestry, *in*: "Plant Cell, Tissue and Organ Culture," J. Reinert and Y.P.S. Bajaj, eds., Springer-Verlag, Berlin (1977).
3. F. Hammerschlag, Peach micropropagation, *in*: "Proc. Conf. on Nursery Production of Fruit plants Through Tissue Culture," U.S.D.A., Maryland (1980).
4. E.N. Crizaldo, Cited *in*: "Tropical Legumes: Resources for the Future," National Academy of Sciences, Washington (1979).
5. H.J. Maydell, Tree and shrub species for agroforestry systems, *Plant Research and Development* 7:44-59 (1978).
6. Anon., "Tropical Legumes: Resources for the Future," National Academy of Sciences, Washington (1979).

7. G. Stenlid, On the physiological effects of phloridzin and some related substances upon higher plants, Physiol. Plant. 21:882-894 (1968).
8. V.I. Kefeli and M. Kutacek, Phenolic substances and their possible role in plant growth regulation, in: "Plant Growth Regulation," P.E. Pilet, ed., Springer-Verlag, Berlin (1977).
9. O.P. Jones, Effect of phloridzin and phloroglucinol on apple shoots, Nature 262:392-393 (1976).
10. O.P. Jones and S.G.S. Hatfield, Root initiation in apple shoots cultured in vitro with auxins and phenolic compounds, J. Hort. Sci. 51:495-499 (1976).
11. D.J. James and I.J. Thurbon, Rapid in vitro rooting of the apple rootstock M-9, J. Hort. Sci. 54:309-311 (1979).
12. A. Hutchinson, C.D. Taper and G.H. Towers, Studies of phlor-idzin in Malus, Can. J. Biochem. Physiol. 37:901-910 (1959).
13. O.P. Jones and M.E. Hopgood, The successful propagation in vitro of two rootstocks of Prunus: the plum rootstock "Pixy", P. insititia and the cherry rootstock F 12/1, P. avium, J. Hort. Sci. 54:63-66 (1979).
14. D.J. James, The role of auxins and phlorogulcinol in adventi-tious root formation in Rubus and Fragaria grown in vitro, J. Hort. Sci. 54(4):273-277 (1979).
15. D.J. James, V.H. Knight and I.J. Thurbon, Micropropagation of red raspberry and the influence of phloroglucinol, Scientia Hort. 12:313-319 (1980).
16. R.N. Basu, T.K. Bose, B.H. Roy and A. Mukhopadhyay, Auxin synergists in the rooting of cuttings, Physiol. Plant. 22:649-652 (1969).
17. B.E. Haissig, Influences of auxins and auxin-synergists on adventitious root primordium initiation and development, N.Z. J. For. Sci. 4:311-323 (1974).
18. O.P. Jones, M.E. Hopgood and D. O'Farrell, Propagation in vitro of M-26 apple rootstocks, J. Hort. Sci. 52:235-238 (1977).
19. W.D. Lane, Regeneration of apple plants from shoot meristem-tips, Plant Sci. Lett. 13:281-285 (1978).
20. A.M. Vieitez and M.L. Vieitez, Culture of chestnut shoots from buds in vitro, J. Hort. Sci. 55:83-84 (1980).
21. K.V. Thimann, The formation and growth of roots, in: "Hormone Action in the Whole Life of plants," University of Massachu-setts Press, Amherst (1977).
22. NBRI Newsletter, Lucknow, India 8(3):27-28 (1981).
23. A.R. Mehta, Physiological aspects of organ differentiation in vitro, in: "Plant Tissue Culture, Genetic Manipulation and Somatic Hybridization of Plant Cells," P.S. Rao, ed., BARC, Bombay (1980).

RECOVERY OF VIRUS FREE PLANTLETS OF CULTIVATED JUTE SPECIES

K. Das, R.R. Sinha and S.K. Sen

Programme in Genetical Research
Botany Department
Bose Institute
Calcutta 700009
INDIA

INTRODUCTION

Meristem culture techniques have helped in the past to generate symptomless or virus free plants of species of over 35 genera (1). As there exists a difference between the rates of plant growth and virus multiplication, it has been possible to culture the young and fast growing tissue, i.e., the meristem of a plant under virus free conditions. Meristem culture techniques date back about three decades (2). These techniques were first used in the fifties in order to free plants of virus (3). Since then it has been used to produce virus-free plants of several crops, including cassava (4), potatoes (5), white clover (6), red clover (7, 8), and alfalfa (8). These techniques have also been adopted for rapid asexual propagation of orchids (9), pea (10) and other crops.

Some of the high fiber yielding cultivars of jute species *Corchorus olitorius* and *C. capsularis* are highly infected with the virus-caused chlorosis disease, resulting in reduced seed set, mottling of leaves, poor fiber quality and frequent plant death. The virus is transmitted through seeds, hence getting rid of the infection often becomes tricky. The present communication deals with the results of our study directed to the use of shoot tip culture for elimination of the pathogen from viral infected basic genotypes of two jute species.

MATERIALS AND METHODS

Explants measuring 2-3 cm in length and containing crown shoot meristems obtained from highly infected field grown plants of cv JRO 524 and cv JRO 978 of C. olitorius and C. capsularis respectively form the experimental material. Explants were surface sterilized by placing them in 0.2% mercuric chloride solution for 5 min., 5 min. in sterile water, 5 min. in 70% ethanol and then 5 changes of waer. Shoot meristems were dissected aseptically using a dissecting microscope (X 30) under a laminar flow chamber. They were cultured in culture tubes or in Erlenmeyer flasks. Growth of plantlets or callus was possible to be induced by incubating the cultures at 27°C $\pm$ 1°C in continuous cool white fluorescent light at an intensity of approximately 4000 lux.

Shoot meristems 0.2-0.8 mm long (measured from the cut surface to the highest point on the shoot tip) and consisting of meristem domes were cultured. Plantlet development was assessed after 6-8 weeks and the regenerated plantlets were either transferred directly to soil or to sand, watered frequently with Hoagland's solution (4) and covered with polythene film for 2 weeks to maintain high humidity.

The basal culture media tested were MS (11) and modified SH (12). The composition of modified SH (MSH) medium was as follows (mg/l): KNO_3, 2300; NH_4NO_3, 500; $MgSO_4.7H_2O$, 400; $NH_4H_2PO_4$, 300; $CaCl_2.2H_2O$, 200; $MnSO_4.H_2O$,5; H_3BO_3, 5; $ZnSO_4.H_2O$, 3; KI, 1; $CuSO_4.5H_2O$, 0.2; $Na_2MoO_4.2H_2O$, 0.1; $COCl_2.6H_2O$, 0.075; $FeSO_4.7H_2O$, 27.8; Na_2 EDTA, 37.8; Inositol, 700; Thiamin, 7; Niacine, 5; Pyridoxin, 0.5; Casein hydrolysate, 250, or coconut milk , 15-20%; Sucrose, 3%; pH 5.8. The growth regulators used were Indole-3-acetic acid (IAA), Naphthalene acetic acid (NAA), Indole-3-butyric acid (IBA), 2,4-dichlorophenoxy acetic acid (2,4-D), Gibberellic acid (GA_3), 6-benzylaminopurine (BAP) and 6-furfurylaminopurine (kinetin). All media were solidified with 0.8% agarose and autoclaved at 121°C for 15 min. after adjusting the pH to 5.8.

RESULTS AND DISCUSSION

Various concentrations of IBA, IAA, 2,4-D and GA_3 were tested with different concentrations of BAP and kinetin in MSH and MS basal medium. Response of MSH medium was observed to be better over the MS basal medium. The salient features of our results are presented in Table 1. It was observed that IBA with kinetin induced direct shoot formation whereas NAA or 2,4-D with kinetin induced callus formation from the meristem tips of both the jute species. Meristems of C. olitorius cv JRO 524, when cultured on medium containing IBA and kinetin both at the level of 0.2 mg/l, regenerated plants directly with little callus , whereas apical meristem tips of C.

Table 1. Influence of Explant Size and Phytohormones on Apical Shoot Meristem Culture of C. olitorius and C. capsularis.

	(mg/l)				
Genotype	IBA 0.2 Kin. 0.2	IBA 0.5 Kin. 0.2	NAA 0.5 Kin.	2,4-D 0.2 Kin. 0.5	GA_3 0.2 Kin. 0.5
C. olitorius					
cv JRO 524					
Dome 0.2-0.4 mm	50(24)16*	NG	50(38)$^+$	50(42)$^{+++}$	NG
Dome 0.4-0.8 mm	50(36)23	NG	50(44)$^+$	50(36)$^{+++}$	50(1)0
C. capsularis					
cv JRC 978					
Dome 0.2-0.4 mm	NG	50(30)22	50(26)$^+$	50(34)$^{++}$	NG
Dome 0.4-0.8 mm	NG	50(40)13	50(34)$^+$	50(40)$^{+++}$	NG

* Total number of meristematic domes inoculated (cultures survived) virus free plants.

$^+$ very slow; $^{++}$ moderate; $^{+++}$ vigorous callus growth; NG no growth.

capsularis cv JRC 978 formed plants on media containing IBA and kinetin at the level of 0.2 mg/l and 0.5 mg/l, respectively. 2,4-D and NAA with kinetin both at the level of 0.5 mg/l induced slow growth of callus which formed shining globular structures after 3 weeks. These calli were then transferred to IBA and BAP containing media. The medium containing IBA and BAP at the level of 0.2 mg/l and 0.5 mg/l respectively induced multiple shoot formation. The green globular structures thus initiated were transformed into leafy structures and then into plantlets within 60 days of culture. The meristematic domes cultured on medium containing 2,4-D and kinetin did not help differentiation in any of the phytohormones tested.

Plants regenerated directly from shoot tips or from callus were devoid of root. After 30.32 days of growth, when these plants were transferred to hormoneless medium, normal root systems developed.

A proper balance of cytokinin and auxin is expected to result in the regeneration of whole plants. C. olitorius and C. capsularis meristems cultured on MSH medium supplemented with various levels of IBA and kinetin regenerated only shoots, whereas presence of NAA with a cytokinin in the medium produced callus only. The amount of endogenous hormones present in the explant could alter the morphogenetic response. Plant regeneration is achieved in the presence of exogenously applied IBA in conjunction with a cytokinin, and the differentiated shoots fail to develop even after prolonged incubation on the same medium. Induction of roots on differentiated shoots could be accomplished by means of the same nutrient medium supplemented with NAA. This indicates that endogenous level of cytokinins synthesized in themeristem approaches the required cytokinin-auxin balance (in association with exogenously applied NAA) to bring about root formation and ultimately whole plant regeneration. Any additional supply of exogenous cytokinins to the culture medium disrupts this balance, thereby resulting in the differentiation of shoot only.

Shoot meristems have a tendency to form callus from basal cut surface. The extent of callus formation depends upon the concentration of phytohormones in the medium. Some of these calli showed multiple shoot formation when the medium was supplemented with the phytohormones necessary for direct shoot formation from shoot meristems. Similar results with Cicer arietinum and Phaseolus vulgaris was observed earlier (13).

The size of the meristem tips used for culture was also proved to be an important factor in elimination of the viruses. It was observed that though the smaller the tip size the lesser was the chance for survival or regeneration, yet the greater was the possibility for getting a virus free plant. Ranges (0.2-0.8 mm) of meristem dome size of both the jute cultivars were cultured. The rate of survival was recorded after 4 weeks of culture. It was found that in the case of the 0.2-0.4 mm size group frequency of virus free plants was about 40-45% via direct plant regeneration and about 50-60% via callus to plant formation.

Experiments were conducted to test the absence or presence of virus in regenerated plants. Etiologically leaves of infected plants have mottle or white patches. Leaves of the presumptive healthy plants obtained through apical meristem cultures were crushed in pestle and mortar and extracts were mixed with sterile distilled water (1:1). This extract was then applied on the upper surfaces of leaves of healthy plants grown in greenhouse. The upper epidermis of the leaves were removed before applying the extracts, and the plants were kept under observation for several weeks. It was expected that the presumptive healthy leaves carry virus particles to their extract if they are not free from the disease. In very few cases we could see that the leaves contained virus particles causing infection of the disease to normal plant.

For elimination of virus particles, size of the meristem tip planted an important role. Usually it was noticed that the smaller the tip size, the greater is the chance for obtaining a plant free of viruses as observed also in the past (14, 15, 16). It may be due to the fact that viruses take time to be in the apical dome and its adjacent tissue. There are instances that virus free plants could be produced from meristem tips reported to be infected later (17).

ACKNOWLEDGEMENT

The present investigation has been carried out under the financial support of a research project (No. HCS/DST/310/76) from the Department of Science and Technology, New Delhi, India.

REFERENCES

1. F. Quak, Meristem culture and virus free plants, in: "Plant Cell, Tissue and Organ Culture," J. Reinert and Y.P.S. Bajaj, eds., Springer-Verlag, Berlin (1977).
2. E.A. Ball, Development in sterile culture of stem tips and subjacent region of Tropaeolum majus L. and of Lupinus alba L., Am. J. Bot. 33:301-308 (1946).
3. G. Morel and C. Martin, Guerison de Dahlias atteints d'une maladic a virus, C.R. Acad. Sci. 235:1324-1325 (1952).
4. K.K. Kartha and O.L. Gamborg, Elimination of cassava mosaic disease by meristem culture, Phytopathology 65:826-828 (1975).
5. F.C. Mellor and R. Stace-Smith, Virus-free potatoes by tissue culture, in: "Plant Cell, Tissue and Organ Culture," J. Reinert and Y.P.S. Bajaj, eds., Springer-Verlag, Berlin (1977).
6. O.W. Barnett, P.B. Gibson and A. Seo, A comparison of heat treatment and meristem tip culture for obtaining virus-free plants of Trifolium repens, Plant Dis. Rep. 59:834-837 (1975).
7. G.C. Phillips and G.B. Collins, Virus symptomfree plants of red clover using meristem culture, Crop Sci. 19:213-216 (1979).
8. V.A. Cheyne and P.J. Dale, Shoot tip culture in forage legumes, Plant Sci. Lett. 19:303-309 (1980).
9. T. Murashige, Plant propagation through tissue cultures, Ann. Rev. Plant Physiol. 25:135-166 (1974).
10. K.K. Kartha, O.L. Gamborg, F. Constabel and J.P. Shyluk, Regeneration of cassava plants from apical meristems, Plant Sci. Lett. 2:107-113 (1974).
11. T. Murashige and F. Skoog, A revised meidum for rapid growth and bioassays with tobacco tissue cultures, Physiol. Plant. 15:473-497 (1962).
12. K. Das, "Application of Plant Cell/Tissue Culture Technique in Cultivated Jute Species of India for Their Possible Use in Crop Husbandry," Ph.D. Thesis, Calcutta University, India (1982).
13. K.K. Kartha, Meristem culture and cryopreservation - methods and application, in: "Plant Tissue Culture," T.A. Thorpe, ed., Academic Press, New York (1981).
14. O.M. Stone, Factors affecting the growth of carnation plants from shoot apices, Ann. Appl. Bot. 52:199-209 (1963).

15. K. Mori, Production of virus-free plants by means of meristem culture, Jap. Agri. Res. Quart. 6:1-7 (1971).
16. P.J. Dale, Meristem tip culture in Lolium multiflorum, J. Exp. Bot. 26:731-736 (1975).

STUDIES ON INDUCTION AND DIFFERENTIATION OF ANDROGENIC CALLUS OF *SOLANUM MELONGENA* L

N. R. Misra, T.M. Varghese, N. Maherchandani and R.K. Jain

Department of Botany
Haryana Agricultural University
Hissar 125004
INDIA

INTRODUCTION

Since the success of Guha and Maheshwari (1) in producing haploid plants in *Datura innoxia*, anther culture and pollen culture techniques have been successfully employed in the production of androgenic haploid plants in several plant species. An attempt has been made by us to produce androgenic haploids of eggplant (*Solanum melongena* L.) and to trace the morphogenetic steps involved. This paper deals with the selection of adequate nutrient medium to induce callus from anthers and the differentiation into shoots and roots.

MATERIALS AND METHODS

Floral buds of *Solanum melongena* L. var. H_4 of various sizes were collected from the experimental farm of Haryana Agricultural University between 8:30 and 9:00 AM in the months of August to September and fixed in FAA. Dehydration and embedding were done by conventional technique using xylene as clearing agent. Microtome sections of 10 μ thickness were cut and stained with safranin and fast green to assess the developmental stages of microsporogenous tissue and to work out a relationship with the size of the bud.

For culturing, the freshly collected buds of appropriate sizes were sterilized with absolute alcohol for ten seconds and the anthers were placed on the following sterilized nutrient media: MS (2); NN (3); and DG (4).

Each of these media was modified by adding 2.0 mg/l of IAA and 1.0 mg/l of kinetin. Two anthers were inoculated in each tube. As DG medium was found to be the best for callus growth, this medium was used for further subculturing.

For differentiation the following media were used:

1. M_I = MS + IAA (4.0 mg/l) + Kinetin (2.56 mg/l)

2. M_{II} = MS + 2,4-D (0.2 mg/l) + Inositol (400 mg/l)

3. M_{III} = NN + IAA (0.1 mg/l)

4. M_{IV} = DG + NAA (0.1 mg/l) + Kinetin (2.0 mg/l)

5. M_V = Blaydes medium (5) + IAA (5.0 mg/l) + Kinetin (0.5 mg/l) + Coconut milk (15% v/v).

RESULTS AND DISCUSSION

An examination of transverse sections of buds of varying sizes, ranging from 0.1 to 2.0 cm long, was made. In a flower bud of size 1.7 cm, the microsporangium contained uninucleated microspores. This was the stage at which the anthers were taken out from bud for inoculation on the medium.

After seven days, the anthers dehisced at the apex and callus came out through this region in all the media. Out of 40 anthers inoculated on each medium, 16 anthers developed calli in MS medium, 10 in NN and 32 in DG medium. Further callus growth was observed in 12 calli on NN medium, 12 on MS medium and 28 on DG medium. Calli formed in MS and NN media were white and compact, whereas in DG medium these were loose and fragile with yellowish white color.

For the maintenance of callus the subculturing was done on DG medium after every 2-3 weeks and callus mass doubled in this period.

Out of five media used for differentiation, greening of the calli was observed in M_I, M_{II} and M_{IV}, but not in M_{III} and M_V medium. After about 10 days of transfer of the callus in this medium, dark green shoot primordia made appearance in small depressions of the callus surface (Figure 1). The primordia were large and translucent. Root initiated from the callus after 15-20 days of shoot formation. A well developed root with a large number of equidistantly placed lateral roots giving feathery appearance was observed. Several plantlets made appearance on the differentiating medium (Figure 2).

Out of the three media used for callus induction, DG medium

Figure 1. Callus with initiating shoot apex. 1X.

Figure 2. Callus with large number of haploid plantlets. 1 1/2X.

with 2.0 mg/l IAA and 1.0 mg/l kinetin was found to be the best for callus growth, although the other two also induced callusing, suggesting that the nutrient requirement for the callus initiation in this plant is different from that of the callus proliferation and growth.

The study also revealed that DG medium with 0.1 mg/l of NAA and 2.0 mg/l of kinetin is most suited for differentiation of callus. It is also evident that higher proportion of auxin to kinetin is required for callus growth whereas for differentiation of the callus, higher concentration of kinetin is required.

As the shoot initiation took place much in advance to that of root initiation, one of the possibilities that can be visualized is that the initiation of shoot changed the auxin-kinetin ratio such that conditions favorable for root initiation were created.

Raina and Iyer (6) also reported the formation of plants from anther culture of Solanum melongena L., but their callus had its origin from the connective tissue. In the present study we observed that there was no proliferation of connective tissue. The callusing was only from microsporogenous tissue. The root tip cells of the differentiation plantlets had twelve chromosomes.

ACKNOWLEDGEMENTS

The authors are grateful to Prof. J.B. Chowdhury, Dept. of Genetics, Haryana Agricultural University for providing certain facilities. Financial assistance by DAE to N.R. Misra and C.S.I.R. to R.K. Jain in the form of J.R.F. is also thankfully acknowledged.

REFERENCES

1. S. Guha and S.C. Maheshwari, In vitro production of embryos from anthers of Datura, Nature 204:497 (1962).
2. T. Murashige and F. Skoog, A revised medium for rapid growth and bioassays with tobacco tissue cultures, Physiol. Plant. 15:473-497 (1962).
3. J.P. Nitsch and C. Nitsch, Haploid plants from pollen grains, Science 163:85 (1969).
4. P.M. Gresshoff and C.H. Doy, Development and differentiation of haploid Lycopersicon esculentum (tomato), Planta 107:161-170 (1972).
5. D.F. Blaydes, Interaction of kinetin and various inhibitors in the growth of soybean tissue, Physiol. Plant. 19:748-753 (1966).
6. S.K. Raina and R.D. Iyer, Differentiation of diploid plants from pollen callus cultures of Solanum melongena L., Z. Pflanzenzuchtg. 70:275-280 (1972).

ANDROGENETIC HAPLOID CALLUS OF TROPICAL LEGUMINOUS TREES

D.N. De and P.V. Lakshmana Rao

Applied Botany Section
Indian Institute of Technology
Kharagpur 721302
INDIA

INTRODUCTION

To date androgenetic haploid calli have been obtained from a number of plants. Apart from the pioneering work of Tulecke (1) on Ginkgo biloba, success is met with only a few trees, e.g. Populus, Camellia, Prunus (2), Hevea (3). However, so far no differentiation of haploid plantlet of trees has been obtained. there exists no report on successful induction of haploid callus of any leguminous tree. The present report deals with the successful induction of androgenetic haploid callus in two tropical leguminous trees, viz. Peltophorum pterocarpum (DC) K. Heyne (Copper pod, yellow poinciana) and Albizia lebbek (L.) Benth. (Siris, East Indian Walnut). Attempts have also been made with Tamarindus indica L. (Tamarind) and Samanea saman (Jacq.) Merril. (Raintree).

EXPERIMENTS AND RESULTS

All the selected trees have a certain physiological aspect common to them, they flower from the hottest and driest period of summer (max. temp. 43°C; mean temp. in April 34°C) to the rainy season (max. temp. 35°C; mean temp. in July 31°C) in India. All the trees attain good height and size and are relatively fast growing. Their pollen grains have a very thick exine and the cytoplasmic content is very dense, rendering them as exceedingly difficult materials for cytological determination of nuclear state. The pollens also show a high osmotic pressure as indicated by the fact that as much as 30% sucrose is necessary to prevent bursting of pollen grains or pollen tubes in solid or liquid medium. The pollens of

Albizia lebbek and *Samanea saman* are aggregated in a number of polyads.

The flowering branches of the young vigorously growing trees are brought to the laboratory in the forenoon and kept for a while in room temperature (28°C-30°C). In view of the extreme difficulty in routine detection of nuclear state of the pollen grains, a correspondence of nuclear state with the length of the flower bud, color of the petal and color of the anther was established and used for selection of the anthers. The surface of the buds was sterilized with 0.1% mercuric chloride followed by repeated washing in sterile distilled water. The anthers were aseptically excised and cultured at 25°C on both solid and liquid medium. For Nitsch (4) solid medium 1.0% agar was added to the liquid medium, and for Murashige and Skoog (5) medium 0.8% agar was added. After a great deal of experimentation, it was recognized that cold treatment, e.g. 3-5°C for 24-72 h, does not influence the response of the anthers. For best response, sucrose concentration should be at least 3.0% in the solid medium. The pollen culture in either solid or liquid MS or Nitsch media was not successful for any of the trees.

Peltophorum pterocarpum

Table 1 summarizes the results on response of the anthers at various stages of development on MS medium supplemented with NAA (2.0 mg/l), IAA (2.0 mg/l) and BA (2.0 mg/l). It is seen that the best callusing is obtained from the anthers cultured either at early uninucleate (90.8%) and late uninucleate (88.4%) pollen grains. In

Table 1. Effect of Developmental Stage of Pollen at Anther Inoculation on Callus Formation after 20 Days.

Color of anther	Stage of microsporogenesis	No. of anthers cultured	No. of anthers producing callus	% of callusing anthers
White	Meiosis	75	0	0
Yellowish white	Tetrads	110	12	8.1
Yellow	Uninucleate	176	160	90.8
Orange yellow	Late uninucleate	130	115	88.4
Deep orange	Binucleate	125	16	12.8

view of the success with anthers containing uninucleate pollen grains, trials were given in MS medium supplemented with various growth substances. Only the positive results are summarized in Table 2. It is seen that the MS solid medium supplemented with NAA (3.0 mg/l), IAA (2.0 mg/l) and BA (2.0 mg/l) yielded the highest number of calli.

The development of the callus from the uninucleate stage was cytologically examined after colchicine treatment (0.05% for 3h) followed by fixation in acetic-ethanal (1:3) and staining with feulgen solution and squashing in a drop of 2.0% carbol fuchsin. The uninucleate pollen grain becomes a multinucleate structure containing 2-4 nuclei in about 8 days. The subsequent cell wall formation and cell division produce a spheroidal or ellipsoidal structure. Sometimes the pollen produces a short filament of 3-5 cells which by subsequent transverse and longitudinal division produces the callus. At about 15 days the anther lobes burst along the length of the suture. Under best conditions of growth, the callus grows vigorously. On repeated transfer and subculturing the callus can be grown indefinitely. The chromosome counts of several calli confirmed their haploid nature (n = 14). After about 15 days, various levels of ploidy were noticed. So far no differentiation of the callus to plantlet has been obtained.

Albizia lebbek

Table 3 summarizes the results on the response of the anthers containing uninucleate pollen grains on MS solid medium supplemented with various growth substances. It is seen that MS medium with 2,4-D (0.5 mg/l), NAA (2.0 mg/l) and kinetin (2.0 mg/l) produced the highest frequency of callusing. The pollens of A. lebbek are organized in 8 polyads in each anther and each polyad consists of 16 or 32 pollen grains arranged in 3 tiers. The formation of the haploid callus involves division of all the pollens of a polyad, so that at the initial stage the spheroidal callus is a mass of cells which are of different genetic make-up. The callus continues to grow through repeated transfers and subculturing and no differentiation has so far been obtained.

Tamarindus indica and Samanea saman

Numerous attempts on various media were made for development of the androgenetic callus of tamarind. Occasionally multicellular embryoids were obtained from microspores cultured in Nitsch liquid medium in the presence of continuous white light (500 lux). The embryoids aborted after about 25 days. Although Samanea is taxonomically very close to Albizia and produces the same type of polyad, it failed to produce under identical conditions.

Table 2. Androgenesis of Anthers of *Peltophorum pterocarpum* on MS Solid Medium Supplemented with Various Growth Substances. (Data based on 200 anthers at uninucleate stage on each medium. Observations on pollen based on 10 anthers from each medium.)

	Growth hormones* (mg/l)										
Medium	2,4-D	K	CM	NAA	BA	IAA	CH	% of callusing anthers	No. of pollen studied	No. of multi-nucleate pollen	No. of multi-cellular pollen
MS	-	-	-	-	-	-	-	-	580	-	-
MS-1	2	-	-	-	-	-	-	30	600	2	2
MS-2	2	2	-	-	-	-	-	42	620	4	3
MS-3	1	-	15%	-	-	-	-	36	600	2	2
MS-4	1	-	-	-	1	1	-	52	560	8	4
MS-5	-	-	-	2	2	-	-	60	680	10	8
MS-6	-	-	-	3	2	2	-	86	720	12	14
MS-7	-	1	-	2	-	-	500	40	610	3	4

* Abbreviations: 2,4-D: 2,4-dichlorophenoxy acetic acid; K: kinetin; CM: coconut milk; NAA: naphthalene acetic acid; BA: benzyl adenine; IAA: indole acetic acid; CH: casein hydrolysate.

Table 3. Androgenesis of Anthers of Albizia lebbek on MS Solid Medium Supplemented with Various Growth Substances. (Data based on 200 anthers of uninucleate stage cultured on each medium at the end of 4 weeks in culture.)

Medium	Growth hormones (mg/l)					% of callusing anthers
	2,4-D	K	NAA	BA	IAA	
MS	-	-	-	-	-	0
MS-a	1	2	-	-	1	10
MS-b	0.5	2	2	-	-	40
MS-c	-	1	1	-	-	4
MS-d	-	-	1	2	2	16
MS-e	-	-	2	2	1	12

DISCUSSION

The successful induction of androgenetic haploid callus of leguminous tropical trees is recorded here, possibly for the first time. The formulation of the culture medium is not principally different from those used for other plants. The unique feature, however, is the lack of response of the anthers to the cold treatment. The cold treatment is not necessary, possibly because the laboratory temperature of 28°C and incubation temperature at 25°C are cool enough for the thermophilic pollens to go for callusing.

The formation of a single callus from a polyad of Albizia is intriguing. In view of the fact that for development of embryoid or callus in pollen culture, a minimum of crowding, of about 10^4 to 10^6 cells/ml is necessary, the polyad offers a natural crowding which must have an adaptive value. Callusing or embryoid formation of polyads may lead to fascinating developmental and morphological patterns. They may result in a callus with a mosaic of genetically different tissues. An admixture of diverse types of tissues may form coalescent embryos. Even cytomixic phenomena involving fusion and exchanges of nuclei of neighboring cells may lead to curious chromosomal variation in addition to polysomaty and haploid mosaic. The possibility of increasing the variability of the species is immense.

It is now widely appreciated that the haploid genome in eukaryotes is rather unstable. In order to obtain a true haploid tree, a definitive method of stabilizing the genome is to be evolved. Although substances like p-fluorophenylalanine have proved to be of dubious value in maintaining haploidy (6), it is a pointer to the usefulness of metabolic inhibitors as genome stabilizers. In our anxiety to get good growth we induce genomic disbalance by growth substances. It is to be accepted that rapid growth is an anathema to differentiation. All haploid plants that have been raised so far are herbaceous (2), and herbaceous plants show relatively lesser order of differentiation. The presence of a single non-adaptive gene, not to speak of a deleterious gene, will mar the possibility of the haploid plantlet acquiring its arboreal habit which demands better differentiation to withstand natural hazards of perennation. Thus production of a haploid plant with arboreous habit is both conceptually and technically challenging.

REFERENCES

1. W. Tulecke, A tissue derived from the pollen of *Ginkgo biloba*, *Science* 117:599-600 (1953).
2. I.K. Vasil, M.R. Ahuja and V. Vasil, Plant tissue cultures in genetics and plant breeding, *Adv. Genetics* 20:127-215 (1979).
3. R. Sachuthananthavale, Cultivation *in vitro* of callus tissue derived from anthers of *Hevea braziliensis*, *Plant Breeding Abstr.* 47:3954 (1977).
4. C. Nitsch, Cultured of isolated microspores, *in*: "Applied and Fundamental Aspects of Plant Cell, Tissue and Organ Culture," J. Reinert and Y.P.S. Bajaj, eds., Springer-Verlag, Berlin (1977).
5. T. Murashige and F. Skoog, A revised meidum for rapid growth and bioassays with tobacco tissue culture, *Physiol. Plant.* 15:473-497 (1962).
6. P.S. Mathews and I.K. Vasil, The dynamics of cell proliferation in haploid and diploid tissues of *Nicotiana tabacum*, *Z. Pflanzenphysiol.* 77:222-236 (1976).

FIELD EXPERIMENTS WITH ANTHER DERIVED LINES OF BARLEY (*HORDEUM VULGARE*) AND RYE (*SECALE CEREALE*)

Bärbel Foroughi-Wehr, Wolfgang Friedt and Gerhard Wenzel

Biologische Bundesanstalt für Land- und Forstwirtschaft
Institut für Resistenzgenetik
D-8059 Grünbach
Federal Republic of Germany

INTRODUCTION

Over the last years, the usefulness of haploids in plant breeding programs has been discussed extensively. The Haploid Method, first developed in model plants, is now also applicable to crop plants, e.g. the cereals. The potential advantages of haploids, i.e. acceleration of breeding programs and simplification of selection, are counterbalanced by various obstacles like possible negative gametophytic selection. However, experimental results are still limited up to now, so that a final evaluation of the Haploid Method is not yet possible. Therefore, we have tested androgenetic barley and rye lines under field conditions for their agronomic performance. Some recent data are reported herein.

PRODUCTION OF ANDROGENETIC RYE AND BARLEY LINES

The plant material first used at Ladenburg for the extraction of haploid rye (*Secale cereale*) lines consisted predominantly of F_1 hybrids between self-incompatible commercial winter rye material and self-compatible descendants from the cross *S. cereale* cv. Heine's Hellkorn X *S. vavilovii* (Experiment 1, see Table 1 - left). The commercial material mainly belonged to the 'Petkus' variety group. In addition, two short straw lines (M II and M III, obtained from H. Kuckuck, Hannover) had been incorporated. The androgenetic self-compatible and incompatible doubled-haploids were afterwards hybridized to a self-compatible inbred line of the 'Carstens' variety group (L 217, obtained from H.H. Geiger, Stuttgart-Hohenheim). Details of propagation of androgenetic material were reported by Hoffmann and Wenzel (1).

Table 1. History of Rye Anther Culture

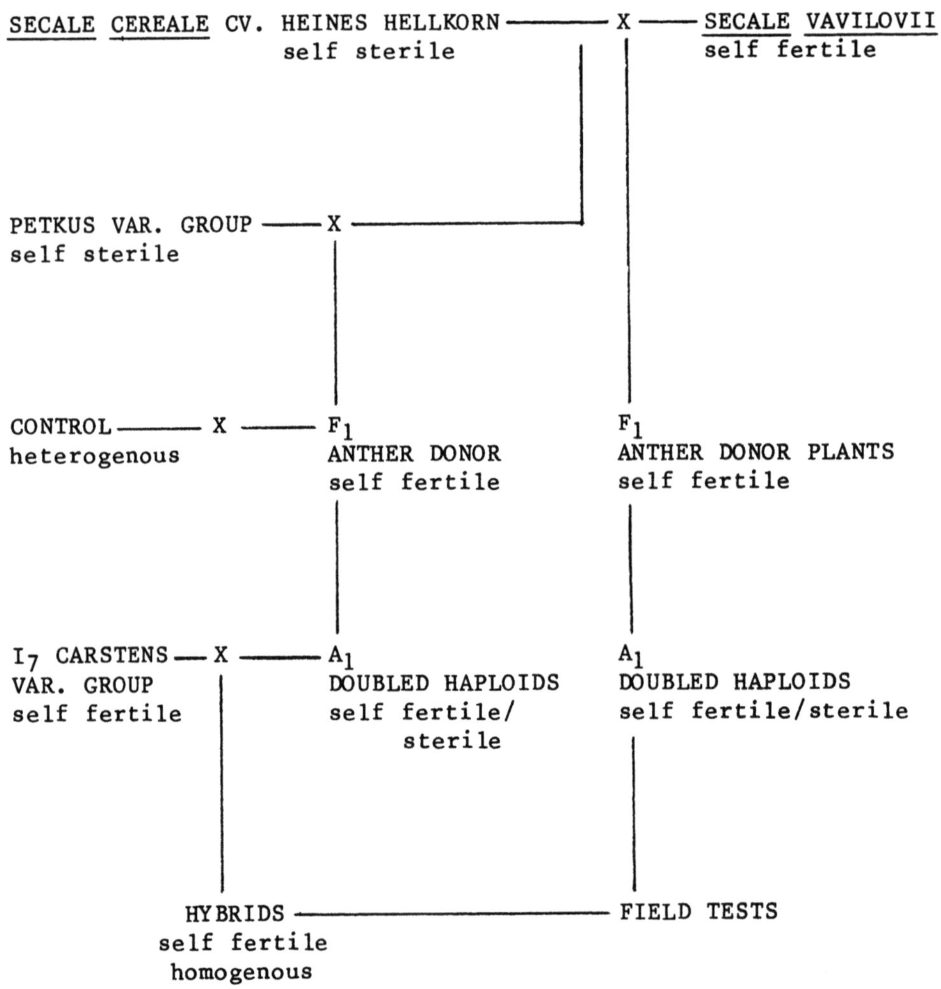

For anther culture Experiment 2, carried on at Grünbach (Table 1 - right), F_1 plants of the hybrid S. cereale cv. Heine's Hellkorn X S. vavilovii have been used. This hybrid is highly sterile, due to the presence of two chromosomal translocations in the heterozygous condition. As a consequence, in meiotic metaphase a sexivalent (VI) in addition to four regular bivalents (II) can be observed, chromosome distribution is disturbed and pollen are usually unfunctional. Details of anther culture and plant propagation are principally the same as reported earlier by Wenzel et al. (2).

For anther culture of spring barley (Hordeum vulgare) F_1 seed of commercial crosses, kindly provided by J. Breun, Herzogenaurach, was used. Anther-donor plants were continuously grown in the greenhouse or out of doors and anthers with uninucleate pollen grains were plated on agar as reported earlier by Foroughi-Wehr et al. (3). From all of the hybrids, except one, green individuals could be regenerated. However, callus formation and plant regeneration patterns showed large differences between crosses. The maximum proportion of green plants was two per 100 anthers plated. The results indicate that genotypic differences with regard to anther culture responsiveness exist.

Chromosome counts in root tip cells of regenerated green plants (A_1 plants) showed that the majority of individuals (ca. 70%) was diploid (2n = 2x = 14) and most of the remainder were haploid (2n = x = 7). Haploid plants were treated with colchicine according to Jensen (4) and thereby doubled-haploid plants were produced artificially. Spontaneous double-haploids usually showed no segregations in successive generations. Seed of A_1 lines was sown into the field for propagation and observation in the A_2 generation (1980). In the A_3 generation (1981) 2,000 viable seeds of each line as well as the original cross parents were sown into 6 m^2 drill-plots at Grünbach. A simple lattice design (13 x 13) was used.

RYE FIELD EXPERIMENTS

Experiment 1 (Ladenburg)

A total of 138 green androgenetic rye lines was recovered. Under optimum conditions, up to 10% of the plated anthers from the hybrid Mut. 520 X M II developed macroscopic structures (1).

Similarly, other hybrids of M II yielded comparatively high regeneration rates. However, even in the best of the hybrids the problem of high frequency of albino plantlets could not be solved satisfactorily. From the 138 androgenetic plants 40 could be grown to maturity. Of these, 15 were completely sterile, 21 could be selfed and four were crossed. From the 25 plants fertilized, seven turned out to be heterozygous and 18 were homozygous. Of the latter as many as 14 were self-fertile while four were self-incompatible.

In the homozygous lines and in some of their hybrids with an inbred line ('Carsten' var. group, Table 1), four characters were examined in detail: grain weight, plant height, spike length and alkylresorcinol content (1).

The mean values of different anther derived lines exceeded the means of the respective control F_2 on both the positive as well as the negative side. Some of the lines exhibited the reduced straw length of one cross parent combined with improved grain weight of the other.

Experiment 2 (Grünbach)

More than 100 green plants have been regenerated from F_1 donor material of the cross *S. cereale* cv. Heine's Hellkorn X *S. vavilovii*, up to now. The average plant regeneration rate was about two per 1,000 anthers plated. The majority of regenerated individuals was spontaneously diploid (2n = 2x = 14), the remainder were - like in barley - mostly haploid (2n = x = 7). In addition tetraploid individuals (2n = 4x = 28) occurred.

Due to chromosomal constitution of the cereale-vavilovii hybrid, as described above, sterile (partially and completely) progenies have to be expected in the segregating generations, in addition to normal self-incompatible genotypes. About one third of all anther derived plants grown up to now exhibited complete sterility. The fertile individuals will be tested for homozygosity in their A_2 generation. Some morphological characters of ten selected A_1 lines are presented in Table 2. The individuals are ordered according to their phenotypic similarity to either *S. vavilovii* (V, top) or *S. cereale* (C, bottom). It is evident that besides the parental phenotypes, all kinds of recombinants were recovered. The latter deviate from either parent with regard to self-fertility, grain weight, fragility of the rachis, as well as other characters not described here. Self-fertility of *S. vavilovii* can easily be combined with the regular grain weight of cultivated rye. Even for wide crosses like this the Haploid Method of producing homozygous diploid lines may be preferable to conventional inbreeding procedures.

BARLEY FIELD EXPERIMENTS

In a replicated trial at Grünbach, 134 doubled-haploid spring barley lines have been tested in 1981 for yield and yield-components as well as disease resistances. In addition, 33 cross parents and two control varieties have been included. The average grain yield

Table 2. Morphological Characters of 10 Diploid Lines out of Anther Culture of the F_1 Hybrid *Secale cereale* cv. Heine Hellkorn X *S. vavilovii*.

A1 PLANT	NO. OF SPIKES PER PLANT	NO. OF FLORETS PER SPIKE	NO. OF SEEDS PER SPIKE	SEED SET % (SELFED)	GRAIN WEIGHT (MG)	RACHIS	TYPE V - C*
R 337	15	45.2	20.0	44.2	10.3	Fragile	V
R 340/1	11	52.9	16.0	30.2	7.3	Fragile	V
R 427/1-1	14	63.3	14.9	23.5	10.9	Firm	V/C
R 675	8	75.7	23.1	30.6	13.9	Fragile	V/C
R 348/1	15	44.8	11.0	24.6	20.6	Fragile	C/V
R 639	15	33.7	2.9	8.6	22.8	Fragile	C/V
R 394	11	58.5	3.6	6.2	19.5	Fragile	C/V
R 446/1-1	14	69.6	0.4	0.6	30.0	Fragile	C/V
R 580	24	66.8	0.2	0.3	23.3	Fragile	C/V
R 469/1	11	57.1	8.5	14.9	26.9	Firm	C

* C = *S. cereale* type; V = *S. vavilovii* type; C/V and V/C = Recombinant types.

of the total trial was rather high (50.7 dt ha^{-1}), and the mean yield of the doubled-haploid (DH) strains was slightly, but not significantly, lower than the mean of all parents (50.2 vs. 50.7 dt ha^{-1}) (Table 3). Variation of yield among DH-lines was also not significantly different from the parental variation as indicated by the respective coefficients of variation (Table 3, C.V.). Similar results were obtained for grain weight and culm length (plant height). Differences in means and C.V.s between DHs and parents were not significant (Table 3). However, in both characters the DH-lines showed a tendency to an improvement as compared to the total of parents. A rather different result was found for heading date (earliness). Here, the DHs showed an increased variation and corresponding reduced mean heading date as indicated in Table 3. The segregation of very short and, therefore, late DH-lines is probably one of the reasons for the reduced mean grain yield of the doubled-haploid progenies.

Much more positive are the results of mildew observations in the field plots during spring and early summer 1981. A number of tested parents showed partial or complete mildew resistance in the field. On average the parents reach a mildew score of 4.7 (1 = fully resistant, 9 = completely susceptible), whereas the DH-lines show a mean value of 3.9 (Table 3). Simultaneously, the doubled-haploid strains exhibit an increased variation with regard to mildew, which clearly is extended to the positive direction, as indicated by the extreme values in Table 3. If the mildew score of each of the lines is compared with both respective parents, then the following can be deduced (5).

20% of the lines were more susceptible than the respective more susceptible parent, 11% showed similar mildew reaction as the more susceptible parent, and 41% of the DH-lines were in this respect similar to the more resistant parent (Table 4). Of special interest are those lines which showed intermediate mildew reactions (14%) or which were even less susceptible than the respective more resistant parent (14%). As a total, 48% of the DH-lines were thereby deviating from both parents and can be considered recombinants. Taking into consideration that recombination was restricted to the F_1 generation, the observed frequency of recombinant lines with regard to only one character - mildew resistance - is surprisingly high. However, it should be expected that the majority is due to recombination of mono- or oligogenically inherited mildew resistances ('major genes').

CONCLUSIONS

Doubled haploid lines of rye and barley show a similar extent of variation as conventionally produced segregating populations. Furthermore there are clear differences between families with regard

Table 3. Variation and Mean Values for Several Characters of Doubled-haploid Spring Barley Lines as Compared to Their Cross Parents, Field Trial, Grunbach 1981.

		MILDEW (1 - 9)	HEADING DATE IN JUNE	CULM LENGTH (CM)	GRAIN WEIGHT (MG)	GRAIN YIELD (DT HA-1)
PARENTS	MIN.	0.8	2.7	62.0	29.9	36.1 (33)
	MAX.	9.3	14.1	92.3	48.2	62.0
	MEAN	4.7	10.4	80.9	41.9	52.7
	C.V.	0.48	0.21	0.11	0.10	0.10
DH (134)	MIN.	0.5	5.4	56.2	29.4	32.9
	MAX	9.1	24.3	98.6	50.5	59.4
	MEAN	3.9	11.8	78.7	42.1	50.2
	C.V.	0.70	0.31	0.12	0.08	0.10
DIFFERENCE	C.V.s	SIGN.	SIGN.	N.S.	N.S.	N.S.

MIN./MAX. = Lowest and highest; MEAN = Mean value; C.V. = Coefficient of variation.

Table 4. Field Reaction to Powdery Mildew (Erysiphe graminis Hordei) of Spring Barley Lines Derived from Anther Culture of F_1 Hybrids, Drill Trial, Grünbach 1981.

	DOUBLED-HAPLOID LINES IN A_3	
MILDEW REACTION	NO.	%
< LOW PARENT*	31	20.3
= LOW PARENT	17	11.1
INTERMEDIATE	22	14.4
= HIGH PARENT	62	40.5
> HIGH PARENT	21	13.7
TOTAL	153	100.0

* Deviations significant at P = 0.05, 2 replications.

to anther culture responsiveness as well as agronomic performance of regenerated DH-lines. In rye, a varying proportion of anther derived lines shows segregation (heterozygosity), whereas in barley DH-lines are almost exclusively true-breeding (homozygous). This enables the safe selection of defined recombinant genotypes in very early generations (F_2, F_3).

REFERENCES

1. E. Hoffmann and G. Wenzel, Selfcompatibility in microspore-derived doubled-haploid rye lines and single grain selection for alkylresorcinol content, Theor. Appl. Genet. 60:129 (1981).
2. G. Wenzel, F. Hoffmann and E. Thomas, Increased induction and chromosome doubling of androgenetic haploid rye, Theor. Appl. Genet. 51:81 (1977).
3. B. Foroughi-Wehr, G. Mix, H. Gaul and H.M. Wilson, Plant production from cultured anthers of Hordeum vulgare L., Z. Pflanzenzuchtg. 77:198 (1976).
4. C.J. Jensen, Barley monoploids and doubled monoploids: Techniques and experience, in: "Barley Genetics III, Proc. Third Intl. Barley Genet. Symp.," H. Gaul, ed., Verlag Karl

Thiemig, Munchen (1976).
5. W. Friedt and B. Foroughi-Wehr, Anther culture of barley (*Hordeum vulgare* L.): Plant regeneration and agronomic performance of homozygous diploid progenies, *in*: "Barley Genetics IV, Proc. Fourth Intl. Barley Genet. Symp.," Edinburgh (in press) (1981).

FACTORS RESPONSIBLE FOR THE PRODUCTION OF SUBPROTOPLASTS IN BRASSICA OLERACEA VAR. CAPITATA

B. Vatsya and S. Bhaskaran

Hindustan Lever Research Centre
Chakala, Andheri East
Bombay 400099
INDIA

INTRODUCTION

The semi-autonomous cytoplasmic organelles have genomes of their own, which are responsible wholly or partially for some traits which a plant breeder might like to incorporate into the nuclear genome background of another species. Thus, at times it may be desirable to produce cytoplasmic hybrids (cybrids) containing the cytoplasm of both but nucleus of only one of the parents (1). Such a combination is attainable by any of the follwoing means: X-irradiation of the protoplasts of one of the parents leading to the degeneration of its nuclei in hybrid cells (2); directional and total elimination of chromosomes of one of the parents in proliferating hybrid cells (3); and fusion of nucleated protoplasts with anucleated protoplasts of the other.

We had reported in an earlier paper (4) the production of anucleated protoplasts (subprotoplasts) in large numbers by increasing the osmolarity of the incubating enzyme mixture during protoplast isolation. Here we give mathematical explanation for subprotoplast formation and a slight modification to the earlier method to get higher yields of subprotoplasts.

MATERIALS AND METHODS

A cultivar of cabbage (*Brassica oleracea* var. *capitata*) named Drum Head was used for this study. Surface sterilized seeds of this cultivar were germinated in agar tubes (0.8% dissolved in tap water) under aseptic conditions at 26°C under 16 h light (3000 lux) provided by white fluorescent tubes. After 10 days, the cotyledonary leaves were placed in ON (5) medium for preculture. These precultured leaves were used for protoplast isolation after 7 and 20 days to see the effect of leaf size on yield of subprotoplasts.

The enzyme mixture consisted of 2% cellulase, 1% macerozyme (both obtained from Kinki Yakult Mfg. Co. Ltd., Japan) and 0.77 M mannitol. The pH of the enzyme solution was adjusted to 5.6. After 6 h of incubation in enzyme mixture at 25°C, the protoplasts were separated from the debris by centrifugation at 85 g for 5 min. The protoplasts from the band at the top of the tube were taken and suspended in 0.77 M mannitol solution. They were centrifuged again at 85 g for 5 min., and protoplasts were collected from the band at the top of the tube. The protoplasts were stained with a few drops of 10^{-3} M acridine orange dye and observed under fluorescent microscope with a yellow filter to observe the nuclei which give characteristic orange/green fluorescence. Protoplasts not exhibiting any fluorescence were taken as anucleated protoplasts or subprotoplasts.

RESULTS AND DISCUSSION

The cotyledonary leaves precultured for 7 days gave a subprotoplast yield of 33% while the subprotoplast yield from the leaves precultured for 20 days was as much as 69%, on incubating in the enzyme mixture containing 0.77 M mannitol as osmoticum for 6 h.

We had shown previously (4) that subprotoplast formation depended on the hypertonicity of the incubating medium. Bradley (6) had suggested that subprotoplast formation depended upon the ratio of length and diameter of the cell. According to this "if a cell behaves as a fluid drop when it is elongated, it will spontaneously separate into two parts when the length (L) is greater than π times the diameter (D) taken at right angles to the long axis of the cell." It is difficult to understand from this how the cell will elongate in a plasmolysate leading to increase in L/D ratio.

However, subprotoplast formation can be explained if we take the ratio of surface area and the volume into consideration. It can be shown mathematically that in a shrinking protoplast (because it is kept in a hypertonic solution) the decrease in surface area and that in volume is not proportional.

$$\text{Volume (V)} = 4/3\pi r^2 \quad \ldots\ldots\ldots\ldots (1)$$

$$\text{Surface area (S)} = 4\pi r^2 \quad \ldots\ldots\ldots\ldots (2)$$

taking logarithm on both sides

$$\log V = \log 4/3 + \log \pi + 3 \log r$$

$$\log S = \log 4 + \log \pi + 2 \log r$$

If $\log V = w$ and $\log S = u$, then:

$$w = \log 4/3 + \log \pi + 3 \log r \quad \text{..........(3)}$$

$$u = \log 4 + \log \pi + 2 \log r \quad \text{...........(4)}$$

On differentiating these equations:

$$\frac{\partial w}{\partial r} = 3(1/r) = 3/r$$

$$\frac{\partial u}{\partial r} = 2(1/r) = 2/r$$

So

$$\left(\frac{\partial w}{\partial r}\right) / \left(\frac{\partial u}{\partial r}\right) = \frac{3/r}{2/r} = 1.5$$

The rate of decrease of volume is 1.5 times more thant he rate of decrease of surface area. This leads to an increase in surface area/volume ratio. Up to a certain stage this is taken care of due to flexibility of the membrane, but after some time the protoplast should acquire a condition where the same volume is surrounded by more surface area. This condition is achieved by subprotoplast fusion.

It can be shown that the surface area of two spheres will be more than the surface area of a single sphere having volume equal to the sum of volumes of the two spheres.

Let

V = fixed volume of a sphere of radius r

V_1 and V_2 vary so that

$$V = V_1 + V_2$$

Let the corresponding surface areas by S, S_1 and S_2 respectively then it can be shown that:

$$S_1 + S_2 > S$$

If

$$V = 4/3\pi r^3, \quad S = 4\pi r^2$$

$$V_1 = 4/3\pi r_1{}^3, \quad S_1 = 4\pi r_1{}^2$$

$$\text{and } V_2 = V - V_1$$

when surface areas are put in terms of volume.

$$S = (3^{2/3}, 4^{1/3}, \pi^{1/3})\ V^{2/3} = a\ V^{2/3}$$

$$S_1 = (3^{2/3}, 4^{1/3}, \pi^{1/3}),\ V_1{}^{2/3} = a\ V_1{}^{2/3}$$

where

$$a = (3^{2/3}, 4^{1/3}, \pi^{1/3})$$

"a" is a positive number.

For $S_1 + S_2 > S$

$$S_1 + S_2 - S = D \text{ should be positive}$$

$$D = a\ [V_1{}^{2/3} + (V - V_1)^{2/3} - V^{2/3}]$$

Let $A = V^{1/3}$ and $A_1 = V_1{}^{1/3}$.

Then D may be written:

$$D = A_1{}^2 - A^2 + (A^3 - A_1{}^3)^{2/3}$$

Let A_1 be expressed in terms of A by:

$$A_1 = kA, \text{ where } k < 1$$

For this it may be shown that:

$$D = A^2\ [k^2 - 1 + (1 - k^3)^{2/3}]$$

here A^2 is a positive quantity.

By expanding the expression $k^2 - 1 + (1 - k^3)^{2/3}$ it can also be shown to be a positive quantity. So D is positive, which means that $S_1 + S_2 > S$.

Therefore, it can be shown on theoretical grounds that a shrinking protoplast in a plasmolysate will divide into two to maintain proper surface area/volume ratio. The explanation may also be applied to other cases where subprotoplast formation has been reported in the tissue kept in plasmolysate.

If it is assumed that one protoplast buds off to give rise to one subprotoplast, then subprotoplast yield of more than 50% should not be expected. Subprotoplast yield of as much as 70%, obtained with the expanded leaves precultured for 20 days, can be explained if more than one subprotoplast is formed from one protoplast. This can be achieved if the protoplast after forming one subprotoplast again forms yet another subprotoplast. This is possible if the protoplast does not attain proper surface area/volume ratio, even after forming one subprotoplast because of its big size and to attain stability undergoes another budding to form a second subprotoplast.

By varying the osmolarity of incubating medium and by manipulating the cell size by preculturing the tissue, the frequency of subprotoplast formation could be controlled. It is desirable, however, to have a pure population of anucleated protoplasts for studies involving exclusive transfer of genetic information contained in cytoplasmic organelles via fusion. Because of the density differences between nucleated and anucleated protoplasts, differential centrifugation should allow separation of these two types of protoplasts. Experiments are underway in our laboratory in this direction.

ACKNOWLEDGEMENTS

The authors are indebted to Mr. Daljit Singh for his help in the mathematical computation. One of us (B.V.) is grateful to the Indian Agricultural Research Institute, New Delhi, for the award of a Senior Fellowship which made this study possible.

REFERENCES

1. S. Bhaskaran, S.K. Mukherjee, R. Majid, N. Gupta and B. Vatsya, *In vitro* genetic manipulations of higher plants, *Curr. Sci.* 50:653-661 (1981).
2. A. Zelcer, D. Aviv and E. Galun, Interspecific transfer of cytoplasmic male sterility by fusion between protoplasts of normal *Nicotiana sylvestris* and X-ray irradiated protoplasts of male sterile *N. tabacum*, *Z. Pflanzenphysiol.* 90:397-407 (1978).
3. K.N. Kao, Chromosomal behaviour in somatic hybrids of soybean-*Nicotiana glauca*, *Mol. Gen. Genet.* 150:225-230 (1977).
4. B. Vatsya and S. Bhaskaran, Production of subprotoplasts in *Brassica oleracea* var. *capitata*: a function of osmolarity of the media, *Plant Sci. Lett.* 23:277-283 (1981).
5. K. Ohyama and J.P. Nitsch, Flowering haploid plants obtained from protoplasts of tobacco leaves, *Plant Cell Physiol.* 10:917-921 (1972).
6. P.M. Bradley, Production of enucleated plant protoplasts of *Allium cepa*, *Plant Sci. Lett.* 13:287-290 (1978).

ISOLATION AND REGENERATION OF THE MESOPHYLL PROTOPLASTS OF BRASSICA JUNCEA CV. PRAKASH

R.K. Aggarwal, D.R. Sharma, H.O. Mehra and R.K. Singh

Department of Genetics
Haryana Agricultural University
Hissar - 125005 (INDIA)

INTRODUCTION

The availability of techniques for isolation and fusion of protoplasts has raised the possibility of their use in overcoming the barriers of hybridization in plant species (1). To achieve this goal it is imperative to work out conditions for the isolation and culture of protoplasts from the parental species. Mesophyll protoplasts of B. Napus (3, 4) and B. Oleracea var. botrytis (Bhartendu Vatsya personal comm.). The work described here deals with the isolation and culture of mesophyll protoplasts of Brassica juncea cv. Prakash.

MATERIALS AND METHODS

For the isolation of protoplasts, fully expanded leaves were taken from field grown plants. Isolation was done by the one-step enzymatic method (5). Leaves were surface sterilized by dipping them in 70% ethyl alcohol for 20 seconds and then treating them with 20-30% 'Polar' liquid bleach along with a few drops of 'Teepol' for 10-15 minutes. After 3 washings with sterile water the lower epidermis of the leaves was removed and leaves were cut into small pieces. About 100 mg of these peeled leaf segments were incubated in 5 ml of enzyme solution (1 or 2% cellulase 'Onozuka R-10' and 1% macerozyme in 0/5M mannitol aqueous solution) in small petri plates at 25±1°C and 30±1°C in the dark for 3-7 h. After incubation the petri plates were gently shaken to liberate the protoplasts and contents were filtered through a sterile layered muslin cloth. This step was followed by centrifugation at 100xg for 3 minutes. The protoplast pellet was washed thrice by resuspending each time in 0.5M mannitol solution. Protoplast counts were made with the help of a haemocytometer. For culturing the protoplasts, final washing was done with the medium employed for culturing them. The washed protoplasts were then cultured in the agar medium of Ohyama and Mitsch (6) ('ON' medium) as well as liquid medium of Kartha et al.(3) in which N-Z amine was replaced by case in hydrolysate. The cultures were kept at 25±1°C in the dark and examined routinely with the help of an inverted microscope. Formation of cell wall was tested by plasmolysing the newly formed cells with 0.8 M sucrose solution.

RESULTS AND DISCUSSION

Protoplast Isolation

Under the incubation conditions used, the protoplast liberation started within 2 h of incubation and continued for several hours. However the highest yield was achieved with the help of 1% macerozyme and 2% cellulase at 30°C in 5 h (2.42x 10^6protoplasts/g fresh weight). The protoplasts were spherical with chloroplasts arranged along the periphery of the plasma membranes. The nucleus was invariably found to be eccentric and placed near the plasma membrane. The size of the protoplasts varied from 20 to 65 μm. As is evident from Table-1, the use of 2% cellulase was superior in all the combinations as compared to 1% concentration. Incubation at lower temperature (25°C) required 7 h for the maximum release of protoplasts whereas higher temperature (30°C) facilitated quicker release and maximum yield was obtained in 5 h as also observed in *Petunia* (7). A sharp decline in protoplast yield at 30°C after 5 h is indicative of deleterious effects of the enzymes on the plasma membranes of the liberated protoplasts. Similar decline in yield of protoplasts at higher temperature has also been observed in case of *Nicotina* sps. (8, 9).

Protoplasts in Culture

Protoplasts appeared to synthesize cell walls within 72-96 h in culture. The constituent cells became slightly oblong (Fig-3) culminating finally in the formation of two daughter cells in a period of about 5-6 days. Sustained divisions were obtained only in a few cases which finally could form micro colonies in solid 'ON' medium (Fig-4, 5). This medium failed to support the growth of these newly formed cell aggregates. On the other hand protoplasts cultured in the liquid medium of Kartha et al. (3) is known to support the growth of *B. Napus* protoplasts showing thereby that these species within this genus exhibit similarities at least for their nutritional requirement. Such observations have quite frequently been made that a medium standardized for the explants of one species also favors cell proliferations in

Table 1. Effect of enzyme concentration, Temperature and incubations period on protoplast isolation in *Brassica juncea* cv. Prakash.

Enzyme solution	Incubation temperature (oC)	Incubation period (hrs.)	Protoplast yield ($X.10^4$ protoplasts/g of leaf tissue
E_1^*	25 ± 1	3.0	11.78 ± 1.95
		5.0	48.03 ± 2.1
		7.0	66.93 ± 4.5
E_2^{**}	25 ± 1	3.0	63.39 ± 4.2
		5.0	128.57 ± 13.85
		7.0	232.12 ± 13.0
E_1	30 ± 1	3.0	8.93 ± 1.05
		5.0	76.0 ± 2.8
		7.0	81.43 ± 6.35
E_2	30 ± 1	3.0	116.53 ± 12.8
		5.0	242.85 ± 9.4
		7.0	88.39 ± 7.0

E_1^* = Cellulase 1% + Macerozyme 1%; E_2^{**} = Cellulase 2% + Macerozyme 1%

other species of the same genus.

The division of protoplasts and consequent formation of colonies in this material shows that even field grown plants can also be used for successful culture of protoplasts unlike in a few other species of this genus (10). Protoplasts isolated from greenhouse grown plants of *B. rapa* have also been reported to be amendable for culture and regeneration of calli.

ACKNOWLEDGEMENT

The financial assistance provided by Council of Scientific and Industrial Research, New Delhi in the form of Junior Research Fellowship to the senior author is fully acknowledged.

REFERENCES

1. I.K. Vasil, M.R. Ahuja, V. Vasil, Plant tissue culture in genetics and plant breeding, Adv. Genet., 20:127-216 (1979).
2. T.H. Ulrich, J.B. Chowdhury, J.M. Widholm, Callus and root formation from mesophyll protoplasts of *Brassica rapa*, Plant Sci. Lett., 19:347-354 (1980).
3. K.K. Kartha, M.R. Michayluk, K.N. Kayo, O.L. Gamborg and F. Constabel, Callus formation and plant regeneration from mesophyll protoplasts of rape plants (*Brassica napus* L. cv. Zephyr), Plant Sci. Lett., 3:265-271 (1974).
4. E. Thomas, F. Hoffmann, I. Potrykus and G. Wenzel, Protoplast regeneration and stem embryogenesis of haploid androgenic rape, Mole. Gen. Genet., 145:245-248 (1976).
5. J.B. Power, E.C. Cocking, A simple method for the isolation of very large numbers of leaf protoplasts by using mixtures of cellulase and pectinase, Biochem. J., III:33-37 (1969).
6. K. Ohyama, J.P. Nitsch, Flowering haploid plants obtained from protoplasts of tobacco leaves, Plant Cell Physiol., 13:229-236 (1972.
7. H. Binding, Regeneration of haploid and diploid plants from protoplasts of *Petunia hybrida*, Z. Pflanzenphysiol., 74:324-356 (1974).
8. D.R. Sharma, Studies on hybridization of somatic cells of plant origin *in vitro*, (Ph.D. Thesis, Division of Genetics, I.A.R.I., New Delhi) (1975).
9. R. Gill, A. Rashid, S.C. Maheshwari, Isolation of mesophyll protoplasts of *Nicotiana rustica* and their regeneration into plants flowering *in vitro*, Physiol. Plant, 47:7-10 (1979).

REGENERATION OF PLANTS FROM PROTOPLASTS OF *SOLANUM MAMMOSUM* L

P.M. Kumar, S.K. Ghose*, S. Sen* and S.K. Sen

Programme in Genetical Research
Department of Botany
Bose Institute
Calcutta 700009
INDIA

* Department of Genetics and Plant Breeding
B.C.K.V.V.
Kalyani
West Bengal
INDIA

INTRODUCTION

The potentials of isolated plant protoplasts for fundamental and applied research studies of crop plants are being appreciated more than ever as discussed by many workers recently (1, 2). *Solanum mammosum* L. is an exotic Solanaceous species and is considered a promising source of solasodine for pharmaceutical industries (3). This communication reports on the success that we attained in isolation and regeneration of cultured cell protoplasts of this medicinal crop plant.

MATERIALS AND METHODS

Protoplasts were isolated from suspension cultured cells of tender stem derived callus of *S. mammosum* L. on SH (4) medium with 2,4-D (1.0 mg/l) and kinetin (0.5 mg/l) as growth substances. Three week old friable calli were utilized to initiate suspension cultures on SH medium supplemented with 2,4-D (0.5 mg/l) and kinetin (0.1 mg/l). Addition of adenine sulphate (4.0 mg/l) and casein hydrolysate (500 mg/l) were tested for the enhancement of cells in culture. Suspensions were agitated continuously in a gyratory shaker at 120 rpm under an illumination of 1500 lux. Subcultures were made every 36-48 h and the growing cells were utilized as the source of protoplasts.

The "one step" method of protoplast isolation (5) was employed. Cells collected by centrifugation were incubated in an enzyme mixture containing 1% Macerozyme, 4% cellulase "Onozuka" R-10 in mannitol as osmoticum (650 mOs/kg.H_2O). pH of the mixture was adjusted to 5.8. One gm/10 ml material were incubated for 3 h in a roller shaker at 25°C $\pm$ 1°C. Protoplasts were purified through repeated and centrifugation following the standard procedure. Viability was tested using the technique of Larkin (6).

Regeneration of the purified protoplasts was tested using the MDA (microdroplet array) technique (7). In order to find out a suitable culture medium twelve established media (8) were tested. Responsive culture conditions were further tested in liquid thin layer or soft agar plating at 2 x 10^4 protoplasts/ml. Petri plates sealed with parafilm were incubated at 25°C $\pm$ 1°C under light intensity of 1500 lux and dark cycle of 16/8 h. Cell wall regeneration was judged through Calcofluor white ST test (9).

RESULTS

Protoplast release was clearly a function of the quality of the cell suspension used. When rapidly growing cultures (subcultured 36-48 h before isolation) of a highly dispersed type were used as the source of the protoplasts, a majority of the cells could be converted to protoplasts. Addition of adenine sulphate (4.0 mg/l) and casein hydrolysate (500 mg/l) to the suspension medium showed enhancement of cell growth. Formulated enzyme mixture could generate production of 80-85% of protoplasts after 3 h of incubation. Viability test indicated 70-75% of protoplasts to be healthy with granules and small vacuoles (Figure 1).

Of the twelve established media tested, five of them responded favorably. Modified SH medium yielded the best result. In SSH medium, cell wall formation was frequently seen followed by few cell divisions. In MSH, MS and NT only cell wall regeneration could be seen. Addition of 2,4-D to the SH medium at the level of 1.0 mg/l caused higher rate of division. Maximum number of cell wall formation as studied through Calcofluor white ST test was observed within 24 h of culture in SH medium with 2,4-D (1.0 mg/l) and kinetin (0.5 mg/l). Myo-inositol was reduced to 100 mg/l and sucrose to 2%. pCPA was eliminated. These cells were seen to undergo the first cell division (Figure 2) after 4-5 days and cell colonies were visible by the end of the third week (Figure 3). Though the initial division frequency was around 35%, final colony formation (sustained division) was less than 15% (Figure 4). These colonies grew faster

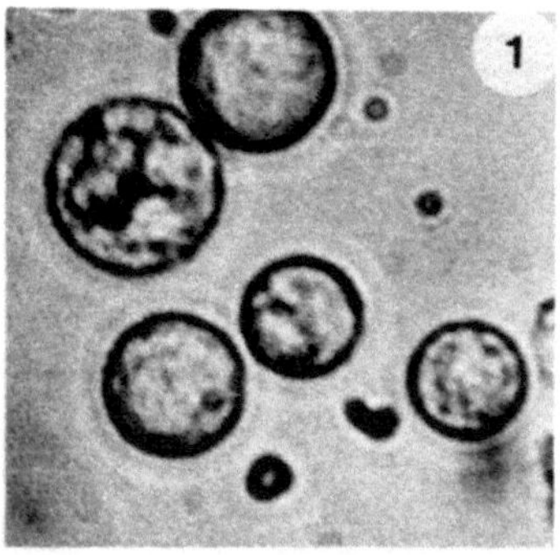

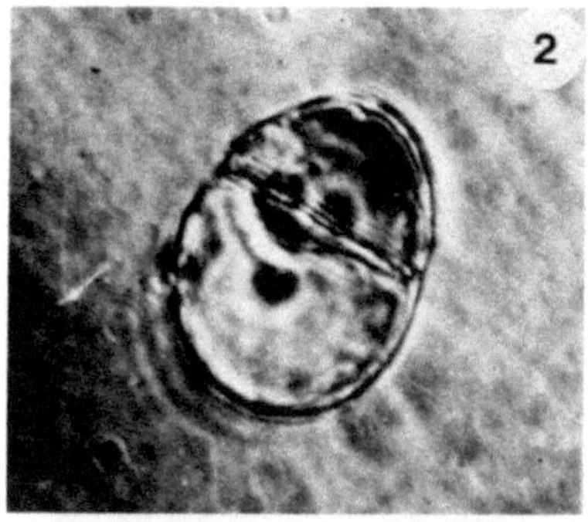

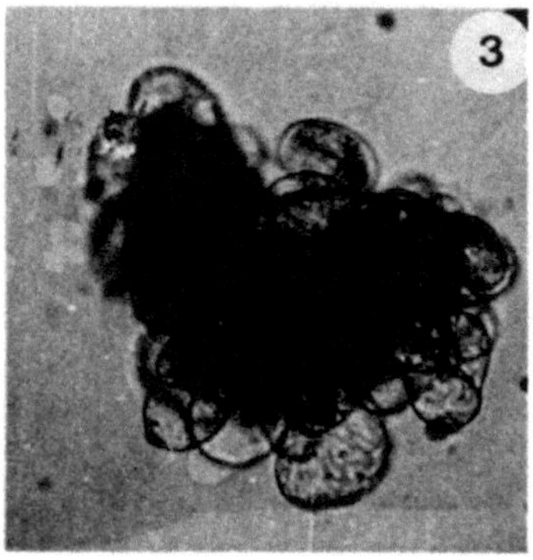

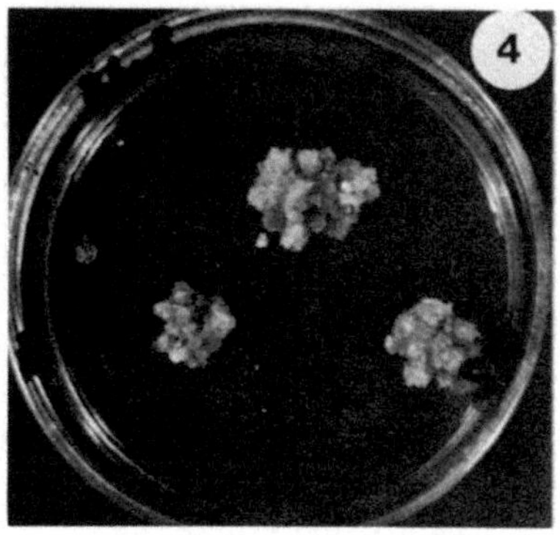

Figure 1. Freshly isolated protoplasts from cultured single cells of *S. mammosum* L.

Figure 2. First division of the protoplast after 4-5 days.

Figure 3. Cell colony after 3 weeks.

Figure 4. Calli developed after fifth week.

in the medium solidified with soft agar.

The growing colonies, when later transferred to MS medium enriched with 0.1 mg/l 2,4-D and 0.2 mg/l of kinetin, showed differentiation of shoots which could be observed within 3 weeks (Figure 5). Though minute hairy roots were seen to be induced, they did not grow well in the same medium. Transferring the material to the same MS medium devoid of hormones induced healthy roots to grow within two weeks. The fully grown plants (Figure 6) were maintained in the normal MS medium with 0.1 mg/l of 2,4-D and 0.2 mg/l of kinetin.

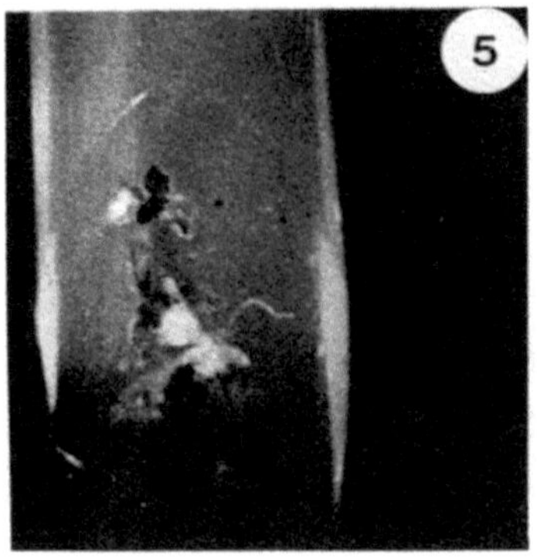

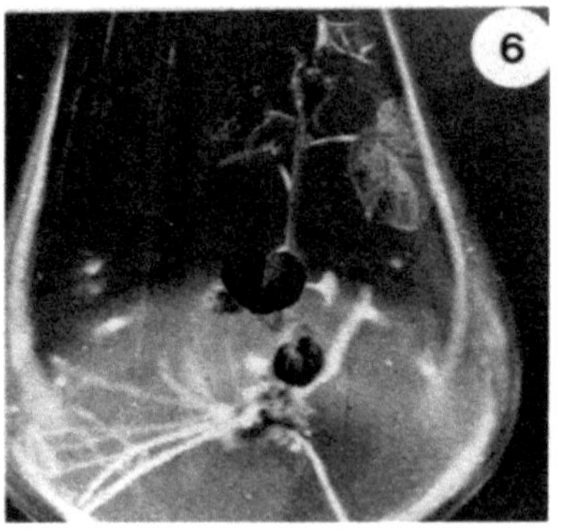

Figure 5. Initiation of shoots in MS medium supplemented with 2,4-D (0.1 mg/1) and kinetin (0.2 mg/1).

Figure 6. Fully grown plant in MS medium.

DISCUSSION

We observed in our study that suspension cultured cells subcultured 36-48 h before isolation yielded maximum number of viable protoplasts. Less vacuolated and thin walled cells contributed to the high yield of protoplasts.

The MDA technique utilized was convenient enough to study and compare the suitability of many culture media at the same condition with regard to the regeneration of protoplasts. The external prerequisites for somatic embryogenesis are uniform. It is necessary that the cultures are in the presence of an auxin (suitably 2,4-D) at least for a certain period and at reduced nitrogen source. The introduction of Murashige and Skoog (10) medium facilitated shoot initiation in the presence of 0.1 mg/1 2,4-D and 0.2 mg/1 kinetin. The necessity of reduced nitrogen (1,650 mg/1 NH_4NO_3) and the low concentration of auxin in the medium were found to play critical roles. The auxin-kinetin balance is critical in regulating the growth and development of embryos. Shoot differentiated calli were transferred to hormone-free medium where root development and growth of healthy shoots could be caused.

ACKNOWLEDGEMENTS

Grants from the Department of Science and Technology, Government of India, to S.K. Sen and a fellowship grant from C.S.I.R., New Delhi to P.M. Kumar are thankfully acknowledged.

REFERENCES

1. Y.P.S. Bajaj, Protoplast isolation, culture and somatic hybridization, in: "Plant Cell, Tissue and Organ Culture," Y.P.S. Bajaj and J. Reinert, eds., Springer-Verlag, Berlin (1977).
2. I.K. Vasil, M.R. Ahuja and V. Vasil, Plant tissue cultures in genetics and plant breeding, Adv. in Genet. 20:127-215 (1979).
3. T. Telek, H. Delpin and E. Cabanillas, Solanum mammosum as a source of solasodine in lowland tropics, Eco. Bot. 31:120-128 (1977).
4. R.U. Schenk and A.C. Hildebrandt, Medium and techniques for induction and growth of monocotyledonous and dicotyledonous plant cell cultures, Can. J. Bot. 50:199-204 (1972).
5. J.B. Power and E.C. Cocking, Isolation of leaf protoplasts. Macromolecule uptake and growth substance response, J. Exp. Bot. 21:64-70 (1970).
6. P.J. Larkin, Purification and viability determinations of plant protoplasts, Planta 128:213-216 (1976).
7. I. Potrykus, C.T. Harms and H. Lorz, Multiple drop array (MDA) technique for the large scale testing of culture media variations in hanging microdrop cultures of single cell system. I. The technique, Plant Sci. Lett. 8:(1979).
8. P.M. Kumar, K. Das, R.R. Sinha, P. Mukherjee and S.K. Sen, Interspecific somatic protoplast fusion products in cultivated jute species, in: "Proc. of Intl. Symp. on Plant Cell Culture in Crop Improvement," K.L. Giles and S.K. Sen, eds., Plenum Press, New York (in press) (1982).
9. T. Nagata and I. Takebe, Cell wall regeneration in and cell division of isolated tobacco mesophyll protoplasts, Planta 92:301-308 (1970).
10. T. Murashige and F. Skoog, A revised medium for rapid growth and bioassays with tobacco tissue culture, Physiol. Plant. 15:473-497 (1962).

INDEX

MIX
Papier aus verantwortungsvollen Quellen
Paper from responsible sources
FSC® C105338

If you have any concerns about our products,
you can contact us on
ProductSafety@springernature.com

In case Publisher is established outside the EU,
the EU authorized representative is:
Springer Nature Customer Service Center GmbH
Europaplatz 3, 69115 Heidelberg, Germany

Printed by Libri Plureos GmbH
in Hamburg, Germany